43,50/75€

THE NATURE OF SYMBIOTIC STARS

ASTROPHYSICS AND SPACE SCIENCE LIBRARY

A SERIES OF BOOKS ON THE RECENT DEVELOPMENTS
OF SPACE SCIENCE AND OF GENERAL GEOPHYSICS AND ASTROPHYSICS
PUBLISHED IN CONNECTION WITH THE JOURNAL
SPACE SCIENCE REVIEWS

Editorial Board

J. E. BLAMONT, *Laboratoire d'Aeronomie, Verrières, France*

R. L. F. BOYD, *University College, London, England*

L. GOLDBERG, *Kitt Peak National Observatory, Tucson, Ariz., U.S.A.*

C. DE JAGER, *University of Utrecht, The Netherlands*

Z. KOPAL, *University of Manchester, England*

G. H. LUDWIG, *NOAA, National Environmental Satellite Service, Suitland, Md., U.S.A.*

R. LÜST, *President Max-Planck-Gesellschaft zur Förderung der Wissenschaften, München, F.R.G.*

B. M. McCORMAC, *Lockheed Palo Alto Research Laboratory, Palo Alto, Calif., U.S.A.*

H. E. NEWELL, *Alexandria, Va., U.S.A.*

L. I. SEDOV, *Academy of Sciences of the U.S.S.R., Moscow, U.S.S.R.*

Z. ŠVESTKA, *University of Utrecht, The Netherlands*

VOLUME 95
PROCEEDINGS

THE NATURE OF SYMBIOTIC STARS

PROCEEDINGS OF IAU COLLOQUIUM No. 70
HELD AT THE OBSERVATOIRE DE HAUTE PROVENCE,
26–28 AUGUST, 1981

Edited by

MICHAEL FRIEDJUNG

Institut d'Astrophysique (CNRS), Paris, France

and

ROBERTO VIOTTI

Istituto Astrofisica Spaziale (CNR), Frascati, Italy

D. REIDEL PUBLISHING COMPANY

DORDRECHT : HOLLAND / BOSTON : U.S.A.
LONDON : ENGLAND

Library of Congress Cataloging in Publication Data

Main entry under title:

The nature of symbiotic stars.

    (Astrophysics and space science library ; v. 95. Proceedings)
    English and French.
    Includes indexes.
    1. Stars, Symbiotic–Congresses. I. Friedjung, M., 1940–
II. Viotti, Roberto, 1939–   III. Series: Astrophysics and space
science library ; v. 95. IV. Series: Astrophysics and space science
library. Proceedings.
QB843.S96F74   623.8   82-3847
ISBN 90-277-1422-3   AACR2

Published by D. Reidel Publishing Company,
P.O. Box 17, 3300 AA Dordrecht, Holland.

Sold and distributed in the U.S.A. and Canada
by Kluwer Boston Inc.,
190 Old Derby Street, Hingham, MA 02043, U.S.A.

In all other countries, sold and distributed
by Kluwer Academic Publishers Group,
P.O. Box 322, 3300 AH Dordrecht, Holland.

D. Reidel Publishing Company is a member of the Kluwer Group.

All Rights Reserved
Copyright © 1982 by D. Reidel Publishing Company, Dordrecht, Holland
No part of the material protected by this copyright notice may be reproduced or
utilized in any form or by any means, electronic or mechanical
including photocopying, recording or by any informational storage and
retrieval system, without written permission from the copyright owner

Printed in The Netherlands

This Volume is dedicated to Andrew David

T H A C K E R A Y

(1910 – 1978)

Pioneer in the Spectroscopy of
Southern Hemisphere Emission Line Stars

TABLE OF CONTENTS

Preface   xi
Scientific Organizing Committee   xiii
List of Participants   xiv
C. FEHRENBACH: Adresse au colloque   xix

INTRODUCTION

J. SAHADE: Fifty years of symbiotic stars   1
A.A. BOYARCHUK: Introductory report on symbiotic stars   11

SESSION I — NEW OBSERVATIONS OF SYMBIOTIC STARS

S. KWOK: Radio observations of symbiotic stars   17
D.A. ALLEN: Infrared studies of symbiotic stars   27
C. EIROA, H. HEFELE, QIAN ZHONG-YU: IR photometry of symbiotic
    stars at Calar Alto observatory   43
Y. ANDRILLAT: Near IR spectra of symbiotic stars   47
Y. ANDRILLAT, L. HOUZIAUX: Photographic infrared spectra of
    symbiotic stars   57
F. CIATTI: Properties of symbiotic stars from studies in the
    optical region   61
N.A. OLIVERSEN, C.M. ANDERSON: Spectra of individual symbiotic
    stars   71
G. MURATORIO, M. FRIEDJUNG: Visual symbiotic spectra obtained with
    the Haute Provence multiphot detector   83
H. NUSSBAUMER: UV line emission of symbiotic stars   85
M.H. SLOVAK, D.L. LAMBERT: Ultraviolet properties of the symbiotic
    stars   103
D.A. ALLEN: X-ray observations of symbiotic stars   115
C.M. ANDERSON, J.P. CASSINELLI, N.A. OLIVERSEN, R.V. MYERS,
    W.T. SANDERS: X-ray detection of the symbiotic star   117
    AG Draconis

SESSION II – DISCUSSION ON INDIVIDUAL STARS

R. VIOTTI, A. GIANGRANDE, O. RICCIARDI, A. CASSATELLA:
Z Andromedae: the prototype? ... 125

M. HACK, P.L. SELVELLI: A review of the properties of the symbiotic star CH Cygni ... 131

D. CHOCHOL, L. HRIC: Recent increase of the activity of the symbiotic star CH Cygni ... 137

V. PIIROLA: Polarimetry of CH Cygni ... 139

A.G. MICHALITSIANOS, M. KAFATOS, R.E. STENCEL, A.A. BOYARCHUK:
UV eclipse observations of CI Cyg ... 141

G.B. BARATTA, A. ALTAMORE, A. CASSATELLA, M. FRIEDJUNG, D. PONZ, O. RICCIARDI: IUE observations of CI Cygni during 1979-1981 ... 145

J. MIKOŁAJEWSKA, M. MIKOŁAJEWSKI: The eclipse of CI Cygni in 1980 on the objective prism spectra ... 147

C. FEHRENBACH: Variation spectrale de CI Cygni en 1975 ... 149

HUANG CHANG-CHUN: Spectral variations of CI Cyg between 1980 and 1981 ... 151

N.A. OLIVERSEN, C.M. ANDERSON, K.H. NORDSIECK: Optical observations of CI Cygni ... 153

A. CASSATELLA: Introductory report on V1016 Cygni ... 157

G. MURATORIO, M. FRIEDJUNG: Recent studies of the spectrum of V1016 Cygni ... 161

J. GRYGAR, D. CHOCHOL: The peculiar symbiotic object V1329 Cygni: single-star versus binary models ... 165

T. IIJIMA: Physical properties of V1329 Cygni ... 169

Y. ANDRILLAT: Variability of HBV 475 in the near infrared ... 173

C. KINDL, H. NUSSBAUMER: Spectral evolution of HBV 475 (= V1329 Cygni) in the ultraviolet ... 175

N.A. OLIVERSEN, C.M. ANDERSON: Introductory remarks on the symbiotic star AG Draconis ... 177

A. ALTAMORE, G.B. BARATTA, A. CASSATELLA, A. GIANGRANDE, D. PONZ, O. RICCIARDI, R. VIOTTI: The ultraviolet spectrum of AG Draconis ... 183

TABLE OF CONTENTS

HUANG CHANG-CHUN: Spectroscopic observations of AG Dra 185

A.G. MICHALITSIANOS, M. KAFATOS: UV time-dependent emission in SY Muscae 191

M.H. SLOVAK: AR Pavonis: the Rosetta stone of the symbiotics 195

* * * : Introductory report on AG Pegasi 201

M. KAFATOS, A.G. MICHALITSIANOS: The peculiar star RX Puppis 203

P.A. WHITELOCK, R.M. CATCHPOLE: Long term trends in the 3.5μ light curve of RX Puppis 207

S. KWOK: HM Sagittae - A most remarkable star 209

C. KINDL, H. NUSSBAUMER: Spectral evolution of HM Sagittae in the ultraviolet 213

P.A. WHITELOCK, R.M. CATCHPOLE, M.W. FEAST: RR Telescopii 215

D. PONZ, A. CASSATELLA, R. VIOTTI: Ultraviolet observations of RR Telescopii 217

M.F. McCARTHY, B.M. LASKER, T.D. KINMAN: V 4049 Sgr 219

T.S. BELYAKINA, R.E. GERSHBERG, Y.S. EFIMOV, V.I. KRASNOBABTSEV, E.P. PAVLENKO, P.P. PETROV, K.K. CHUVAEV, V.I. SHENAVRIN: The Kuwano peculiar object (PU Vulpeculae) 221

SESSION III - INTERPRETATION

A.A. BOYARCHUK: Determination of the term Symbiotic Star 225

M.J. PLAVEC: The symbiotics as binary stars 231

M. FRIEDJUNG: Models for symbiotic stars in the light of the data 253

M. KAFATOS: Symbiotic star UV emission and theoretical models 269

SESSION IV - EVOLUTIONARY CONSIDERATIONS

B. RUDAK: The evolutionary status of symbiotic stars 275

A.V. TUTUKOV, L.R. YUNGELSON: On the model of symbiotic stars 283

J.P. SWINGS: Concluding remarks 297

Subject index 303

Star index 307

PREFACE

Many aspects of symbiotic stars have long puzzled astronomers. For instance while most students of the subject have considered them binary, many have at different times supported single star models. The nature of their outbursts is uncertain, while the dividing line between symbiotic stars and novae is unclear. In any case doubts can even be raised as to whether a class of "Symbiotic Stars" really exists.

Much new data has been obtained in recent years, in particular from the study of radiation outside the visual region. Many symbiotic stars have been studied in the UV with IUE since 1978, while X-rays were detected in a few cases with the Einstein satellite. There have been a number of infrared and radio studies, and the number of known symbiotic stars has also considerably increased. Furthermore theoretical ideas have in recent years been considerably enriched by concepts of stellar winds, and accretion phenomena in binaries including accretion disks. It was therefore extremely opportune and timely to hold the first international meeting exclusively devoted to these stars, so as to consider the new results from such a wide range of observations in different spectral regions, and the conclusions which can be drawn for possible models as well as theories of the nature and structure of symbiotic stars.

After a session devoted to new observations in different spectral regions, a session was spent considering some individual stars. This was because there are considerable differences between different stars, and one needs to understand the physics of each star as an individual. It was only after such an examination that a session was devoted to interpretation (classification and possible models). The meeting ended with a session on evolution and internal structure.

Different themes were introduced by invited papers. There were no formal contributed papers, because we wanted to leave as much time as possible for free discussion. However some short contributions summarizing some recent results were presented, particularly on individual stars.

We have tried as much as possible to fully reproduce the discussion, where space allows us to do this. However there are some gaps, in particular where participants did not complete the sheets on which their comments were to be written. It is for this reason that some of the more numerous remarks made on the last day during the examination of classification and models are unfortunately not reproduced. For instance one participant spoke of God having created both man and the devil.

Most participants considered that most symbiotic stars are binary (90 % according to one), but their physics was the subject of lively controversy. Some thought that one should speak of a "Symbiotic Phenomenon" rather than of "Symbiotic Stars". It may be useful to hold another meeting in a few years, to see whether progress has been made on resolving the different problems.

The meeting was dedicated to the memory of the late Dr. A.D. Thackeray, who in many years of thorough painstaking work provided many basic results concerning southern hemisphere emission line objects, including notably the symbiotic stars AR Pav and RR Tel. Without such a basis, later studies would have lacked secure foundations on which to build. One of us (MF) is fortunate to have known Dr. Thackeray personally, and appreciated his quiet unassuming dedication.

We wish to thank the other members of the scientific organizing committee D.A. Allen, A.A. Boyarchuk, H. Nussbaumer, M.J. Plavec, B. Rudak, J. Sahade, J.P. Swings, and A.V. Tutukov for their active help in the scientific preparation of the meeting. J. Sahade kindly provided us with a historical introduction on the symbiotic stars. The light curves sent us by J. Mattei were very useful during the discussions on the individual stars.

We are very grateful to the members of the local organizing committee Y. Andrillat and Ch. Fehrenbach for making perfect arrangements at the Haute Provence Observatory. In addition a large amount of work was performed for us by the technicians and administrative staff of the observatory; we need in particular to thank the secretaries and the kitchen staff. The beauty of the surroundings and the social events, provided an excellent framework for our scientific discussions. All of us very much appreciated the wonderful concert given by the violinist Nicolas Risler in the dome of the 193 cm telescope.

We are greatly indebted to many people and organizations for their financial support: Dr. Petit as scientific director of the earth, ocean, and space sector of the french "Centre National de la Recherche Scientifique", Mr. Delorme as president of the Conseil Géneral des Alpes de Haute Provence, and the Executive Committee of the International Astronomical Union. In addition the directors of the Institut Geografique National, Drs. Valin and Serres allowed some of the participants to be lodged in a hostel of the Institute. We finally thank the presidents of IAU Commissions 27, 29 and 42 for sponsoring the colloquium.

It may be noted that many contributions are based on observations by the International Ultraviolet Explorer (IUE) collected at the Goddard Space Flight Center of the National Aeronautics and Space Administration, and at the Villafranca Satellite Tracking Station of the European Space Agency. Finally Marina Mele provided some very appropriate illustrations of our "scientific" discussions.

Michael Friedjung

Roberto Viotti

SCIENTIFIC ORGANIZING COMMITTEE

D. A. ALLEN, Anglo Australian Observatory, Australia

A. A. BOYARCHUK, Crimean Astrophysical Observatory, U.S.S.R.

M. FRIEDJUNG, Institut d'Astrophysique, Paris, France

H. NUSSBAUMER, Institute of Astronomy, ETH Zentrum, Zürich, Switzerland

M. PLAVEC, University of California, Los Angeles, U.S.A.

B. RUDAK, Nicholas Copernicus Astronomical Center, Warsaw, Poland

J. SAHADE, Instituto de Astronomía y Física del Espacio, Buenos Aires, Argentina

J. P. SWINGS, Institut d'Astrophysique, Liège, Belgium

A. V. TUTUKOV, USSR Academy of Sciences, Moscow, U.S.S.R.

R. VIOTTI (Chairman), Instituto Astrofisica Spaziale, Frascati, Italy

LIST OF PARTICIPANTS

ALTAMORE, A., Roma, Italy (19)

ANDRILLAT, H., Montpellier, France (11)

ANDRILLAT, Y., O.H.P., France (23)

BALLEREAU, D., Meudon, France (15)

BENSAMMAR, S., Meudon, France (12)

BOYARCHUK, A.A., Crimea, U.S.S.R. (14)

CASSATELLA, A., Villafranca, Spain (5)

CIATTI, F., Asiago, Italy (20)

CHOCHOL, D.C., Skalnate Pleso, Czechoslovakia (2)

FEHRENBACH, Ch., O.H.P., France (34)

FIEDEROVA, A.V., Moscow, U.S.S.R. (7)

FRIEDJUNG, M., Paris, France (35)

GOY, G., Geneva, Switzerland (1)

GRAVINA, R., Lyon, France

GRYGAR, J., Rez, Czechoslovakia (4)

HACK, M., Trieste, Italy (31)

HOUZIAUX, L., Liège, Belgium (23)

HUANG, C.C., Nanking, China (13)

IIJIMA, T., Nagoya, Japan

KAFATOS, M., Fairfax, U.S.A. (27)

KEYES, C.D., Los Angeles, U.S.A. (24)

KINDL, C., Zürich, Switzerland (18)

KWOK, S., Ottawa, Canada (17)

McCARTHY, M.F., Vatican City State (9)

MARTEL, M.T., Lyon, France (16)

LIST OF PARTICIPANTS

MICHALITSIANOS, A., Greenbelt, U.S.A. (26)

MIKOLAJEWSKA, J., Torun, Poland (6)

MURATORIO, G., Marseille, France (22)

NUSSBAUMER, H., Zürich, Switzerland (30)

OLIVERSEN, N.A., Madison, U.S.A. (29)

PATRIARCHI, P., Villafranca, Spain (8)

PLAVEC, M.J., Los Angeles, U.S.A. (36)

RUDAK, B., Warsaw, Poland (3)

SIENKIEWICZ, R., Warsaw, Poland (10)

SLOVAK, M.H., Austin, U.S.A. (28)

SWINGS, J.P., Liège, Belgium (21)

VIOTTI, R., Frascati, Italy (32)

WHITELOCK, P.A., Cape, South Africa (25)

1. Goy, G.
2. Chochol, D.C.
3. Rudak, B.
4. Grygar, J.
5. Cassatella, A.
6. Mikolajewska, J.
7. Fiederova, A.V.
8. Patriarchi, P.
9. McCarthy, M.F.
10. Sienkiewicz, R.
11. Andrillat, H.
12. Bensammar, S.
13. Huang, C.C.
14. Boyarchuk, A.A.
15. Ballereau, D.
16. Martel, M.T.
17. Kwok, S.
18. Kindl, Ch.
19. Altamore, A.
20. Ciatti, F.
21. Swings, J.P.
22. Muratorio, G.
23. Houziaux, L.
24. Keyes, Ch,D.
25. Whitelock, P.A.
26. Michalitsianos, A.
27. Kafatos, M.
28. Slovak, M.
29. Oliversen, N.
30. Nussbaumer, H.
31. Hack, M.
32. Viotti, R.
33. Andrillat, Y.
34. Fehrenbach, Ch.
35. Friedjung, M.
36. Plavec, M.J.

ADRESSE AU COLLOQUE

Mes chers Collègues,

C'est pour moi un grand plaisir de souhaiter la bienvenue aux parteciants à ce 70ème Colloque de l'Union Astronomique Internationale sur "La Nature des Etoiles Symbiotiques".

L'Observatoire de Haute Provence, du Centre National de la Recherche Scientifique, accueille avec plaisir les quarante parteciants venus de 13 pays et qui sont tous des spécialistes des problèmes posés par ces étoiles.

Ces étoiles sont difficiles à comprendre et leur nom même de "symbiotique" provient de la première théorie faite pour les expliquer: l'association d'une étoile chaude at d'une étoile froide. Cette théorie est-elle toujours valable, ou bien avons-nous de nouvelles idées révolutionaires?

De nouvelles et importantes observations ont été faites au cours des dernières années, notamment dans l'ultraviolet spatial, dans l'infrarouge, dans le domaine des rayons X, et même dans le domaine visible. Il faut interpréter tous ces documents.

Ceci est le but principal de la réunion de ces spécialistes venant de toutes les parties du monde: des Etats-Unis et du Canada, de la Chine et du Japon, de l'Afrique du Sud et naturellement de l'Europe où sont représentés l'URSS, la Pologne, la Tchécoslovaquie, sans parler de nos proches voisins de l'Europe.

Pourquoi ce colloque a-t-il lieu à l'Observatoire de Haute Provence? Une des raisons est que des nombreux astronomes ont travaillé sur ce sujet. Je rappellerai le mémoire du Professeur J. Dufay, du Professeur Tcheng Mao Lin, de Mlle Marie Bloch qui ont fait de nombreuses publications sur ce sujet.

Ce travail est continué par les astronomes français qui observent ici, soit comme visiteurs, soit comme résidents. Madame Andrillat et moi-même collaborons très activement avec nos collègues belges, M. Pol Swings, puis MM. L. Houziaux, J.P. Swings et J.M. Vreux, avec nos collègues italiens de Trieste, avec M. C.C. Huang de Nankin et M. A. Woszczyk de Torun.

Ce colloque est dédié à la mémoire de notre ami A.D. Thackeray, disparu si tragiquement il y quelques années, mais permettez-moi d'associer à son nom aussi ceux de J. Dufay, Tcheng Mao Lin et Marie Bloch.

                                                Charles Fehrenbach

                                              Directeur de l'Observatoire de Haute Provence

INTRODUCTION

FIFTY YEARS OF SYMBIOTIC STARS

Jorge Sahade
Instituto de Astronomía y Física del Espacio,
Buenos Aires, Argentina
Member of the Carrera del Investigador Científico,
CONICET, Argentina

Fifty years ago, Merrill and Humason wrote a note that was published the following year in the <u>Publications of the Astronomical Society of the Pacific</u> (Merrill and Humason, 1932), where they called attention to the existence of a group of stars —a very small group, then, with only AX Persei, RW Hydrae and CI Cygni, and "possibly" T Coronae Borealis and R Aquarii as members— characterized by the fact that their spectra display titanium oxide absorption bands together with emissions of He II 4686, [O III] 4363 and other nebular lines. The stars in the group were later called "symbiotic stars" by Merrill, on the ocassion of a paper on BF Cygni that he presented before the American Astronomical Society in 1941 (cf. Merrill, 1958), and their spectra were described, also by Merrill, as "combination spectra".

IAU Colloquium No. 70 comes, therefore, at the right time to celebrate such a significant anniversary in the investigation of symbiotic stars. These have been dealt with in a number of colloquia and symposia, but always as a chapter of a more general subject, and this meeting is the first one ever devoted exclusively to discuss them. Consequently, it provides an unvaluable opportunity to assess our present knowledge in the field and its implications and to plan lines for future research. We already have available a large amount of information over a wide wavelength range and, in addition, the space astronomical observations have opened up new possibilities of understanding phenomena connected with the structure of extended envelopes in stars. So, an exchange of ideas and discussions on the problematics of symbiotic stars at this time should prove to be most useful and to have far-reaching effects in our understanding of the symbiotic stars.

Z Andromedae has been always considered to be the prototype of the group because it was the first member whose spectrum was studied in detail. Such a study was undertaken by H.H. Plaskett (1928) at the Dominion Astrophysical Observatory, Victoria, Canada, and published in 1928. Plaskett identified the high excitation lines and concluded that the spectrum originates in an extended atmosphere where the pressure is lower than that of the solar chromosphere. A few years later, Merrill detected, on Plaskett's prints, the molecular (TiO) absorption bands.

The remark by Merrill and Humason led Merrill to start a series of investigations (Merrill, 1932, 1933, 1934, 1941, 1943, 1944, 1947, 1948, 1950a, 1950b) aimed at gathering information on the spectral and radial velocity behavior of a number of symbiotic stars. Merrill's pioneering work on the field, which was done at the Mount Wilson Observatory over a period of several years, has provided much of our knowledge on the spectral changes that symbiotic stars undergo in parallel with changes in their brightness, and has yielded the picture of the broad correlation between light and spectrum that has been stated as follows,

when the star is faint, a giant M spectrum is prominent;

when the star brightens, an early type shell spectrum develops and the continuum dominates the photographic region and covers the M-type spectrum;

when the star declines in brightness, the shell spectrum weakens and emission lines of progressively increasing excitation and forbidden lines develop.

In connection with the light variability of symbiotic stars we should recall that the photometric work done at Harvard suggested that the symbiotic objects are semiregular, long period variables with periods of the order of one to two years. The names of Mrs. Fleming, the Gaposchkins and Mrs. Mayall, among others, are associated with Harvard important published photometric results.

The coming into being of the McDonald Observatory in 1939 with a quartz prism spectrograph attached to the 208-cm reflecting telescope, that permitted the extension of our knowledge of the spectrum farther into the violet than hitherto possible, gave rise to a series of papers by Swings and Struve that were produced in the early 40's and were devoted to the study of peculiar stars. In these investigations we find a large and valuable contribution to the field of symbiotic stars.

It is interesting to quote from the 1940-41 Annual Report of the Yerkes Observatory the following paragraph that states the reasons underlying Swings and Struve's series of papers on peculiar stars (Swings and Struve, 1935, 1940, 1941a, 1941b, 1942a, 1942b, 1943a, 1943b, 1945; Struve, 1940). In that report, Struve (1942) wrote that "one of the pressing problems of astrophysics is to explain the origin and support of extensive gaseous envelopes surrounding otherwise normal stars. This problem is related to that of supporting the solar chromosphere, and its solution is required before we can be certain that we fully understand the structure and the physical properties of the outer layers of a star".

On the other hand, to Merrill (1958) "these bizarre objects

present challenging problems. In addition to their intrinsic interest as peculiar individuals, there is another reason for studying them. The apparently anomalous phenomena which are so conspicuous and so easily open to study in these stars may perhaps be exaggerated or pathological examples of features, which, scarcely noticed, occur in a minor degree in many other stars. Symbiotic stars may thus be strategic objects in which to study phenomena actually of fairly common occurrence. For example, it is possible that studies of symbiotic stars may eventually extend our comprehension of phenomena in normal dwarf stars of type G, e.g., the sun with its mysterious corona".

So, Merrill, as well as Swings and Struve, were attracted by the symbiotic objects not only because these stars were so peculiar in their spectrum, but largely because it was hoped that their study would throw light upon the problems of extended atmospheres in stars. And there is no need to stress again how greatly the three scientists have contributed to our knowledge of the symbiotic stars in the photographic region.

Now a crucial question comes up, namely, which are the criteria that would permit us to decide whether or not an object is a symbiotic star. Originally, Merrill's designation was supposed to single out a group of objects characterized, as we have already said, by the combination, in their spectra, of features of a low temperature object with features that require high excitation conditions. This characterization was certainly not enough because it led to non-homogeneous lists of objects when attempts to produce catalogues of symbiotic stars were made.

In a review paper, Boyarchuk (1969) suggested that an additional criterium be added, namely, that the brightness of the object be variable with an amplitude up to 3 magnitudes and with a period of several years; furthermore, he pointed out that the late-type component should actually be an M giant. Boyarchuk's criterium would leave out recurrent novae such as RS Ophiuchi and slow novae like RR Telescopii, which are generally considered to belong to the group.

The degree of excitation in the extended envelope reaches, in some of the objects, a level that is reminiscent of that of the solar corona. For instance, [Fe XIV] and [A XI] have been observed in RS Oph and in T CrB, while [K XI], [Ca XIII] and [Ni XV] were found in RS Oph. In other objects, however, the excitation reaches much lower levels and one detects lines of [Fe III], [O III] and [Ne III] for instance.

More recently, Allen (1979) has suggested that the criteria for membership be stated as follows: 1) The object must appear stellar; 2a) Emission from ions of greater than 55 eV ionization potential (i.e. He II emission) must at some time have been present; evidence for stellar spectral type G or later must also exist; 2b) In the absence of convincing evidence of a late-type star, the ionization potential represented must at some time have exceeded 100 eV (e.g. [Fe VII] emission).

We need to analyze and discuss thoroughly the criteria that define the symbiotic stars. Then, we will be able to decide whether or not stars like WY Velorum or 17 Leporis or AX Monocerotis, which until rather recently were considered as symbiotic objects, or, as a matter of fact, any other star, actually do belong to the group.

The first catalogue of symbiotic stars, or stars with combination spectra, was a short list included as Table II in Merrill and Burwell's (1933) <u>Catalogue of Be and Ae Stars</u>. Further cataloguing attempts were later due to Bidelman (1954), Payne-Gaposchkin (1957), Boyarchuk (1969), Wackerling (1970) and Allen (1979); the latter list containing 115 entries, 3 of them, Magellanic Cloud objects.

Among the catalogued objects we find recurrent novae like T CrB and RS Oph, definitely binary systems like those in Table 1, taken from Sahade and Wood's (1978) book on interacting binaries, and slow novae like RR Tel, V1016 Cygni, RT Serpentis and HM Sagittae.

Table 1
Symbiotic Stars that are Binary Systems

Star	Period(days)	Spectrum	$f(\mathfrak{m})$
17 Leporis	260	M1 III + B9	0.24
AX Monocerotis	232	gK + B3nn	3.0
T Coronae Borealis	230	M3 III + sd Be	
AR Pavonis	605	M3 III + sd	
AG Pegasi	820	M3 III + ...	0.014

As far back as in 1934, Hogg (1934) suggested that in symbiotic objects we are dealing with binary systems that combine "normal, possibly somewhat variable M giant, and a variable, very high temperature dwarf of the visual magnitude of about +2, which excites a nebular envelope... fainter than normal planetaries". This interpretation is the most generally accepted one and places the group within the evolutionary framework of close pairs (Plavec, 1973; Paczyński and Rudak, 1980). At any rate, since all novae are close binary systems (cf. Sahade and Wood, 1978), it would seem that, to the objects in Table 1, we should add the slow novae and the recurrent novae that are symbiotic objects.

The observations of AR Pavonis that were worked out by Thackeray and Hutchings (1974) suggested that the M star fills its Roche lobe and that there is a stream of matter flowing towards the hot member of the pair. This picture is in line with what we know about mass outflow in interacting binaries but is at variance with Hutchings, Cowley and Redman's (1975) model for AG Pegasi. The picture for AR Pav also agrees with the one advanced by Kuiper (1940, 1941) when he proposed his theoretical interpretation of β Lyrae and pointed out that T CrB, Z And, AX Per, CI Cyg, WY Geminorum probably are cases of instability at the Lagrangian point L_1 of binaries that combine "a giant extending up to L_1, and a dwarf".

If the close binary interpretation of symbiotic stars would hold, then these would represent a certain stage in the evolution of a particular group of double stars and such a stage would be characterized by the particular combination of objects that we have already mentioned. The interaction of these components would give rise to outbursts and to the spectral changes that are observed. Boyarchuk (1966, 1967) tried to assign some figures to the model by attempting to reproduce the spectrum of Z And at different times and concluded that the three sources that contribute to the continuous spectrum are an M giant, a hot companion with a temperature of the order of 10^5 K and a nebula characterized by an excitation temperature of some $17000°$ and an electron density of $n_e \gtrsim 10^6$ cm^{-3}. Boyarchuk also showed that the hot component is responsible for the very large variation in brightness of Z And, and that the companions to the M giants in symbiotic stars should be located below the main sequence in the HR diagram.

IUE observations of symbiotic stars (cf. Sahade and Brandi, 1981) have been partly planned so as to ascertain whether the ultraviolet spectra would confirm or disprove the notion that all symbiotic stars are binaries. So far, the results seem to confirm that we are dealing with binary systems.

One of the symbiotic slow nova objects that have been more thoroughly followed in their behavior as a nova, is RR Tel. Thackeray has published several papers (Thackeray, 1950, 1953, 1955, 1959, 1977; Thackeray and Webster, 1974) and also reproductions of spectra that depict the changes since its outburst in 1946. At the Córdoba Observatory, Landi Dessy and myself have also collected a large amount of material of RR Tel that starts at about the same time as Thackeray's observations. We have, then, available the most valuable set of spectrograms that describe the evolution of a slow nova spectrum that results from the interaction of the two components of the binary system.

Another problem that has been considered since an early date refers to the kind of objects that result from the evolution of the symbiotic stars. Several astronomers have, at one time or the other, suggested that symbiotic stars develop into planetary nebulae. The rationale for the idea lies in the fact that the observations suggest, as we have already mentioned, that there is a sort of a nebula associated with the object or objects (cf. Aller, 1954; Boyarchuk, 1969) and because there are cases which seem to be intermediate between a symbiotic star and a planetary nebula. As we pointed out (Sahade, 1976), to investigate the possibility of the symbiotic stars turning into planetary nebulae "imply to investigate the nature of the central stars of the latter objects". And this field, about ten years old, is not easy because we are then dealing with faint objects; at any rate, information is being slowly gathered and, eventually, we may be able to step on more solid ground regarding this subject.

A few papers have proposed, on different basis, that the symbiotic stars can be sorted out in two different groups. Thus,

Ilovaisky and Wallerstein (1968) suggest two groups depending on whether the excitation of the emission spectrum arises in shock dissipation or in the radiation field;

Webster and Allen (1975) [see also Allen, 1979] find objects with infrared excess that suggests the presence of hot circumstellar dust clouds, and objects without dust emission; moreover,

Sahade and Brandi (1981) have proposed two groups based on the characteristics of the far ultraviolet spectrum: those that display strong emissions of highly ionized species would be in one group, and those that show only very few or no emissions would be in a second group.

Finally, Paczyński and Rudak (1980) have discussed, on the assumption that all of them are binaries, two types of symbiotic stars, in the framework of binary star evolution: type I would correspond to the cases where "the luminosity is produced in a stably burning hydrogen shell", and type II would correspond to the cases where "hydrogen burning proceeds through shell flashes".

The thing is that there is no one-to-one correlation between the groups of the different proposals or suggestions. So, perhaps the apparent observational groupings result from phase effects or perhaps there is no physical connection whatsoever between the criteria used for the classifications. This is another point that may become clarified with further observations at different phases in the spectral evolution of the objects.

I should like to finish this introduction by pointing out the change in the attitude of the astronomers towards the symbiotic stars which is most illustrative of how trends change in Astronomy. When the existence of the group was brought out, its members were considered very bizarre objects, difficult to handle and to understand properly. The contributions by Merrill, Swings and Struve were very important to open up the field and build up information and were complemented by those of other scientists like Mao Lin and Marie Bloch, who observed at Haute Provence, and many others, particularly A.D. Thackeray, in South Africa, and A.A. Boyarchuk, in the Soviet Union.

Research on the symbiotic stars acquired new impetus when it was thought that they could be understood in terms of close binary evolution. But still when the first proposals for the IUE were evaluated, the only proposal for observing the symbiotic stars that was submitted deserved a second category qualification. However, before the first IUE observing runs were over, the proposals containing symbiotic objects

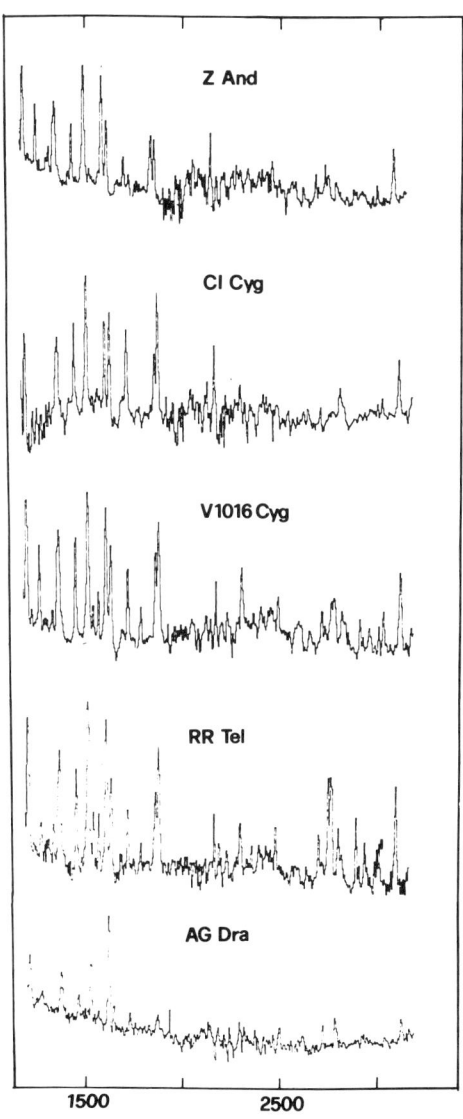

The low resolution ultraviolet spectra of selected symbiotic stars obtained with the International Ultraviolet Explorer (IUE).
Log of dereddened fluxes are given according to $E(B-V) = 0.35$ (Z And), 0.40 (CI Cyg), 0.28 (V1016 Cyg), 0.10 (RR Tel), and 0.06 (AG Dra). The spectrum of CI Cyg was obtained in August 1980 just after eclipse. AG Dra was observed in June 1979 during a minimum phase. (Courtesy of Angelo Cassatella).

began pouring in and now the number of images that have been secured is amazingly large.

So, it would seem that we are now entering a new era in the investigation of symbiotic stars and IAU Colloquium No. 70 will undoubtedly be instrumental in suggesting the pathways to follow for the writing of the new chapter, fifty years after our group of objects was brought to the limelight.

REFERENCES

Allen, D.A. 1979, Changing Trends in Variable Star Research, eds. F.M. Bateson, J. Smak and I.H. Urch (Hamilton, New Zealand: U. of Waikoto), p.
Aller, L.H. 1954, Pub. Dominion Astrophys. Obs. Victoria 9, 321.
Bidelman, W.P. 1954, Astrophys. J. Suppl. 1, 175.
Boyarchuk, A.A. 1966, Astr. Zu. 43, 976.
Boyarchuk, A.A. 1967, Astr. Zu. 44, 12.
Boyarchuk, A.A. 1969, Non-Periodic Phenomena in Variable Stars, ed. L. Detre (Budapest: Academic Press), p. 395.
Hogg, F.S. 1934, Pub. American Astron. Soc. 8, 14.
Hutchings, J.B., Cowley, A.P. and Redman, R.O. 1975, Astrophys. J. 201, 404.
Ilovaisky, S.A. and Wallerstein, G. 1968, Pub. Astron. Soc. Pacific 80, 155.
Kuiper, G.P. 1940, Pub. American Astron. Soc. 10, 57.
Kuiper, G.P. 1941, Astrophys. J. 93, 133.
Merrill, P.W. 1932, Astrophys. J. 75, 413.
Merrill, P.W. 1933, Astrophys. J. 77, 44.
Merrill, P.W. 1934, Astrophys. J. 81, 312.
Merrill, P.W. 1941, Pub. Astron. Soc. Pacific 53, 121.
Merrill, P.W. 1943, Astrophys. J. 98, 336.
Merrill, P.W. 1944, Astrophys. J. 99, 15.
Merrill, P.W. 1947, Astrophys. J. 105, 120.
Merrill, P.W. 1948, Astrophys. J. 107, 317.
Merrill, P.W. 1950a, Astrophys. J. 111, 484.
Merrill, P.W. 1950b, Astrophys. J. 112, 514.
Merrill, P.W. 1958, Étoiles à raies d'émission (Cointe-Sclessin: Institut d'Astrophysique), p. 436.
Merrill, P.W. and Burwell, C.G. 1933, Astrophys. J. 78, 87.
Merrill, P.W. and Humason, M.L. 1932, Pub. Astron. Soc. Pacific 44, 56.
Paczyński, B. and Rudak, B. 1980, Astron. Astrophys. 82, 349.
Payne-Gaposchkin, C. 1957, Galactic Novae, eds. J.H. Oort, M.G. Minnaert and H.C. van de Hulst (Amsterdam: North-Holland Pub. Co.).
Plaskett, H.H. 1928, Dominion Astrophys. Obs. Victoria 4, 119.
Plavec, M. 1973, Extended Atmospheres and Circumstellar Matter in Spectroscopic Binary Systems, ed. A.H. Batten (Dordrecht: Reidel), p. 216.
Sahade, J. 1976, Mém. Soc. R. Sci. Liège, 6e série, 9, 303.

Sahade, J. and Brandi, E. 1981, The Universe at Ultraviolet Wavelengths, ed. R.D. Chapman (NASA CP-2171), p. 451.
Sahade, J. and Wood, F.B. 1978, Interacting Binary Stars (Pergamon Press), p. 114.
Struve, O. 1940, Proc. Nat. Ac. Sci. 26, 458.
Struve, O. 1942, Pub. American Astron. Soc. 10, 201.
Swings, P. and Struve, O. 1935, Astrophys. J. 71, 312.
Swings, P. and Struve, O. 1940, Astrophys. J. 91, 546.
Swings, P. and Struve, O. 1941a, Astrophys. J. 93, 356.
Swings, P. and Struve, O. 1941b, Astrophys. J. 94, 291.
Swings, P. and Struve, O. 1942a, Astrophys. J. 95, 152.
Swings, P. and Struve, O. 1942b, Astrophys. J. 96, 257.
Swings, P. and Struve, O. 1943a, Astrophys. J. 97, 194.
Swings, P. and Struve, O. 1943b, Astrophys. J. 98, 91.
Swings, P. and Struve, O. 1945, Astrophys. J. 101, 224.
Thackeray, A.D. 1950, Mon. Not. R. Astron. Soc. 110, 45.
Thackeray, A.D. 1953, Mon. Not. R. Astron. Soc. 113, 211.
Thackeray, A.D. 1955, Mon. Not. R. Astron. Soc. 115, 236.
Thackeray, A.D. 1959, Observatory 79, 11.
Thackeray, A.D. 1977, Mem. R. Astron. Soc. 83, 1.
Thackeray, A.D. and Hutchings, J.B. 1974, Mon. Not. R. Astron. Soc. 167, 319.
Thackeray, A.D. and Webster, B.L. 1974, Mon. Not. R. Astron. Soc. 168, 101.
Wackerling, L.R. 1970, Mem. R. Astron. Soc. 73, 153.
Webster, B.L. and Allen, D.A. 1975, Mon. Not. R. Astron. Soc. 171, 171.

INTRODUCTORY REPORT ON SYMBIOTIC STARS

A. A. Boyarchuk
Crimean Astrophysical Observatory
U.S.S.R. Academy of Science

The term "Symbiotic Star" was first introduced by P.W. Merrill (1958) and now is used for designation of astronomical objects, whose spectra represent a combination of absorption features of a low temperature star with emission lines of high excitation. At first the group of these objects contained four stars: Z And, BF Cyg, CI Cyg and AX Per (Merrill 1933). Many symbiotic stars were discovered as results of different sky surveys in the last decade. The infrared observations have confermed the presence of a cool component in many emission line objects even if the visual observations do not give evidence for the latter. In the catalogue published by Boyarchuk in 1969 the number of symbiotic stars was 21 as well as 16 probable ones. Recently, Allen (1979) has published a catalogue containing more than 100 objects with complex spectra.

More than ten reviews of Symbiotic Stars were published. The problemes of symbiotic stars were discussed during several colloquia. All of this means that symbiotic stars are important objects of astronomical investigations. Many reports on different problems of Symbiotic Stars will be made at this Colloquium, and I should like to point out only a few aspects, which are rather new and I suppose are important for the understanding of Symbiotic Stars.

New observations give more support for a binary hypothesis of the nature of Symbiotic Stars. Here one should note that infrared observations have shown the presence of cool components in many cases. The observation of several stars in the spectral region 1200-3200 A, which were carried out with IUE, shows that hot components of Symbiotic Stars have a temperature of more than 40000°K.
It was known that many Symbiotic Stars show periodic variations of radial velocities. Recently, eclipses were discovered for at least two stars: AR Pav (Mayall 1937) and CI Cyg (Belyakina 1976).

Of course the direct or indirect evidence of binary nature cannot be given by us for each Symbiotic Star. They need more careful observation.

The IUE observations give more information on the nebula. The semi-forbidden lines give a possibility to determine the physical conditions in the dense parts of the nebula. The electron density varies from 10^{11} to 10^{6} cm^{-3} and at the same time the temperature varies from 40000°K to 15000°K. The spectral index of the radio emission means that the density of the nebula varies as r^{-2}. All of this means that the nebulae of Symbiotic Stars are very inhomogeneous.

Some progress has been made in the theory. The accretion in the binary system has been investigated. Bath (1977) considered accretion-heated white dwarfs or main sequence stars. In these cases a mass transfer of 10^{-3} to 10^{-6} M$_\odot$yr^{-1} is needed in order to explain the light curves. Tutukov and Yungelson (1976) and Paczynski and Rudak (1980) have studied the model of a binary consisting of a red giant and of an accreting hot degenerate CO-dwarf. In this case nuclear reactions are the main source of the energy. A mass transfer of 10^{-7} M$_\odot$yr^{-1} is needed for an explanation of light curves of Symbiotic Stars. In both cases the high temperature must be deep in an accretion disk. The observations of soft X-ray radiation seem to confirm this. The temperature of the X-ray radiation is of the order of a few hundred thousand degrees.
Unfortunately single star models of Symbiotic Stars have not been investigated enough. Especially the relative intensities of emission lines and absorption features have not been considered in any single star model.

Thus Symbiotic Stars are very complicated objects where different physical processes exist. The explanation of the nature of Symbiotic Stars will help us to understand the nature of other types of non-stable stars. I hope that this Colloquium will be important for the investigation of Symbiotic Stars.

REFERENCES

Allen, D.A.: 1979, Proc. IAU Coll. No.46 "Changing Trends in Variable Star Research", edited by F.M. Bàteson, J. Smak, I.H. Urch, University of Waikato, Hamilton, p. 125.
Bath, G.T.: 1977, Mon. Not. R. astr. Soc. **178**, 203
Belyakina, T.S.: 1976, Inf. Bull. Var. Stars No. 1169
Boyarchuk, A.A.: 1969, "Non-Periodic Phenomena in Variable Stars", ed. L. Detre, Academic Press, Budapest, p.395
Merrill, P.W.: 1933, Astrophys. J. **77**, 44
Merrill, P.W.: 1958, "Etoiles à Raies d'Emission", 8th Coll. d'Astrophys.

Liège, p. 436
Paczynski, B., Rudak, B.: 1980, Astron. Astrophys. _72_, 349
Tutukov, A.V., Yungelson, L.R.: 1976, Astrophysics _12_, 531

DISCUSSION FOLLOWING BOYARCHUK

Hack: I have a question about the stars for which you have presented a binary model. What is the evidence for binarity? Are orbital radial velocity curves available? Or is there good evidence of eclipses? How reliable is the observational evidence for binarity?

Boyarchuk: The orbital radial velocity curves were published for BF Cyg, RW Hya, AG Peg and R Aqr. The light curve of CI Cyg shows eclipses. The energy distribution over wide spectral regions for BF Cyg, CI Cyg, AG Peg, Z And and AG Dra also gives arguments in favour of the binarity of these stars.

Friedjung: Though I believe that symbiotic stars are binary, there is a major problem in the interpretation of the radial velocity variations of some of them. A few cases agree very well with a binary conception, but others are harder to interpret in this framework.

Nussbaumer: You gave radial velocities for some symbiotic stars. There seemed to be more or less periodic changes when lines of NIII and HeII were measured, whereas the radial velocities appeared rather constant in the FeII lines. Have you further comments on these differences?

Boyarchuk: I believe that the FeII lines are formed in the outer parts of the cool giant's atmosphere, and therefore the displacements represent the cool component's motion. If the cool component is a normal red giant, its mass is equal to 3–5 $M_\odot$. In this case we shall expect orbital velocities of less than 10 km s^{-1}, for typical periods of about two years. It is necessary to have very accurate measurements of radial velocities in order to obtain a conclusion about the periodic variation in the case of FeII lines.

Slovak: Recent VLA observations at 6 cm by Ghigo and Cohen (1981, Ap.J. _245_, 998) resolved an asymmetric nebula around AG Peg of size 0.2–0.3 arcsec. This size is 2–3 times larger than predicted by a spherically symmetric single component wind model (Wright and Barlow 1975, M.N. _170_, 41) and indicates the complexity of the wind(s) interacting in the system.

Viotti: Let me express a number of "desiderata". I believe that important points necessary for understanding the nature of the symbiotic stars are:
(1) the luminosity classification of the cool spectrum component,
(2) the radii of the M I-II-III ... stars,
(3) radial velocity measurements for long time scales (i.e. many 'periods'),
(4) search for "regular" variability of the cool spectral component, expecially near the energy maximum ($\sim 1\ \mu$),
(5) what is the meaning of the radial velocity curves derived from the emission lines? do they reflect an orbital motion or something else?

McCarthy: I have a comment on item 4 of your list of desiderata. I agree that this field is most important for future photometric and spectroscopic studies. I wish to recall that J. Stebbins and G. Kron stated that in their experience <u>all</u> M giants showed variations of at least 0.1 or 0.2 magnitudes, and that we must suppose M stars to be variable until proven not variable.

SESSION I

NEW OBSERVATIONS OF SYMBIOTIC STARS

Chairmen: R. Viotti and M. Plavec

Introductory reports on:

RADIO (S. Kwok)

INFRARED (D.A. Allen)

NEAR INFARED (Y. Andrillat)

OPTICAL (F. Ciatti)

ULTRAVIOLET (H. Nussbaumer, and M.H. Slovak and D.L. Lambert)

X-RAY (D.A. Allen, and C.M. Anderson, J.P. Cassinelli, N.A. Oliversen, R.W. Myers and W.T. Sanders)

Symbiotic Stars: Cool Star with a Hot Envelope?

RADIO OBSERVATIONS OF SYMBIOTIC STARS

Sun Kwok
Herzberg Institute of Astrophysics
National Research Council of Canada
Ottawa, Canada K1A 0R6

I. INTRODUCTION

One of the defining characteristics of symbiotic stars is the presence of nebular emission lines, which indicates the existence of a circumstellar nebula. Since the excitation and ionization of atomic emission lines depend on the temperature and density of the emission region, line observations (optical or ultraviolet) therefore only selectively probe part of the circumstellar nebule. Radio observations being less sensitive to "hot spots", reflect the global character of the ionized region and are more suitable for determining the density structure of the nebula. Recent advances in the technique of aperture synthesis allow observations to be made with high resolution ($\sim 0\overset{"}{.}1$) and direct mapping of nebular structure is now a possibility. Since the circumstellar nebula is most likely to have originated from the stellar components of the symbiotic system, a better understanding of the nebula may provide important clues to the nature of symbiotic stars.

II. RADIO SURVEYS OF SYMBIOTIC STARS

The search for radio emission from symbiotic stars began in the early seventies. Early detection include V1329 Cygni (Altenhoff and Wendker 1973), V1016 Cygni (Purton, Feldman and Marsh 1973), and R Aqr (Gregory and Seaquist 1974). Systematic surveys have been carried out by Bath and Wallerstein (1976, 15 objects), Gregory and Seaquist (cf. Feldman and Kwok 1979, 17 objects), Wright and Allen (1978, 91 objects) and Kwok (1982, 41 objects). The emission line star survey by Purton et al. (1981) also contained a number of symbiotic objects.

Among the 112 stars catalogued by Allen (1979), approximately 10% are detected in radio wavelengths. The low detection rate is not entirely surprising. If we adopt a binary model for symbiotic stars where mass is being transferred from the cool to the hot component, it can be shown that the binary separation cannot exceed approximately twice the radius of the cool component (assuming a mass ratio of 3).

This implies that the mass-transfer region is no greater than several $\times 10^{13}$ cm; or at a distance of 500 pc, an angular size of $\sim 10^{-2}$ arc seconds. No thermal radio emission can be expected to be detected using present-day instruments.

It is interesting to note that most of the detected objects are not "classical" symbiotic stars but more suitably described as slow novae. This, coincidently, agrees with the separation of symbiotic stars into two classes as described by Paczynski and Rudak (1981), who distinguished between symbiotic stars with quasi-periodic variations (type I) and those with abrupt optical brightenings which last for years (type II). A difference in accretion rates is suggested to be responsible for the separation of the two types. Such distinctions were also noted by Allen (1979), who designated symbiotic stars with dust-like infrared excesses as type D and those without type S. Approximately 20% of the objects in Allen's (1979) catalogue is type D and these correlate well with the radio emitters. Webster and Allen (1975) also noted that [OIII] forbidden lines are stronger in type D objects and this is explained by Allen (1979) that type S object have denser envlopes and the forbidden lines are collisionally de-excited. Mira-like variability is also found to be more common in type D objects (Feast, Robertson and Catchpole 1977), suggesting that the cool component is a Mira. Most of the slow novae are classified as type D.

The brightest objects in the radio are all slow-novae which have undergone recent eruptions: RR Tel (1945), V1016 Cygni (1964), and HM Sge (1975). This seems to suggest that radio emissions are related to an ejection process, resembling the case of classical novae. For this reason, I shall briefly discuss the radio properties of novae and attempt to establish possible links between novae and symbiotics.

III. WIND MODEL OF NOVAE

Since the mechanism of radio emission is believed to be thermal free-free emission (with the exception of RX Pup, Seaquist 1977) radio-emitting slow novae must have circumstellar nebulae hundreds of times larger than those of symbiotic stars. Classical novae have long been known to be radio sources (Hjellming 1974) and one is tempted to seek a connection between novae and slow novae. In fact, Bath (1977) has suggested that symbiotic stars, slow novae and classical novae may all operate under similar physical mechanisms. Disc accretion by the compact companion at rates close to the Eddington limit produces an ultraviolet continuum source which produces a radiatively-driven wind. Variation in the wind mass-loss rate leads to a shifting optically-thick surface which could result in light variation observed in symbiotic stars. He suggests that the difference between novae and slow novae may just be the rate of the decrease in the mass-loss rate.

Bath and Shaviv (1976) and Gallagher and Starfield (1976) have already demonstrated that many observed properties of classical novae (including the visible light curves) can be explained by such a wind model. Since the free-free opacity is higher in radio wavelengths, the optically thick surface of the wind is in fact larger in the radio and the nova envelope can remain optically thick for a longer period of time than in the optical. It would be interesting to test the wind model of novae and compare its predictions with the observed radio "light curves". Furthermore, there is increasing evidence that fast stellar winds ($v \sim 2000$ km s^{-1}) are also present in slow novae (Andrillat and Swings 1982), suggesting that the wind model may also be applicable to slow novae.

Let us assume that after the optical outburst, mass loss rate from the hot companion decreases as a power law:

$$\dot{M}(t) = \dot{M}_o (t_o/t)^\alpha \tag{1}$$

Note that the $\alpha=0$ case corresponds to a pure stellar wind situation and $\alpha=\infty$ corresponds to a sudden ejection. For $\alpha>1$, the total mass of the nova envelope is finite:

$$\Delta M = \int_{t_o}^{\infty} \dot{M}(t) dt = \frac{\dot{M}_o t_o}{\alpha - 1} \tag{2}$$

The density distribution in the nova envelope at any instant is given by:

$$\rho(r,t) = \frac{\dot{M}_o}{4\pi v r^2} \left(\frac{vt_o}{vt-r+r_o}\right)^\alpha \tag{3}$$

where v is the wind velocity and r_o the base radius of the wind. From equation (3), free-free emission from the nova envelope can be calculated at any epoch. Figure 1 shows the radio "light curves" of V1500 Cygni in 3 frequencies (Hjellming et al. 1979). Theoretical curves are fitted with the following parameters: $\Delta M = 5 \times 10^{-5} M_\odot$, $\alpha = 2$, $t_o = 1$yr and $r_o = 10^{11}$ cm. In the rising parts of the light curves where the whole envelope is optically thick the radio flux densities are those given by an expanding blackbody:

$$S_\nu = \frac{B_\nu}{D^2} \pi [v(t-t_o) + r_o]^2 \tag{4}$$

where B_ν is the Planck function. At later time intervals, the optically-thick surface will shrink from the physical edge of the envelope and the spectrum will have an intermediate spectral index (<2 but >0) until the whole envelope becomes optically thin and the declining stage begins.

We can see that qualitative features of the radio light curves can be reproduced. Although the model and its associated parameters may not be unique, it offers a good physical explanation to the radio behavior of classical novae.

IV. INDIVIDUAL OBJECTS

I shall now discuss the radio properties of individuals objects where spectral and/or temporal information is available. Attempts will also be made to interpret the observations in the framework of the above model.

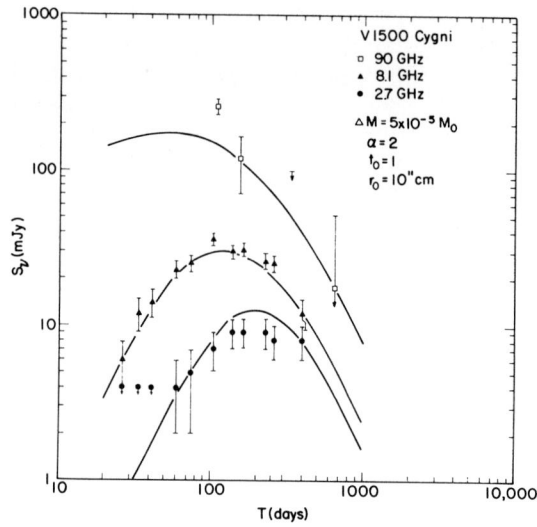

Figure 1. *Model fitting to the radio light curves of V1500 Cygni. Radio data is taken from Hjellming et al. (1979).*

A. V1016 Cygni

The radio spectrum of V1016 Cygni is probably the best measured among all symbiotic stars. The optically-thick part of the spectrum has a spectial index of 0.8 and shows no sign of variation over several years. The $\lambda 2.8$ cm flux density has been monitored continuously since V1016 Cygni's initial detection in 1973 but only marginal evidence of variation is found (Purton *et al.* 1981). Since there is an 8-year gap between the optical outburst and the first radio measurement, we cannot completely rule out the possibility that it has undergone a rise and fall an is the case of classical novae. Nevertheless, it definitely demonstrates the existence of a quiescent component which is incompatible with a rapidly expanding blackbody. Also, the observed radio emission cannot be due to a fast wind from the compact component because the observed flux densities require a mass loss rate $>10^{-3} M_\odot$ yr^{-1} if v>1000 km s^{-1}.

The value of the spectral index, as well as the frequency dependence of the angular sizes (0."35 at $\lambda 6$ cm and 0."19 at $\lambda 2$ cm, Newell and Hjellming 1981), however, indicate that the optically-thick radio emission arise from a wind-like situation. There is evidence that the cool component of V1016 Cygni is an M giant and the presence of the

9.7 μm silicate feature suggests that mass is being lost by the M-giant. When the hot component was "turned on" in 1965, the M-giant wind would have become ionized. Assuming a mass loss rate of $\sim 10^{-5}$ $M_\odot$ yr^{-1} and an ejection velocity of ~ 10 km s^{-1}, the level of free-free emission from the ionized M-giant wind is comparable to the observed value.

Although the above picture adequately explains the quiescent radio emission from V1016 Cygni, the nebula of V1016 Cygni may have a more complicated structure, as evident by the high-resolution (HPBW$\sim$0''.07) λ1.3 cm VLA map obtained by Newell and Hjellming (1981). The map shows a bright rim of size 0''.3×0''.5 with a cavity inside. Weaker emissions (halo) can be seen outside the rim structure. One may identify the halo as the M-giant wind (optically thin at λ1.3 cm) and the bright rim as the result of the interaction between the M-giant wind and fast wind (v$\sim$1400 km s^{-1}, Andrillat et al. 1982) from the hot component, similar to the interacting-winds model proposed for planetary nebulae by Kwok, Purton, and FitzGerald (1978). If this interpretation is correct then the radio emission should become optically thin at progressively lower frequencies (Kwok and Purton 1979). Newell and Hjellming suggest that this is in fact happening: the λ1.3 cm flux density they measured in 1981 is lower than earlier measurements reported by Purton et al. (1981) and the turn-over frequency might have moved to 20 GHz in 1981 from 40 GHz a few years ago.

In summary, there are probably two stellar winds in V1016 Cygni (one from each stellar component) and their interaction may result in the shell structure seen by Newell and Hjellming (1981). The shell is probably moving at relatively low velocity, for after 16 years from the initiation of the new fast wind, the M-giant wind is still the dominant contributor to the radio emissions below 10 GHz. Figure 2 shows a schematic diagram of the interacting-winds model.

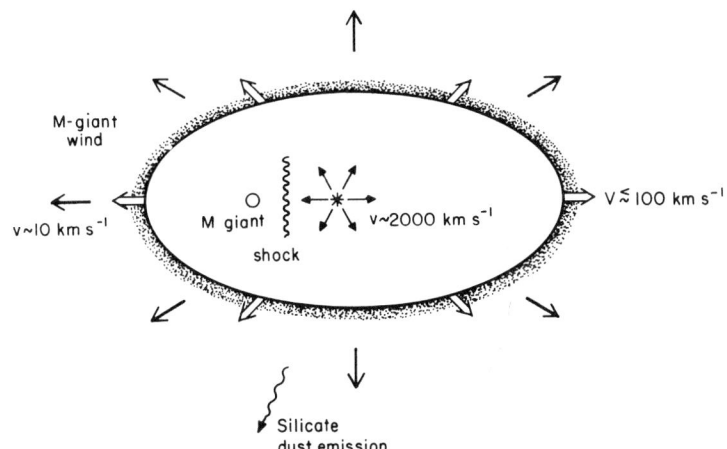

Figure 2. *A schematic diagram of the interacting-winds model for slow novae.*

B. HM Sge

HM Sge had an optical outburst in 1975 and is in many respects similar to V1016 Cygni. Its infrared spectrum shows a strong silicate feature, suggesting the presence of an M-giant wind. A Wolf-Rayet feature of $\sim$2000 km s^{-1} (Wallerstein 1978, Allen 1980) has been seen, which can be attributed to a fast wind originating from the compact companion. HM Sge is easily resolved by the VLA. Observations made in May 1981 by Kwok, Bignell and Purton show the source to be elongated with a position angle of $\sim$40°. The angular sizes (1981) are also found to be dependent on frequency, with major axes of 1".5, 0".43 and 0".39 for 1.4, 5 and 15 GHz respectively.

In contrast to V1016 Cygni the radio spectrum of HM Sge has been undergoing rapid evolution. Figure 3 shows that the 10.6 GHz flux density has been increasing almost linearly at 15 mJy/yr since 1977. Figure 4 shows the spectra of HM Sge in 1977 and 1980. This continued brightening suggests a resemblance to the optically-thick phase of classical novae but the time scale is much longer. Since the spectrum is optically-thick to at least 15 GHz, the expansion velocity of the optically-thick surface can be estimated from the change in flux density. The derived value is of the order of 100 km s^{-1}, far less than the wind velocity of the hot component.

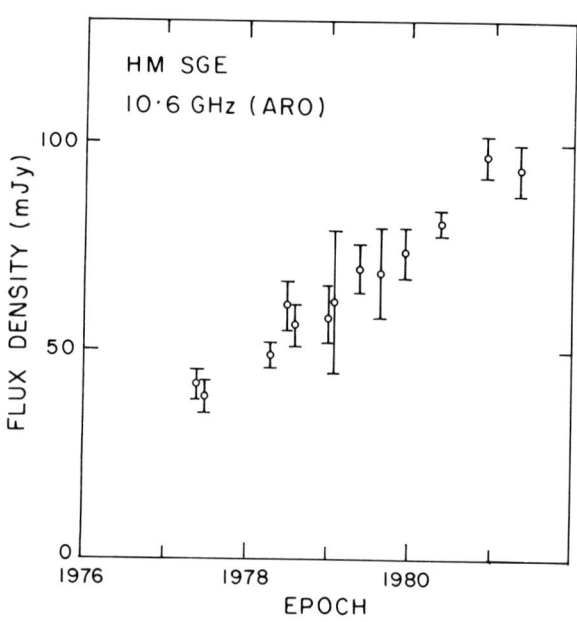

Figure 3. $\lambda 2.8$ cm light curve of HM Sge (Purton, Kwok and Feldman 1982).

It is clear the nova model of §III cannot adequately explain the spectral evolution of HM Sge. If one takes into account the presence of the M-giant wind and adopts the interacting-winds picture of Figure 2, then the optically-thick emission could have originated from $\sim$100 km s^{-1} expanding shell, which is also increasing in mass as more material in the M-giant wind is being swept up. The observed elongated

structure could also
be explained by the
asymmetric expansion
predicted by the
model.

C. V1329 Cygni

V1329 Cygni
(HBV 475) had an
outburst in 1966.
Its [OIII] emission
is much weaker than
V1016 Cygni,
indicating a denser
nebula. Radio
emission (12±4 mJy
at 10.7 GHz) was
first detected by
Altenhoff and Wendker
in 1973. Between
1973 and 1978,
Purton *et al.*
(1981) report an
average flux
density of 14±5 mJy
at 10.6 GHz. In
1978, Hjellming (1981)
measured a flux
density of 5 mJy at
5 GHz using the VLA.
However, a
subsequent
measurement by

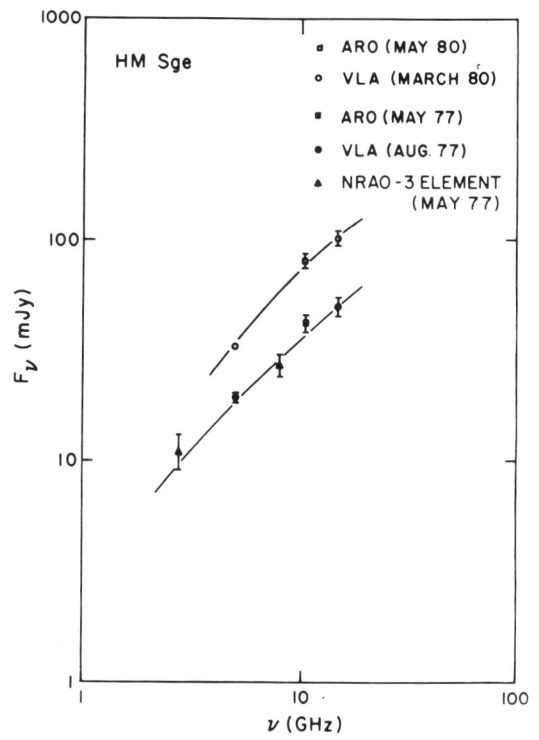

Figure 4. Spectra of HM Sge in 1977 and 1980.

Kwok, Purton and Keenan (1981) found the source to have disappeared
below an upper limit of 1 mJy. This suggests that V1329 Cygni is on its
optically-thin decline phase (similar to novae) and a fast wind
(Tamura 1981) from the compact companion may be responsible for the
formation of the nebula.

D. AG Peg

AG Peg had an optical outburst in 1850 and has been on a steady
visible decline since. Gallagher *et al.* (1979) suggest that the
bolometric luminosity has remained constant and the visible light curve
is best explained by a shrinking photosphere due to a very slowly
decreasing wind. Evidence for a 2000 km s^{-1} wind has been found by
Keyes and Plavec (1980). If the radio emission (13±2 mJy at 10.6 GHz)
found by Gregory, Kwok and Seaquist (1977) arises from this wind then
the mass loss rate is 1.4×10^{-5} (D/kpc)$^{3/2}$ $M_\odot$ yr^{-1}. Recent measurements
by Ghigo and Cohen (1981) at 5 GHz are consistent with a wind-like
spectrum.

E. R Aqr

R Aqr is distinct from the other symbiotic stars for it possesses an extensive nebulosity. The nebula is expanding and is believed to have been ejected 630 years ago. Radio emission from R Aqr has been detected by Gregory and Seaquist (1974) and it is found to be variable on a short time scale, possibly associated with ejection events. VLA observations of R Aqr obtained by Spoka et al. (1981) show an angular size of $\sim 0\rlap{.}''3$. Since this is much smaller than the optical nebula, the radio emission might have resulted from a recent ejection.

F. RR Tel

RR Tel is classified as a type D object and probably contains a Mira variable (Feast, Robertson and Catchpole 1977). Wright and Allen (1978) determine the radio spectral index to be ~ 0.6. Unfortunately not enough temporal coverage is available to determine its possible variations.

V. SUMMARY

The fact that most symbiotic starts do not have detectable radio emission is consistent with the hypothesis that the mass transfer is via Roche lobe and the optical and ultraviolet emission lines originate from the mass transfer region. However, in a number of slow novae, stellar winds from either the cool or the hot components can generate an extended circumstellar envelope from which radio emission arises. It is interesting to note that when an M-giant wind is present, there is no need for the cool component to fill the Roche lobe, and accretion can occur via the M-giant wind. The evolution of the radio spectra of some slow novae is best explained by the interaction of two stellar winds, and this is supported by the line profile analysis by Willson and Wallerstein (1981). It is clear from radio observations that in classical novae the hot-star wind dominates and is normal symbiotic systems wind is not an important element. Slow novae, being intermediate objects, may have their origins tied to the presence of a stellar wind from the cool component.

REFERENCES

Allen, D.A. 1979 in *Changing Trends in Variable Star Research*, ed.
    F.M. Bateson, J. Smak and I.H. Urch, p.125.
Allen, D.A., 1980, *Mon. Not. Roy. Astron. Soc.*, **190**, 75.
Altenhoff, W.J. and Wendka, H.J. 1973, *Nature*, **241**, 37.
Andrillat, Y., Ciatti, F. and Swings, J.P. 1982, *Astrophys. Space Sci.*
    in press.
Bath, G.T. 1977, *Mon. Not. Roy. Astron. Soc.*, **178**, 203.
Bath, G.T. and Shaviv, G. 1976, *Mon. Not. Roy. Astron. Soc.*, **175**, 305.
Bath, G.T. and Wallerstein, G. 1976, *Publ. Astron. Soc. Pacific*, 88, 759.
Feast, M.W., Robertson, B.S.C. and Catchpole, R.M. 1977, *Mon. Not. Roy. Astron. Soc.*, **178**, 499.

Feldman, P.A. and Kwok, S. 1979, *J. Roy. Astron. Soc. Canada*, 73, 271.
Gallagher, J.S. and Starfield, S.G. 1976 *Mon. Not. Roy. Astron. Soc.* 176, 53.
Gallagher, J.S., Holm, A.V., Anderson, C.M. and Webbink, R.F. 1979, *Astrophys. J.*, 229, 994.
Ghigo, F.D. and Cohen, N.L. 1981, *Astrophys. J.*, 245, 988.
Gregory, P.C. and Seaquist, E.R. 1974, *Nature*, 247, 532.
Gregory, P.C., Kwok, S. and Seaquist, E.R. 1977, *Astrophys. J.*, 211 429.
Hjellming, R.M. 1974, in *Galactic and Extra-Galactic Radio Astronomy*, ed. G.L. Verschum and K.I. Kellerman, Springer-Verlag, p.159.
Hjellming, R.M. 1981, paper presented at North American Workshop on Symbiotic Stars, Boulder.
Hjellming, R.M., Wade, C.M. Van denberg, N.R. and Newell, R.T. 1979, *Astron. J.*, 84, 1619.
Keyes, C.D. and Plavec, M.J. 1980, in *Close Binary Stars*, ed. M.J. Plavec, D.M. Popper and R.K. Ulrich, p.535.
Kwok, S. 1982, in preparation.
Kwok, S. and Purton, C.R. 1979, *Astrophys. J.*, 229, 187.
Kwok, S., Purton, C.R. and FitzGerald, P.M. 1978, *Astrophys. J. (Letters)*, 219, L125.
Kwok, S., Purton, C.R. and Keenan, D.W. 1981, *Astrophys. J.*, in press.
Newell, R.T. and Hjellming, R.M. 1981, paper presented at the North American Workshop on Symbiotic Stars, Boulder.
Paczynski, B. and Rudak, *Astron. Astrophys.*, 82, 349.
Purton, C.R., Feldman, P.A. and Marsh, K.A. 1973, *Nature*, 245, 5.
Purton, C.R., Kwok, S. and Feldman, P.A. 1982, in preparation.
Purton, C.R., Feldman, P.A., Marsh, K.A., Wright, A.E. and Allen, D.A. 1981, *Mon. Not. Roy. Astron. Soc.*, in press.
Seaquist, E.R. 1977, *Astrophys. J.*, 211, 547.
Spoka, R.J., Dwek, E., Zuckerman, B., Michalitsianos, A. and Hobbs, R. 1981, paper presented at the North American Workshop on Symbiotic Stars, Boulder.
Tamura, S. 1981, preprint.
Wallerstein, G. 1978, *Publ. Astron. Soc. Pacific*, 90, 36.
Webster, B.L. and Allen, D.A. 1975, *Mon. Not. Roy. Astron. Soc.*, 171, 171.
Willson, L.A. and Wallerstein, G. 1981, paper presented at the North American Workshop on Symbiotic Stars, Boulder.
Wright, A.E. and Allen, D.A. 1978, *Mon. Not. Roy. Astron. Soc.*, 184, 893.

DISCUSSION ON RADIO OBSERVATIONS

Cassatella: There is no sign of such high velocities in the ultraviolet IUE spectra of V1016 Cyg and HM Sge or probably in the optical. This seems to be in contradiction with the high expansion velocity (about 1400 km s^{-1}) derived from the radio observations, unless these UV lines are not formed around the compact object.

Kwok: The 1400 km s^{-1} expansion velocity is taken from the H$_\alpha$ line width observed by Swings and Andrillat.

Friedjung: Firstly I object strongly to the term "slow novae" used for objects like V1016 Cyg that are rather different from ordinary novae. I also have a question. Can you give an upper limit to the mass loss rate from ordinary symbiotic stars, which do not show radio emission? All normal red stars have winds, and it would be useful to have upper limits to the mass loss rate.

Kwok: Not all normal red stars have winds. From infrared data of Gehrz and Woolf (1971 Ap.J. 165, 285) only red giants with spectral type later than M2-3 show wind characteristics. However for a fully ionized wind to be detectable in the radio (1mJy at 5 GHz), $\dot{M}/v$ has to be $> 10^{-8}$ $(M_\odot yr^{-1})/(km\ s^{-1})$ assuming $D \approx 1$ kpc.

Kafatos: Is the shock arising from the colliding winds stationary, and what is the associated temperature? I would like also to comment that in the IUE range P Cygni profiles are generally not seen either in the hot lines or in the cool (e.g. MgII) lines of symbiotic spectra. Some exceptions are known (e.g. RX Pup), but they are rare.

Kwok: The equilibrium expansion velocity of the shock front depends on the relative strength of the two winds (see Kwok et al 1978 Ap.J. 219, L125).

Keyes: AG Peg is the exception to the correlation between radio emission and Allen D-type characteristics. Please comment on this.

Kwok: Radio emission from AG Peg is consistent with a wind of $\sim 10^{-8}$ $M_\odot yr^{-1}$ at a speed of ~ 2000 km s^{-1}. Therefore it can be explained by a nova-like decaying wind model. The radio emitting nebula is probably not formed by an M-giant wind.

INFRARED STUDIES OF SYMBIOTIC STARS

David A. Allen
Anglo-Australian Observatory

(read by J. P. Swings)

ABSTRACT

Infrared photometry and spectroscopy of symbiotic stars is reviewed. It is shown that at wavelengths beyond 1μm these systems are generally dominated by the cool star's photosphere and, indeed, are indistinguishable from ordinary late-type giants. About 25% of symbiotic stars exhibit additional emission due to circumstellar dust. Most of the dusty systems probably involve Mira variables, the dust forming in the atmospheres of the Miras. In a few cases the dust is much cooler and the cool component hotter; the dust must then form in distant gas shielded from the hot component, perhaps by an acccretion disk.

Spectroscopy at 2μm can be used to spectral type the cool components, even in the presence of some dust emission. Distances may thereby be estimated, though with some uncertainty.

Spectroscopy at longer wavelengths reveals information about the dust itself. In most cases this dust appears to include silicate grains, which form in the oxygen-rich envelope of an M star. In the case of HD 330036, however, different emission features are found which suggest a carbon-rich environment.

EARLY DAYS

It was Jean Pierre Swings who first suggested to me that symbiotic stars would merit study at infrared wavelengths. Our paper (Swings and Allen 1972) demonstrated that these systems are dominated in the 1-4μm region by the cool giant star, in most cases known to be present from optical work. The subject did not seem to merit further attention at the time, our interests then lying with stars shrouded in circumstellar dust. The only such object amongst the symbiotic stars was RX Pup, and this was soon shown by Sanduleak and Stephenson (1973) to be in a low-excitation state. It seemed reasonable to argue that the high excitation of the classical symbiotic stars prevented the formation of dust, and that RX Pup was proof that dust was ready to form within such systems whenever the destructive ultraviolet source was (by some means) extinguished.

Within the year Swings was following up my infrared photometry in the southern hemisphere by optical spectroscopy of objects exhibiting dust emission between 1 and 4µm. We had confidently expected to find more low-excitation Fe II and [Fe II] emission stars by this means. However, he wrote "I haven't found [Fe II] yet, but [Fe VI]". Swings' spectra did not at the time reveal the symbiotic nature of the [Fe VI-VII] stars, so the fallacy of my argument remained unproved.

In the mean time, Glass and Webster (1973) were applying infrared photometry to some southern symbiotic stars, with results that differed from our (mostly northern) survey. In particular, they found a sizeable infrared excess in RR Tel. It was not at first clear how to explain the infrared data on RR Tel; it must be remembered that at that time no optical evidence had accrued to suggest the presence of a cool star in the system. We now know that a combination of an M giant and dust emission produces the energy distribution found by Glass and Webster.

From these beginnings it was but a small step for Webster and Allen (1975) to show that indeed symbiotic stars divide into two groups - those in which the 1-4µm continuum shows only the presence of a cool star (type S), and those in which dust emission dominates (type D).

THE TWO TYPES

This outwardly esoteric classification appears to be of some significance: several correlations exist between it and the optical and radio properties. Specifically, D-type symbiotic stars generally show evidence for more extended, ionized gaseous envelopes which provide weak radio emission together with an environment of low-enough electron density to allow a rich spectrum of forbidden lines to form. Additionally, where forbidden-line spectra are seen in S-type symbiotic stars, the lines tend to be only of high excitation, suggesting that many S-type objects are density bounded, whilst D-type objects are radiation bounded, at least in some parts.

Additionally, there seems to be a significant reduction in the proportion of D-type symbiotic stars near the galactic centre. This may, however, reflect observational selection. Most symbiotic stars in that part of the sky are near the limit of the objective-prism surveys by which they have been discovered. The additional effect of circumstellar absorption expected in the D-types may render such stars undetectable.

Finally, correlations with spectral type and variability will be discussed below.

TWO-MICRON SPECTRA

In most of the D-type symbiotic stars the presence of an M star is not easily inferred from the optical data. Skeptics thus argued (and sometimes still do) that these systems might not after all be symbiotic. To overcome these arguments, two-micron spectroscopy has proved powerful. Within the 2.0-2.5μm atmospheric window lie not only the steam bands, which depress the continua of M giants towards both ends of the window, but also the CO band heads. The CO spectral break at 2.3μm is a certain indicator of the presence of a cool star. By spectroscopy at these wavelengths, cool giants were shown to be present in the D-type objects He 2-38 and RR Tel (Allen *et al* 1978), and RX Pup (Barton, Phillips and Allen 1979).

Two-micron spectroscopy has since been used (Allen 1980a) to spectral type the cool components of symbiotic systems. In only the stars He 2-104, H1-36 and W16-312 is there doubt about the existence of a cool giant. The distribution of spectral types differs markedly from that in the field (Allen 1980a), in the sense that the symbiotic stars are biassed heavily towards the coolest M stars. If the symbiotic stars are interacting binaries, this observation is naturally accounted for by the greater propensity for mass loss (and hence mass transfer) amongst such late giants. Unfortunately, a precise luminosity classification cannot be established from the present data, so attempts to derive distances are somewhat at risk. Better results would probably be obtained by spectroscopy in the 0.7-1.0μm region. From the presently determined distances it would seem that we are sampling as far afield as the galactic centre in many cases. This allows a preliminary estimate that there are 10^3 symbiotic stars in the Galaxy.

Amongst stars of known spectral type, circumstellar dust is found primarily in those of type G or later than M6. There is no instance of a D-type system of spectral type from K0 to M2.

VARIABILITY

Hyland has extensive but unpublished data on RX Pup which show the star to vary like a Mira. However, it was Feast, Robertson and Catchpole (1977) who first pointed out the variability by up to 2 magnitudes of the D-type symbiotics (including RX Pup) at infrared wavelengths. Before them, Harvey (1975) had suggested a periodicity of about 450 days in the D-type object V1016 Cygni, and subsequently variations in either the infrared or the optical red continuum have been seen in several other D-types. To date no corresponding change has been recorded in any S-type system, though intercomparison of various observers' data suggests changes of order 0.3 mag in some instances, and possibly more in AX Per.

It would be reasonable to infer that most D-type symbiotic stars involve Mira variables. Indeed, such stars possess a penchant for shrouding themselves in dust. However, it must be cautioned that the infrared observations are not yet so extensive as to allow an unequivocal classification of the type of variability. Perhaps the only convincing example is RR Tel, for which a pre-outburst period is known. A series of observations being accrued at the South African Astronomical Observatory (Feast, private communication) appears entirely consistent with the same period persisting, now almost 40 years after the outburst.

For the present it is sufficient to note that the D-type symbiotic stars appear (from the correlations already noted) to have shed more extensive circumstellar shells than the S types, and that the expulsion of material is more likely from variable (expecially Mira) stars.

DUST TEMPERATURES

There has been extraordinarily little work at longer infrared wavelengths. It must be admitted that the majority of symbiotic systems will be faint at, say, 10μm unless they possess thermal emission from dust. Nonetheless, there would be some interest in determining whether the simple classification scheme based on photometry at short wavelengths persists to 10μm. This is especially the case in view of the correlations between the infrared classification and other properties. At the time of my compilation of a catalogue of symbiotic stars (Allen 1979), about half had been classified only on the basis of their H-K (1.65-2.2μm) colours. The gradual accumulation of data at longer wavelengths has improved the situation, and in some cases the classification has been revised. I have secured deep 3.8μm photometry in a number of symbiotic stars, and longer-wavelength data on a few S-type systems have been presented by Bopp (1981) and, long ago, by Woolf (1973). For the present I discount Woolf's unconfirmed observation of a 20μm rise in Z And. Observations out to 10 or 20μm have been made by a number of authors of the more easily studied D-types.

The available photometry suggests the following range of dust temperatures:

(i) The classic D-type systems such as RR Tel, He 2-38 radiate a significant amount of energy (possibly over 50%) as thermal dust emission. The hottest dust is at about 800-1000K, but in most cases there appears to be some additional contribution from cooler material.

(ii) A small number of D-type systems is known in which the dust temperature is perhaps half as great. In these there is little or no indication of dust emission at 2.2μm, but a large colour index is seen from 2.2 to 3.8μm. Without exception these are the 'yellow symbiotic stars' defined by Glass and Webster (1973) - i.e. those which have cool components of type F-G. I propose that these be distinguished as type D'.

(iii) A very small amount of silicate-like dust may be present around some S-type objects. The only examples of this are R Aqr (Stein *et al* 1969), and CH Cygni in which Bopp (1981) found a weak

silicate feature at 10 and 20μm. Both are very late-type stars with
relatively low-excitation emission spectra. The temperature of the
silicate dust in these stars is unknown. In CH Cyg it is superimposed
on a continuum which is slightly too red for the 2800K star in the
system, and which may represent a very weak black-body dust component.

We now see that not only is there a bimodal distribution of spectral
types for which dust emission occurs, but that the dust emission is
itself distinguished between the two ranges of spectral type. Apparently
dust in the G-type stars is of a different nature, or lies in a different
location relative to the system.

A COMPENDIUM OF OBSERVATIONS

Table 1 lists the available photometry of symbiotic stars at J, H,
K, L, L' and N. The effective wavelengths of the first five filters
are 1.20, 1.65, 2.20, 3.5 and 3.8μm. The data at N represent a broad
10μm filter, though in many cases the quoted value is derived somewhat
loosely from narrow-band data. It should be noted that the present
definition of the J filter is that used at the Anglo-Australian Observatory
and the observatories in Hawaii: it appears to be indistinguishable
from Johnson's originally defined J filter (Johnson 1965), despite the
different quoted effective wavelengths. The following transformations
have been applied to other J data:

$(J-H)_{AAO} = 1.09 (J-H)$ for Cal Tech, Kitt Peak photometry
$(J-H)_{AAO} = 1.07 (J-H)$ for South African photometry

The objects included in Table 1 are, in the main, listed in Allen
(1979). A few have been added:

AS 289 and AS 316, inadvertently omitted from the original. The
1950 coordinates of these stars are, respectively, 18 09 34.7 -11 40 55,
18 39 33.4 -21 20 46;

BI Crucis (Henize and Carlson 1980);

An innominate object discussed by Carter and Feast (1979);

EG Andromedae, shown to be of high excitation by IUE spectroscopy
(Stencel and Sahade 1980);

UV Aurigae, HD 149427, CH Cygni and R Aquarii, which are of lower
excitation but are probably closely related objects.

In addition to the photometry, spectral types of the cool components have been listed. Those derived at 2μm are used only where no
adequate optical determination exists.

TABLE 1   INFRARED DATA ON SYMBIOTIC STARS

Object	IR Type	K	J-H	H-K	K-L	K-L'	N	Refs	Spectral Type	Refs
EG And	S	2.4 to 2.7		0.10	0.1		2.2	B,IRC,SA	M2	W2
SMC S18	D	11.09	0.36	1.00				U	G?	Fig.1
AX Per	S	5.6 to 6.5	0.87	0.25	0.42			Sz	M5	Bk
M1-2	D'	9.81		0.28	1.6		4.0	A1,CB,SA,GKS	G2	O
UV Aur	S	2.13			0.3		0	IRC,W	N	NB
LMC -	D	12.8	0.6	1.8				A6		
LMC S63	S	11.4		0.3				A6	R	A4
Wray 157	D'	9.40	0.73	0.35		1.19		U	G	A5
RX Pup	D	2.1 to 3.1	1.2	1.1	1.4	1.8		FRC,GW,SA,U	M5	A5
Hen 160	S	7.48	1.02	0.41		0.39		U	M7	A5
AS 201	D'	9.93	0.29	0.33		0.98	4.9	CB,U	G	A5
He 2-38	D	4.0 to 5.6	1.4	0.9	1.2	1.5		AG1,FRC,U	M	A4
SS 29	S	10.6 var	0.91	0.28		0.1		U	G	U
SY Mus	S	4.68	1.07	0.32	0.18			FRC,GW,U	M2	SS
BI Cru	D	4.7 to 5.2	1.6	1.4	1.4			A2,U	M	A2
He 2-87	S	5.98	1.60	0.70	0.32			A1,AG1	M7	A5
Hen 828	S	7.12	1.06	0.35				U	M6	A5
SS 38	D	5.7 to 6.5	2.0	1.6		0.2		U	M	A5
Hen 863	S	8.51	0.91	0.19		2.3		U	K4	A5
Hen 905	S	8.47	1.06	0.34				U	K4	A5
RW Hya	S	4.70	0.99	0.17	0.16			FRC,GW,SA	M2	M
Hen 916	S	7.86	1.17	0.32				U	M6	A5
He 2-104	D	6.80	2.04	1.76	2.06			A1,AG1,U		
He 2-106	D	5.5 var	3.6	1.9	2.1			A1,AG1	M	A4
BD-21°3873	S	7.20	0.83	0.19		0.18		U	G	A5

INFRARED STUDIES OF SYMBIOTIC STARS

Object	IR Type	K	J-H	H-K	K-L	K-L'	N	Refs	Spectral Type	Refs
He 2-127	D	7.9 to 8.3	1.2	0.8		1.1		U	M7	A5
Hen 1092	S	7.67	1.09	0.29				AG1,GW	K5	A5
Hen 1103	S	8.36	0.97	0.31				U	M0	A5
HD 330036	D'	7.57	1.00	0.54	1.7		0.7	AG1,GW,U	F-G	U,We
T CrB	S	4.82	0.99	0.23	0.14		4.2	FG1,GKL	M3	K
AG Dra	S	6.4		0.08	0.2		6	B,SA	K1	Bk,R
He 2-147	D	4.3	1.32	0.72	0.9			A1,AG1,U	M8	A5
UKS-Ce 1	S	12.6		0.1:				U	R	LA
Wray 1470	S	7.81	0.98	0.34				U	M4	SS
He 2-171	D	6.3 to 7.2	1.8	1.5	1.8		2.2	AG1,U	M	A4
Hen 1213	S	6.72	1.08	0.29		·0.19		U	K4	A5
He 2-173	S	6.78	1.48	0.32				A1,AG1	M	A4
HD 149427	D'	10.33	0.10	0.40		1.23		AG1,GW,U	A-F	We
He 2-176	D	5.6 var	1.5	0.70	0.6			A1,AG1	M7	A5
Hen 1242	S	6.06	0.90	0.33	0.16			A1,AG1,AS,GW	M6	A5
AS 210	D	6.7	1.8	1.4				U	G?	A4
HK Sco	S	7.96	1.13	0.29				U	M1	A5
CL Sco	S	7.85	0.93	0.22				U	K5	A5
V455 Sco	S	5.92	1.19	0.45	0.33			AG1	M6	A5
Hen 1341	S	7.58	0.98	0.34				U	M0	A5
Hen 1342	S	8.43	1.01	0.31				U	M2	A5
AS 221	S	7.60	1.27	0.57				AG1,U	M4	A5
H2-5	confused								M	A4
Th 3-7	S	8.05		0.50				AG2	M	A5
Th 3-17	S	8.18		0.20				AG2	M3	A5
Th 3-18	S	8.03	1.3	0.35				AG1	M2	A5
Hen 1410	D?	8.41	0.8	0.67				AG1	M3	A5
V2116 Oph	S	8.10	1.66	0.75				GF	M6	A5,DMB
Th 3-30	S	8.30	1.34	0.51				U	K5	A5
Th 3-31	S	7.57	1.28	0.46				AG1	M5	A5

Object	IR Type	K	J-H	H-K	K-L	K-L'	N	Refs	Spectral Type	Refs
M1-21	S	7.21	1.22	0.42				AG1	M2	A5
Pt-1	S	8.58	1.26	0.42				U	M1	TP
RT Ser	S	7.00	1.11	0.42	0.23			FG1,U	M6	A5
AE Ara	S	6.26	1.02	0.29	0.4			AG1	M2	A5
SS 96	S	6.39	1.35	0.46	0.55			AG2	M2	A5
UU Ser	S	9.12	1.04	0.39				U	M	A4
AS 239	D	7.5 to 8.2	1.6	0.9	1.0			FG2	M8	A5
(SSM)	S	8.28	1.34	0.51				CF	M	CF
Hen 1481	D	7.85	1.2	0.51				AG2	M7	A5
H1-36	D	8.06		2.2	2.53		0.6	AG2,U		
W16-312	D	8.02	2.4	1.8				U		
RS Oph	S	6.58	0.70	0.35	0.4		4.8	FG1,GKL,Sz,SA	M0	SS
AS 245	S	7.19		0.50	0.50			U	M6	A5
He 2-294	S	8.40	1.29	0.41				U	M3	A5
B1 3-14	S	8.74	1.22	0.54				U	M6	A5
B1 L	S	7.81	1.46	0.38				AG2	M6	A5
AS 255	S	8.43	0.99	0.30				U	K3	A5
V2416 Sgr	S	4.55	1.49	0.57	0.35			A3	M5	A5
SS 117	S	7.07	1.18	0.50				AG2	M6	A5
Ap 1-8	S	7.9	1.4	0.4				AG2	M0	A5
SS 122	D?	6.63	1.29	0.69				U	M7	A5
AS 270	S	5.53	1.26	0.47				U	M1	A5
H2-38	D	6.67	0.89	0.90	1.02		3.5	AG2,U	M8	A5
SS 129	S	8.03	1.03	0.27				U	K	A4
V615 Sgr	S	7.58	1.03	0.30				U	M	A4
Hen 1591	D	8.96	0.86	0.72		1.54		AG2,U	G	A5
AS 276	S	8.05	0.97	0.29		0.35		U	M4	A5
Ap 1-9	S	8.8		0.3				AG1	K4	A5
AS 281	S	6.95	1.09	0.31				AG1	M5	A5
V2506 Sgr	S	8.42		0.5				AG1	M	A4

Object	IR Type	K	J-H	H-K	K-L	K-L'	N	Refs	Spectral Type	Refs
SS 141	S	8.97	0.99	0.29				U	M	A4
AS 289	S	5.03	1.23	0.54				U	M3	SS
Y CrA	S	6.54	0.98	0.35		0.42		U	M5	A5
V2756 Sgr	S	7.76	0.98	0.27				AG1	M2	A5
CnMy 17	S	7.50	1.14	0.28				AG1	M3	A5
YY Her	S	7.87		0.15				SA	M2	H
He 2-374	S	6.43	1.23	0.44				U	M	A4
AS 296	S	4.50		0.34	0.34			SA	M5	SS
AS 295B	confused							U	M	HH
AR Pav	S	7.16	0.99	0.25				GW	M3	TH
Hen 1674	S	7.67		0.27				U	M5	A5
He 2-390	D	7.66			2.2		2.2	AG1,U	M	A4
V3804 Sgr	S	7.30		0.45				U	M6	A5
V443 Her	S	5.34		0.29	0.14			SA	M3	TG
AS 304	S	7.60		0.33				U	M4	A5
V2601 Sgr	S	8.03		0.27				U	M5	SS
AS 316	S	7.76		0.31				U	M	U
MWC 960	S	7.84		0.19				U	M0	A5
AS 327	S	8.52		0.16				SA	M	A4
FN Sgr	S	7.85	1.06	0.24				GW,SA	M4	A5
Pe 2-16	S	8.09		0.50				AG1	M5	A5
V919 Sgr	S	7.20		0.24				U	M1	A5
CM Aql	S	7.64	1.10	0.45				U	Var	A4,A5,H2
AS 338	S	7.54	0.95	0.35	0.1			A3	M5	A5
BF Cyg	S	6.33	0.87	0.31	0.47			Sz,SA	M5	Bk
CH Cyg	S	-0.6 to -0.8	0.96	0.39	0.5		-2.6	B,GMS,IRC,LPV,Sz,SA	M6	W1
Hen 1761	S	5.55	1.10	0.17	0.17			GW	M3	A5,T
HM Sge	D	3.6 to 4.4	2.0	1.8	1.9		-1.5	B,DHM,M',P	M	DHM
AS 360	S	7.06	1.03	0.39				SA,U	M6	A5
CI Cyg	S	4.45	0.89	0.30	0.41			Sz,SA	M5	Bk

	IR								Spectral	
Object	Type	K	J-H	H-K	K-L	K-L'	N	Refs	Type	Refs
V1016 Cyg	D	4.8 to 6.0		1.8	2.3		-0.2	A1,B,H,SA,U	M3	Bk
RR Tel	D	3.7 to 4.7	1.0	0.8	1.2		0.4	FRC,G,GW,U	M5	A5
He 2-467	S	9.41	0.82	0.23		0.11		A3,U	G	L
V1329 Cyg	S	6.88	1.17	0.48	0.3			A3	M4	AH
CD-43° 14304	S	7.60	0.88	0.19	0.0:			U	K3	A5
V407 Cyg									M	Bk
AG Peg	S	3.91	0.96	0.25	0.23			FRC,GW,Mz,Sz,SA	M3	Bk,CS
Z And	S	5.00	0.91	0.24	0.3			Sz,SA,W	M2	Bk
R Aqr	D	-1.2			0.5		-3.8	IRC	M7	M

References in table

A1 Allen (1973)
A2 Allen (1974a)
A3 Allen (1974b)
A4 Allen (1979)
A5 Allen (1980a)
A6 Allen (1980b)
AG1 Allen and Glass (1974)
AG2 Allen and Glass (1975)
AH Andrillat and Houziaux (1976)
B   Bopp (1981)
Bk Boyarchuk (1970)
CB Cohen and Barlow (1974)
CF Carter and Feast (1979)
CS Cowley and Stencel (1973)
DHM Davidson, Humphreys and Merrill (1978)
DMB Davidsen, Malina and Bowyer (1977)
FG1 Feast and Glass (1974)
FG2 Feast and Glass (1980)
FRC Feast, Robertson and Catchpole (1977)
G   Gehrz et al (1973)

GF	Glass and Feast (1973)
GKL	Geisel, Kleinmann and Low (1970)
GKS	Gillett, Knacke and Stein (1971)
GMS	Gillett, Merrill and Stein (1971)
GW	Glass and Webster (1973)
H	Harvey (1975)
H1	Herbig (1950)
H2	Herbig (1960)
HH	Herbig and Hoffleit (1975)
IRC	Neugebauer and Leighton (1969)
K	Kraft (1958)
L	Lutz et al (1976)
LA	Longmore and Allen (1977)
LVP	Luud, Vennik and Pehk (1978)
M1	Merrill (1950a)
M2	Merrill (1950b)
M'	Merrill (1977)
Mz	Mendoza (1972)
NB	Nassau and Blanco (1954)
O	O'Dell (1966)
P	Puetter et al (1978)
R	Roman (1953)
S	Stein et al (1969)
SA	Swings and Allen (1972)
SS	Sanduleak and Stephenson (1973)
Sz	Szkody (1977)
T	Thackeray (1954)
TG	Tifft and Greenstein (1958)
TH	Thackeray and Hutchings (1974)
TP	Torres-Peimbert, Recillas-Cruz and Peimbert (1980)
U	Unpublished data, D.A. Allen
W	Woolf (1973)
W1	Wilson (1942)
W2	Wilson (1950)
We	Webster (1966)

DISCUSSION OF THE PHOTOMETRY

The following emission mechanisms might be expected to be active in this spectral region:

(i) The photosphere of the cool giant

(ii) Any dust at temperatures of a few hundred Kelvin

(iii) Free-free emission

(iv) Such esoterica as cyclotron radiation from a magnetic accretion column.

The last of these is ignored: there is no evidence for a contribution from any such mechanisms. The possibility of such effects in an interacting binary environment should, however, be borne in mind. Free-free emission is also unlikely to be a significant contributor, as may be estimated from the intensity of Hβ or from the radio flux and spectrum (by a gross extrapolation). However, optically thin free-free emission may contribute a little at 10μm (e.g. in CH Cyg ?), for its intensity relative to the cool star increases roughly as the square of wavelength.

Figure 1 is a plot of the J-H/H-K colour indices for symbiotic stars, together with various diagnostic lines for cool giants with and without dust. It can be seen that the observations agree well with expectations. Note, however, that the Magellanic Cloud stars have colours suggestive of hotter stars with cooler dust than their galactic equivalents.

In theory the JHK photometry, used in conjunction with a good spectral type, could yield an estimate of the interstellar extinction for S-type systems. It is likely, however, that the derived value is too unreliable to be of value.

SPECTRA AT 3 AND 10 MICRONS

The best-known infrared spectral feature in cool stars is the 'silicate bump', a broad emission band centred near 10μm and common in oxygen-rich systems. This is most reliably identified at spectral resolutions of 50 or more, and has been recorded in R Aquarii (Stein et al 1969), HM Sagittae (Puetter et al 1978) and V1016 Cygni (Aitken et al 1980). Photometry through narrow-band filters can also suggest the presence of the silicate band, as in the case of CH Cygni by Bopp (1981). A recent spectroscopic survey by Allen, Aitken and Roche (in preparation) shows most D-type symbiotic stars to have silicate emission, and thus to be oxygen-rich.

In carbon-rich systems the silicate feature is not expected to form. Rather, emission due to silicon carbide is expected around 11μm.

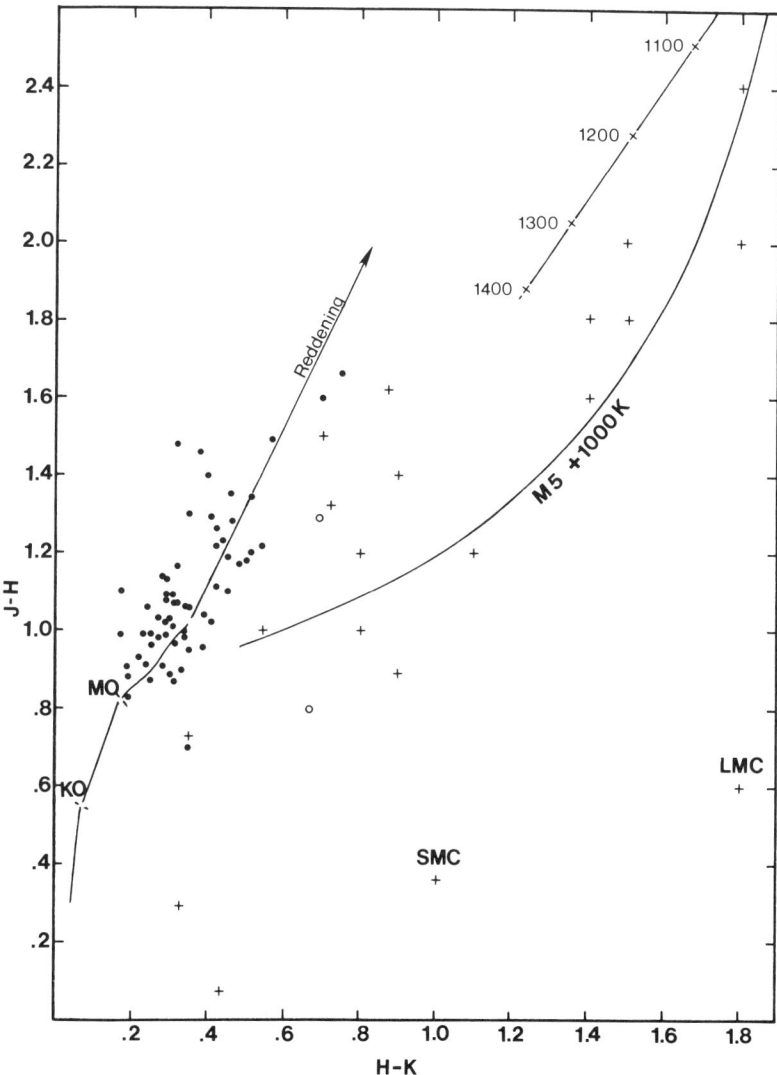

Figure 1. The J-H,H-K colour-colour diagram for symbiotic stars. The S-type systems (dots) are readily explained by M giants with a range of interstellar reddening. The giant sequence is represented by the irregular curve. Crosses indicate D-type systems. If the dust were the dominant emission mechanism, the data should fall on the black-body line, which is marked with temperatures in Kelvin. Combinations of cool star and dust map out a range of locations on the diagram; the combination of an M5 giant with a 1000K dust shell is shown. The four D-type stars near the foot of the diagram must have cool components hotter than 5000K. The Magellanic Cloud symbiotic stars are included in this group. Open circles are probable D-types for which longer-wavelength photometry is needed.

This material probably accounts for the structure in Woolf's (1973) energy distribution for UV Aur.

Distinguishing itself from all others studied, the star HD 330036 was reported by Allen (1981) to have emission in the 3.3-3.5μm region. The emission resembles that in NGC 7027 and other H II regions, and is believed to originate in carbon-rich dust grains lying at the interface of ionized and neutral gas (Aitken et al 1979; Dwek et al 1980; Sellgren 1981). The emission bands at 7.7 and 11.3μm which usually accompany 3.3μm features are also seen (Allen, Aitken and Roche).

This is the first indication that any of the G-type (yellow) symbiotic stars might have an oxygen/carbon ratio less than unity. It also suggests the presence of neutral material in the HD 330036 system.

In the symbiotic stars which have M giant components, the high black-body temperatures are consistent with the formation of dust in the extended atmosphere, where it may be shielded from the destructive radiation of the hot component. The coolness of the dust in the G-type symbiotics (infrared type D'), where the star itself is hotter, indicates that the dust lies at greater distances from the cool component. Believing these to be mass-loss systems implies that the dust forms in these more remote corners. It is hard to envisage the formation of dust in the hostile radiation field of the hot component. Moreover, in HD 330036 the infrared emission bands imply that neutral, dust-laden gas exists. Perhaps the most attractive way to provide the necessary shielding from the hot star is for the latter to be enveloped in an optically thick disk. If this is the case, one cannot discount the possibility that the G-type spectrum is produced by the disk itself. It should be noted that in some cases the G star has been given a supergiant luminosity classification. One is reminded, too, of the B supergiant spectrum in RX Puppis which was mimicked by an optically thick disk or wind (Barton, Phillips and Allen 1979; Klutz, Simonetto and Swings 1978). If in this interpretation one argues that an M giant is still present, as for RX Puppis, the absence of dust as hot as 1000K must also be explained.

CONCLUSIONS

At wavelengths beyond 1μm, symbiotic stars do not appear symbiotic. They are seen to be normal cool giant stars. If these giants are of very late spectral type, and especially if they are Mira variables, they (like giants in the field) usually enshroud themselves in silicate-rich circumstellar dust, and consequently become extremely rubescent. It is difficult to escape the conclusion that symbiotic systems contain normal cool giants, and indeed that the giants are unaffected by even the thousandfold optical brightening of a slow-nova outburst.

One subset of the symbiotic stars, those containing G-type cool components, are distinguished from the remainder by having cooler dust, sometimes carbon rich, which must lie quite remote from the G star.

ACKNOWLEDGEMENTS

I'd like to thank June Holt for typing this, and all the staff of the AAO who have directly or indirectly contributed to the many observations noted herein which were made with the 3.9 m Anglo-Australian Telescope.

REFERENCES

Aitken, D.K., Roche, P.F., and Spenser, P.M.: 1980, Monthly Notices Roy. Astron. Soc. 193, pp.207-212.
Aitken, D.K., Roche, P.F., Spenser, P.M., and Jones, B.: 1979, Astron. Astrophys. 76, pp.60-64.
Allen, D.A.: 1973, Monthly Notices Roy. Astron. Soc. 161, pp.145-166.
Allen, D.A.: 1974a, Inf. Bull. Var. Stars, 911.
Allen, D.A.: 1974b, Monthly Notices Roy. Astron. Soc. 168, pp.1-13.
Allen, D.A.: 1979, Int. Astron. Union. Colloq. 46, pp.125-149.
Allen, D.A.: 1980a, Monthly Notices Roy. Astron. Soc. 192, pp.521-530.
Allen, D.A.: 1980b, Astrophys. Letters 20, pp.131-133.
Allen, D.A.: 1981, Proc. N. Amer. Workshop Symbiotic Stars, in press.
Allen, D.A., Beattie, D.H., Lee, T.J., Stewart, J.M., and Williams, P.M.: 1978, Monthly Notices Roy. Astron. Soc. 182, pp.57P-60P.
Allen, D.A., and Glass, I.S.: 1974, Monthly Notices Roy. Astron. Soc. 167, pp.337-350.
Allen, D.A., and Glass, I.S.: 1975, Monthly Notices Roy. Astron. Soc. 170, pp.579-587.
Andrillat, Y., and Houziaux, L.: 1976, Astron. Astrophys. 52, pp.119-121.
Barton, J.R., Phillips, B.A., and Allen, D.A.: 1979, Monthly Notices Roy. Astron. Soc. 187, pp.813-816.
Bopp, B.W.: 1981, Proc. N. Amer. Workshop Symbiotic Stars, in press.
Boyarchuk, A.A.: 1970, Chapter 3 in "Eruptive Stars", Eds. Boyarchuk, A.A., and Gershberg, R.E., Academy of Sciences, Moscow.
Carter, B.S., and Feast, M.W.: 1979, Inf. Bull. Var. Stars, 1714.
Cohen, M., and Barlow, M.J.: 1974, Astrophys. J. 193, pp.401-418.
Cowley, A.P., and Stencel, R.E.: 1973, Astrophys. J. 184, pp.687-692.
Davidsen, A., Malina, R., and Bowyer, S.: 1977, Astrophys. J. 211, pp.866-871.
Davidson, K., Humphreys, R.M., and Merrill, K.M.: 1978, Astrophys. J. 220, pp.239-244.
Dwek, E., Sellgren, K., Soifer, B.T., and Werner, M.W.: 1980, Astrophys. J. 238, pp.140-147.
Feast, M.W., and Glass, I.S.: 1974, Monthly Notices Roy. Astron. Soc. 167, pp.81-85.
Feast, M.W., and Glass, I.S.: 1980, Observatory, 100, p.208.
Feast, M.W., Robertson, B.S.C., and Catchpole, R.M.: 1977, Monthly Notices Roy. Astron. Soc. 179, pp.499-508.
Gehrz, R.D., Ney, E.P., Becklin, E.E., and Neugebauer, G.: 1973, Astrophys. Letters 13, pp.89-93.
Geisel, S.L., Kleinmann, D.E., and Low, F.J.: 1970, Astrophys. J. 161, pp.L101-L104.
Gillett, F.C., Knacke, R.F., and Stein, W.A.: 1971, Astrophys. J. 163, pp.L57-L59.

Gillett, F.C., Merrill, K.M., and Stein, W.A.: 1971, Astrophys. J. 164, pp.83-90.
Glass, I.S., and Feast, M.W.: 1973, Nature Phys. Sci. 245, pp.39-40.
Glass, I.S., and Webster, B.L.: 1973, Monthly Notices Roy. Astron. Soc. 165, pp.77-89.
Harvey, P.M.: 1975, Astrophys. J. 188, pp.95-96.
Henize, K.G., and Carlson, E.D.: 1980, Publ. Astron. Soc. Pacific 92, pp.479-483.
Herbig, G.H.: 1950, Publ. Astron. Soc. Pacific 62, pp.211-215.
Herbig, G.H.: 1960, Astrophys. J. 131, pp.632-637.
Herbig, G.H., and Hoffleit, D.: 1975, Astrophys. J. 202, pp.L41-L45.
Johnson, H.L.: 1965, Comm. Lunar Planet. Lab. 3, pp.73-77.
Klutz, M., Simonetto, O., and Swings, J.P.: 1978, Astron. Astrophys. 66, pp.283-288.
Kraft, R.P.: 1958, Astrophys. J. 127, pp.625-641.
Longmore, A.J., and Allen, D.A.: 1977, Astrophys. Letters 18, pp.159-162.
Lutz, J.H., Lutz, T.E., Kaler, J.B., Osterbrock, D.E., and Gregory, S.A.: 1976, Astrophys. J. 203, pp.481-484.
Luud, L., Vennik, J., and Pehk, M.: 1978, Soviet Astron. Letters 4, pp.46-48.
Mendoza, E.E.: 1972, Bol. Obs. Tonantzintla Tacubaya, 38, pp.211-214.
Merrill, K.M.: 1977, Int. Astron. Union Circ., 3088.
Merrill, P.W.: 1950a, Astrophys. J. 111, pp.484-494.
Merrill, P.W.: 1950b, Astrophys. J. 112, pp.514-519.
Nassau, J.J., and Blanco, V.M.: 1954, Astrophys. J. 120, pp.129-138.
Neugebauer, G., and Leighton, R.B.: 1969, "Two micron sky survey", NASA SP-3047.
O'Dell, C.R.: 1966, Astrophys. J. 145, pp.487-495.
Puetter, R.C., Russell, R.W., Soifer, B.T., and Willner, S.P.: 1978, Astrophys. J. 223, pp.L93-L95.
Roman, N.G.: 1953, Astrophys. J. 117, pp.467-468.
Sanduleak, N., and Stephenson, C.B.: 1973, Astrophys. J. 185, pp.899-913.
Sellgren, K.: 1981, Astrophys. J. 245, pp.138-147.
Stein, W.A., Gaustad, J.E., Gillett, F.C., and Knacke, R.F.: 1969, Astrophys. J. 155, pp.L3-L7.
Stencel, R.E., and Sahade, J.: 1980, Astrophys. J. 238, pp.929-934.
Swings, J.P., and Allen, D.A.: 1972, Publ. Astron. Soc. Pacific, 84, pp.523-527.
Szkody, P.: 1977, Astrophys. J. 217, pp.140-150.
Thackeray, A.D.: 1954, Observatory, 74, pp.258-259.
Thackeray, A.D., and Hutchings, J.B.: 1974, Monthly Notices Roy. Astron. Soc. 167, pp.319-336.
Tifft, W.G., and Greenstein, J.L.: 1958, Astrophys. J. 127, pp.160-171.
Torres-Peimbert, S., Recillas-Cruz, E., and Peimbert, M.: 1980, Rev. Mexicana Astron. Astrof. 5, pp.51-58.
Webster, B.L.: 1966, Publ. Astron. Soc. Pacific, 78, pp.136-142.
Webster, B.L., and Allen, D.A.: 1975, Monthly Notices Roy. Astron. Soc. 171, pp.171-180.
Wilson, R.E.: 1942, Astrophys. J. 96, pp.371-381.
Wilson, R.E.: 1950, Publ. Astron. Soc. Pacific, 62, pp.14-15.
Woolf, N.J.: 1973, Astrophys. J. 185, pp.229-237.

IR PHOTOMETRY OF SYMBIOTIC STARS AT CALAR ALTO OBSERVATORY

C. Eiroa, H. Hefele, Qian Zhong-yu
Max-Planck-Institut für Astronomie, Heidelberg, FRG
and
Centro Astronomico Hispano-Aleman, Almeria, Spain

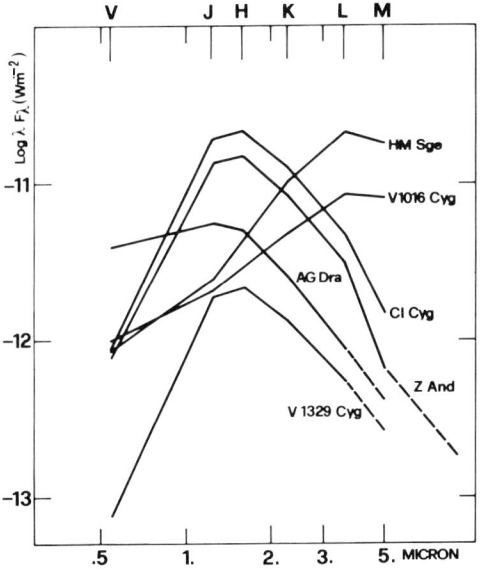

Infrared photometry of symbiotic stars was made with the 2.2 m telescope of the Calar Alto Observatory on July 1981, and are reported in the figure.

Three stars (Z And, CI Cyg, V1329 Cyg) were observed during a phase of minimum activity. We note that they display nearly the same 0.55 to 5 u energy spectrum, which is attributed to the late type star with a colour temperature close to 3000K.

AG Dra shows a similar spectrum from 1.67 u to 5 u, where the cool spectrum dominates, while shortwards there is a large contribution of the "nebular" component, which increased during the recent outburst. The H and K magnitudes are close to those of Swings and Allen (1972) contrary to the result of Bopp (1981).

V1016 Cyg and HM Sge both present a very strong IR flux attributed to dust emission. The IR magnitudes of V1016 Cyg are within the range of variability previously reported. HM Sge is about 0.5-0.9 mag fainter than in June 1977 (Davidson et al. 1978) in spite of the fact that its present visual luminosity is about 0.5 mag brighter.

REFERENCES

Bopp, B.W.: 1981, North American Workshop on Symbiotic Stars, Ed. R.E. Stencel, NBS and University of Colorado, Boulder, p.11

Davidson, K., Humphreys, R.M., Merrill, K.M.: 1978, Astrophys.J. 220, 239
Swings, J.P., Allen, D.A.: 1972, Pub.Astr.Soc.Pacific 84, 523

DISCUSSION ON INFRARED OBSERVATIONS

Whitelock: It is notable that Allen's IR spectral types of the cool components of symbiotic systems are significantly and systematically later than those determined at shorter wavelengths, as remarked upon by Allen in an earlier paper. This is of importance when considering the galactic population distribution of symbiotic stars.
Allen's classification of the cool component of AR Pav is particularly disturbing. He types it as M6III from which he derives a distance of 10 kpc. This is at odds with the optical photometry (Menzies et al, in press) which suggests that the distance of 3.8 kpc derived by Thackeray and Hutchings is more appropriate.

McCarthy: I suggest that all of us who have occasion to use the important distinction between D and S types among the symbiotic objects, try to imitate the care used by Allen in his text, that is to speak of "D Type Symbiotic Stars" and "S Type Symbiotic Star", to avoid any confusion with the classical stellar temperature sequence, epsecially for S type stars.

Slovak: Andriews (1974 MN 167, 635) estimated $E(B-V) < 0.1$ for AR Pav from studying stars in the nearby field. Yet the IUE spectra, using the 2200 A feature, indicates $E(B-V) \approx 0.30$. Hence a large portion of the reddening may be circumstellar, as opposed to interstellar.

Keyes: AR Pav is well out of the galactic plane, so we would not expect any appreciably large reddening than that observed, even if the object were at great distance.

Michalitsianos: Concerning the 2200 A feature, there may be a significantly different extinction law for material local to the system. How one corrects for $E(B-V)$ separating the interstellar component from the local absorption is not clear for symbiotic stars that appear heavily reddened.

Houziaux: Keyes and Plavec determined a value of $E(B-V)=0.12$ for AG Peg (The Universe in the Ultraviolet Wavelengths, NASA, p.443). Is this determination made from the strength of the 2200 A feature? Was this feature assumed to be entirely of interstellar origin? It would be interesting to know if it is found that the $E(B-V)$'s determined from such ultraviolet data are always smaller than the $E(B-V)$'s determined from continuum fluxes in the visible, and if they agree with the $E(B-V)$'s determined from the emission lines ratios. On the other hand, when a star shows a strong infrared excess due to dust emission, it is hard to believe that the stellar ultraviolet radiation around 2200 A is not absorbed in the dust envelope.

Nussbaumer: For V1016 Cyg, which displays an intense dust emission, we have determined the reddening in three different ways: 1) from the NeV lines at 1575 and 2973 A; 2) from the HeII recombination lines; 3) from the 2200 A depression. The three methods agree with each other. They also agree with the result obtained from the Balmer lines of H.

Cassatella: About the near infrared molecular absorption bands, I think it is certainly important to have a means of distinguishing between infrared molecular bands produced in the atmosphere of a cool giant, and those produced for example in a molecular cloud in front of it.

Kafatos: I would like to emphasize, and I think Dr. Nussbaumer will say more about it, that there is some evidence that the chemical abundances (particularly C and Si) in the nebular regions of even S-type symbiotics are different than cosmic abundances. In particular the depletion of C and Si may be interpreted as the presence of dust, although it is difficult to understand how dust could survive in the UV rich environment of the symbiotic nebula.

NEAR IR SPECTRA OF SYMBIOTIC STARS

Y. Andrillat
Observatoire de Haute Provence
04870 St Michel l'Observatoire (France)

It is very well known that symbiotic stars are displaying at the same time spectral characteristics of hot stars, cool stars (principally of M type), nebulae and circumstellar dust envelopes.
The investigation of the near IR is particularly interesting because one finds there important features of these different objects.

HOT STARS

In the near infrared region up to 1μ, the helium lines : He I λ 10830, He II λ 10123, and the oxygen lines OI $\lambda\lambda$ 7772-8446 are very useful to construct theoretical models.

- Helium lines :

the quantitative observations are very crucial because they assign important constraints to the law for ionization distribution in the models.
He I λ 10830 ($2s^3S - 2p^3P$) - upper metastable level - is produced only by a highly excited state, and only at temperatures as high as 20 000°K.
This lines has been found as a strong emission in WR stars and other early type emission line stars : Of, principally in the supergiants (Vreux, Andrillat 1979), Be stars (Andrillat, Swings 1976).
In the later type, it appears in absorption or in emission but with a weak strength (W = 100 - 1000 mA in a few G and early K stars).
It is not present with an appreciable intensity in the M stars (Vaughan, Zirin 1968).

He II λ 10123 : This line is the first member of the Pickering serie (n = 4 ⟶ n = 5).
It has been observed to be in emission in WR, in absorption or in emission in O and Of stars (Andrillat, Vreux 1975 and Vreux, Andrillat 1979). Observations of this transition are extremely important to develop our understanding of helium line formation and the ratio I(10123) / I(4686) permits in particular to decide which mechanism produces He II λ 4686 emission line (collisional desexcitation, pumping...).

- Oxygene lines:

O I λ 8446 ($3^3S^o - 3^3P$) is observed in emission in Be stars, novae and generally its intensity is very high compare to the one of the λ 7772 line.
In many cases, this anomalous strength is explained by a fluorescence mechanism due to Lβ absorption by O I atoms : for example, this process has been confirmed from observations of the other IR transitions λ 10295-302 in the spectrum of Nova Sct 1975 (Andrillat et al. 1975).

The λ 7772 line ($3^5S^o - 3^5P$). This triplet appears in emission in novae and in a few Be stars (Andrillat, Houziaux 1967).
It cannot be excited by a fluorescence mechanism. It is visible in absorption in stars of the other types (A to M). It is an excellent criterion of luminosity (Parsons 1974).
Its intensity is temperature dependent because the high excitation potential (9.15eV) of the metastable $3s^5S$ lower level. It is particularly strong in Supergiants where the extended atmospheres allow low collisional desexcitation rates.

COOL STARS

In the near infrared, there are many characteristic TiO and VO bands which permit to classify the late type stars, principally the M stars. Several works on spectral classification have been done by different authors using a low dispersion who have described the appearance of the TiO and VO bands at various spectral types (Mavridis 1966 - McCarthy and al. 1966 - Wing 1966 - Albers 1974).
All their classifications are based upon the intensity of the principal bands of :
TiO : $\lambda\lambda$ 7054-7126-7589-8194-8300-8432-8859
VO : $\lambda\lambda$ 7900-10500.
The table 1 gives schematically the principal characteristics of the different spectral types of giant M stars.

Table 1

Spectral type	First appearance	
	Ti O	VO
M0	7054-7126	
M2	7589	
M4	8432	
M5	8194-8859	7900
M6	8300	

a complex of VO bands around λ 10500 is very strong in the latest spectral types M8-M10.

It is important to note, among these bands, some are blended with some

telluric bands situated at :
O_2 molecular bands : $\lambda 6870$ (B) - $\lambda 7593$ (A)
H_2O "   "   : $\lambda 7190$ (a) - $\lambda 8227$ (Z) - $\lambda 9060$ (γ) - $\lambda 9420$ (ρ)
However, it is possible to recognize the molecular bands, because the modification due to the blending with the telluric bands are generally very typical.

Near infrared spectral features permit also to determine the luminosity criteria (Albers 1974 - Treanor,McCarthy 1966 - White,Wing 1978).

- Supergiant M stars
The CN bands $\lambda\lambda 7800-8000$ are enhanced in the spectra of M stars of high luminosity (Sharpless 1956). In the supergiant M stars, the $\lambda\lambda$ 8308-8330 features due to the blend of titanium and iron lines are very well visible.

- Giant M stars
In luminosity class III, the CN bands $\lambda\lambda$ 7800-8000 are extremely weak or absent. The Ca II triplet ($\lambda\lambda$ 8498-8543-8663) is strong.

- Dwarf M stars
The Ca H bands near $\lambda 6385$ and $\lambda 6880$ are prominent (Ohmen 1936).
The Na I doublet ($\lambda\lambda$ 8183-8195) is clearly present.
The sodium lines and the Ca H bands are good luminosity criteria for the dwarf stars over the range M0-M5.

NEBULAR LINES

In the near IR, one finds several interesting forbidden lines which are present in novae (nebular phase), Be [ ] stars, planetary nebulae and which are used to determine the physical parameters (T_e, N_e)
- [S III] $\lambda\lambda 9532-9069$, the strongest lines in PN
- [S II] $\lambda\lambda 10284-10370$ which permit access to reddening problems by comparison with the blue doublet ($\lambda\lambda$ 4068-4076).
- [C I] $\lambda\lambda$ 8727-9823, [S I] $\lambda 7726$, [O II] $\lambda\lambda 7319-7331$, [Cl III] $\lambda\lambda 8196-8580$, [Cl IV] $\lambda\lambda 7530-846$, [A III] $\lambda\lambda 7136-7751$, [A IV] $\lambda\lambda 7237-7263$, [A V] $\lambda 7006$
- Some lines of [Fe II] $\lambda\lambda 7155-7452$, [Fe IV] $\lambda\lambda 7189-7221-7923$ have been found in Be [ ] (Swings 1969).

Taking into account the precedent remarks, it appears that the near infrared study will bring interesting informations in particular concerning the spectral classification and the luminosity class of the "cool spectrum". As far as the "hot spectrum" is concerned, it will be possible to precise the formation mechanisms of the helium and oxygen. Nevertheless, there are very few obervations in this region.

SPECTRA OF THE SYMBIOTIC STARS BETWEEN $\lambda\lambda$ 8000-11000

A first glance permits to distinguish 2 kinds of near infrared

spectra of the symbiotic stars:
- Relatively poor spectra which correspond to "old and classical symbiotics" very well known from several decades (RY Sct, EG And, CH Cyg, AG Dra, AG Peg...). (Fig.1)
- Spectra very rich in emission lines and molecular bands corresponding to stars in evolution which had a recent outburst (RX Pup, V1016 Cyg (1964), HM Sge (1975), HBV 475 = V1329 Cyg (1969)). (Fig.2).
In July and August 1981, we have obtained the spectra of these objects (Haute Provence Observatory - 193 cm Telescope - Spectro ROUCAS + Image Tube - spectral range $\lambda\lambda$ 8000-11000 - dispersion 230 A.mm^{-1}).

RY Sct

The spectrum of RY Sct shows [Fe III] emission lines which are currently found in the spectra of some novae and a few symbiotic stars (BF Cyg for example in the blue visible spectrum).
The spectra obtained by Cowley and Hutchings (1976) do not confirm a possible relation with these objects. These authors proposed a model consisting of 2 massive stars embedded in an unusual nebulosity showing strong [Fe III] emissions. RY Sct has been identified as a radio binary source.
- Our spectrum obtained the 3.7.1981 (Fig.1) does not present the characteristics of symbiotic stars. The dominent features are the strong $\lambda\lambda$9532-9069, [S III] and λ10830 He I emission lines. No molecular bands are visible.

EG And = HD 4174 = BD 39°167

From 0.8 to 1.1μ , the spectrum of August 11, 1981 shows only one emission : He I λ10830. (Fig.1).
- Wilson (1950) describes the blue spectrum of October 12, 1949, but he does not identify this element.
- The presence of He I λ10830 permits to assign a temperature to the hot companion : $T \geqslant 20\ 000°$, which is in good agreement with the value $20\ 000° < T < 40\ 000°$ indicated by Smith (1980).
The IR observations by Swings and Allen (1972) indicate an absence of any large scale IR excess, as in the case of AG Peg.
Smith suggests that analogous processes are occuring in EG And and AG Peg and gives $20\ 000° < T < 40\ 000°$.
-Smith proposes a binary model consisting of a M2 III semi regular variable and a hot secondary one, surrounded by a common envelope.
- In the near IR, we do not observe molecular bands : the absence of TiO λ8432 band is compatible with a M2-M4 type star.
- Moreover, we observe a strong Ca II triplet without the CN band 7900 in absorption which characterizes a giant star.
Thus, our IR observations confirm the spectral type proposed by Smith. This author observes changes in the equivalent width, radial velocity and line profile of Hα . He founds a good correlation with a pulsation period of 470 days.
It was interesting to know which phase the He I λ10830 emission corresponds to. In this aim , we have observed the Hα profile pratically

at the same date (2 days after the observation of He I $\lambda 10830$).
We have used a spectrum recording system by analogic TV mounted on the
Echelle spectrograph of the 152 cm Telescope. The dispersion was
55 A.mm^{-1} and the spectral range $\lambda\lambda$ 6250-6925.
Hα is a single emission without structure. The radial velocity of the
peak is - 44 km.s^{-1} and the equivalent width : 2,4 A. These values
correspond to the 0,80 phase (Smith 1980).

CH Cyg

During the last recent outburst (sept.1977), we observed CH Cyg in the
near IR region. Spectral features were visible (Andrillat, Faraggiana
1977):
- many strong TiO molecular bands, in particular $\lambda\lambda$7054-7126-8344-8432-
8859, these 3 later bands being characteristic of M6,5 type stars or
even of a later type.
This result is compatible with the one obtained by Morris (1977) which
attributed a M7 type from the blue spectrum.
- 2 absorption lines of the Ca II triplet were visible, the third was
blended with a TiO band. The presence of these lines indicated that
the M7 star is a giant.
The observations by Smith and Bopp (1977) in the spectral range 5800-
8700 show a 2 components emission profile of Hα, the blue one being
several times stronger than the red one.
These authors did not note any emission features visible in the red.
Then, it seems that the He I lines, 5875-6678-7065, are absent : in
the near infrared, we have not observed the He I λ 10830 emission line.
In July 1981 (Fig.1): we have again observed CH Cyg in the spectral
range 0,8-1,1 μ
- The TiO molecular bands have disappeared
- It is the same for the Ca II triplet
- Only P7 λ 10049 is visible as a broad absorption. P13 is also visible.

CI Cyg

In the spectral range 0,8-1,1 μ, the spectrum of CI Cyg is poor.
However, we have observed some changes between 1974 and 1981.
- 5.8.1974 $\begin{cases} \text{He I } \lambda \text{ 10830 is a strong emission} \\ \text{O I } \lambda \text{ 8446 is a moderate emission} \end{cases}$
- 26.8.1975 $\begin{cases} \text{He I } \lambda \text{ 10830 is absent} \\ \text{O I } \lambda \text{ 8446 ? is not visible but our spectrum is overexposed} \end{cases}$
- 4.7.1981 (Fig.1) : the spectrum is analogous to that of the 5.8.1974.
At this date, C.C.Huang has observed CI Cyg in the blue region : the
He I spectrum is strong. The He II λ 4686 is strong but λ 3203 is absent.
We do not observe He II λ 10123 line. These 2 lines have the same
upper level and it would be interesting to determine the ratio of their
intensities in order to determine an intrinsec possible reddening.
It is the case for AG Dra.

AG Dra

It is very well known that the intensity of the H and He emission lines

are variable.
In the spectral range 0,8-1,1μ, the spectrum observed on July 4,1981 (Fig.1) is characterized by :
- a strong continuum
- intense emission lines of He I λ10830 and He II λ10123
At the same date, C.C.Huang has observed the UV recombination lines of He II λ3203 and also the He II λ4686 line.
The P7 λ10049 Paschen line is a weaker emission than He II λ10123.
- O Iλ8446 is also present in emission.
- No molecular bands are visible.
This remark permits to assign a spectral type earlier than M2 (we have no observations below 0,8μ).
The result is in good agreement with the K3 III classification proposed by Doroshenko and Nikolov (1967) from a study of the energy distribution in the continuum observed in 1961-1962.
During this period, these authors noted also the presence of emission lines : He II λ4686, and He I λ3889, this later being analogous to that of He I λ10830 (metastable level).

AG Peg

In the near IR, the continuum is very strong (Fig.1).
- He I λ10830 is a strong emission
- He II λ10123 is very well visible and its intensity is comparable to that of P7 λ10049.
It clearly appears that the He II λ10123 line is broader than P7.
This broadening confirms the presence of a WR companion star in the spectroscopic binary AG Peg which consists of a normal M3 III star and a hot WN6 star (Boyarchuk 1968).
- The absence of the TiO λ 8432 molecular band and the presence of the strong TiO $\lambda\lambda$ 7054-7126 (Andrillat, Houziaux 1982) and the moderate 7589 bands are in good agreement with the type M3.
- The presence of the strong Ca II triplet in absorption indicates the class luminosity is III. Finally, the spectral IR features confirm the classification proposed by Boyarchuk.
- O I λ 8446 is also present in emission.
Swings and Allen (1972) found no indication of an infrared excess in AG Peg (H-K = 0,23 mag).
In the case where dust is present : H-K $\geq$ 0,4.
However, Boyarchuk found that AG Peg has a circumstellar envelope with Te $\sim$ 17 000K).Perhaps, the hot component provides sufficient energy to heat the circumstellar material to such high temperatures as large scale grain formation cannot occur (Smith 1980).
The temperature of the hot component T = 40 000° has been determined by Gallagher and al. 1979).

RX Pup

From 1940, this star has shown some spectral changes
- 1940 : the spectrum presented an obviously symbiotic character but without any evidence of a late type companion (Swings,Struve 1941).

- From 1972 to 1975 : the high excitation emission lines have disappeared or have a weak strength.
The spectrum in the spectral range λλ 3658-8542 is similar to that of a Be [ ] star (Swings, Klutz 1976).
Moreover, Swings and Allen (1972) discovered a large infrared excess which they attributed to a circumstellar dust shell.
- Between march and december 1979, Klutz and Swings (1981) observe, between λλ 3558 and 5010, an important spectral evolution with an increase of the excitation degree : broad and strong emission lines of He II, N III, O III, [O III] are present. The N III λ 4640 and He II λ 4686 lines intensities are similar to the ones found in WN7-8 stars. The spectrum is comparable to that of 1940.
- 30 november 1980: In the near IR (λλ 5875-10850), one spectrum has been performed at the 3,60 m ESO telescope with the Boller Chivens Spectrograph equipped with a RETICON (dispersion 228 A.mm^{-1}) (Fig.3).
We are observing (Swings 1981) an important increase of the low excitation emission lines, principally :
- He I lines λλ5875, 6678, 7065 : we observe also the λ 10830 line.
- O I λ 8446 which is very strong while λ 7772 is a moderate emission. λ8446 is probably excited by a fluorescence mechanism from Lβ.
In 1979, λ 8446 is a moderate emission and λ 7772 is in absorption (Swings, Klutz 1976).
The Ca II triplet is a strong emission.
Paschen lines are visible up to P17.
Many Fe II lines are present : they are indicated by points (Fig.3).
For the first time, molecular absorption bands are clearly detected : TiO λλ7589-8432 are strong.
These bands permit to attribute a spectral type later than M5 to the cool component.
Moreover, it seems that the λ7900 VO band is present. In this case, the spectral type is M6.
Thus, this observation in the near IR confirms the symbiotic character of RX Pup and precises the spectral type M5 or M6 for the cool companion.

V1016 Cyg, HM Sge

In the near IR, the spectra of these 2 stars are similar and characterized by many strong emission lines (Fig.2).
The He I λ10830 line is dominent.
The Paschen lines are visible up to P19.
He II λ10123 is strong.
O I λ 8446 is present but stronger in V1016 Cyg than HM Sge.
The sulfur forbidden lines are also intense.
Important spectral variations have been observed for these 2 stars. They will be described in the papers concerning the individual stars.

HBV 475

The spectrum obtained on August 8, 1981 shows a strong continuum cut by intense molecular TiO absorption bands (Fig.2).

O I λ8446 is very faint because this line is blend with the strong TiO λ8432 which prevents from seing the Ca II triplet line.
Spectral variations of the object will be given in the section relative to the individual stars.

RERERENCES

Albers,H.: 1974, Astrophys.J.189, 463.
Andrillat, Y., Houziaux,L.: 1967, J.Obs.50, 107.
Andrillat, Y., Fehrenbach, Ch., Vreux,J.M.: 1975, IAU Circ.n°2798
Andrillat, Y., Vreux, J.M.: 1975, Astron.Astrophys.41, 133.
Andrillat, Y., Swings,J.P.: 1976, Astrophys.J.204,L123.
Andrillat, Y., Houziaux, L.: 1982, IAU Coll.70 "The nature of symbiotic stars", this volume, p. 57.
Boyarchuk, A.A.: 1968, Soviet Astronomy 11, 818.
Cowley, A.P., Hutchings, J.B.: 1976, PASP 88, 456.
Doroshenko, V.T., Nikolov, N.S.:1967, Soviet Astronomy 11, 453.
Faraggiana, R., Andrillat, Y.: 1977, IAU Circ.3102.
Gallagher, J.S., Holm, A.V., Anderson, C.M., Webbink, R.F.: 1979, Astrophys.J.229, 994.
Klutz, M., Swings, J.P.: 1981, Astron.Astrophys.96, 406.
McCarthy, M.F., Treanor, P.J., Kent Ford, W.: 1966, Coll.on late type stars, Ed.M.Hack, 100.
Mavridis, L.N.: 1966, Coll.on late type stars, Ed.M.Hack, 420.
Morris, S.C.: 1977, IAU Circ.3101.
Ohman, Y.: 1936, Stockholm Annaler Bd.12 n°3.
Parsons, S.B.: 1964, Astrophys.J.140, 853.
Sharpless, S.: 1956, Astrophys.J.124, 342.
Smith, S.E.: 1980, Astrophys.J.237, 831.
Smith, S.E., Bopp, B.W.: 1977, IAU Circ.3113.
Swings, J.P.: 1969, Thèse Université de Liège.
Swings, J.P., Allen, D.A.: 1972, PASP 84, 523.
Swings, J.P., Klutz, M.: 1976, Astron.Astrophys.46, 303.
Swings, J.P.: 1981, IAU Circ.3560.
Swings, P., Struve, O.: 1941, Astrophys.J.94, 291.
Treanor, J.J., McCarthy, M.F.: 1966, Coll.on late type stars, Ed.M.Hack, 109.
Vaughan, A.H., Zirin, H.: 1968, Astrophys.J.152, 123.
Vreux, J.M., Andrillat, Y.: 1979, Astron.Astrophys.75, 93.
Vreux, J.M., Andrillat, Y.: 1979 , Astron.Astrophys.76, 221.
White, N.M., Wing, R.F.: 1978, Astrophys.J.222, 209.
Wilson, R.E.: 1950, PASP 62, 14.
Wing, R.F.: 1966, Coll.on late type stars, Ed.M.Hack, 231.

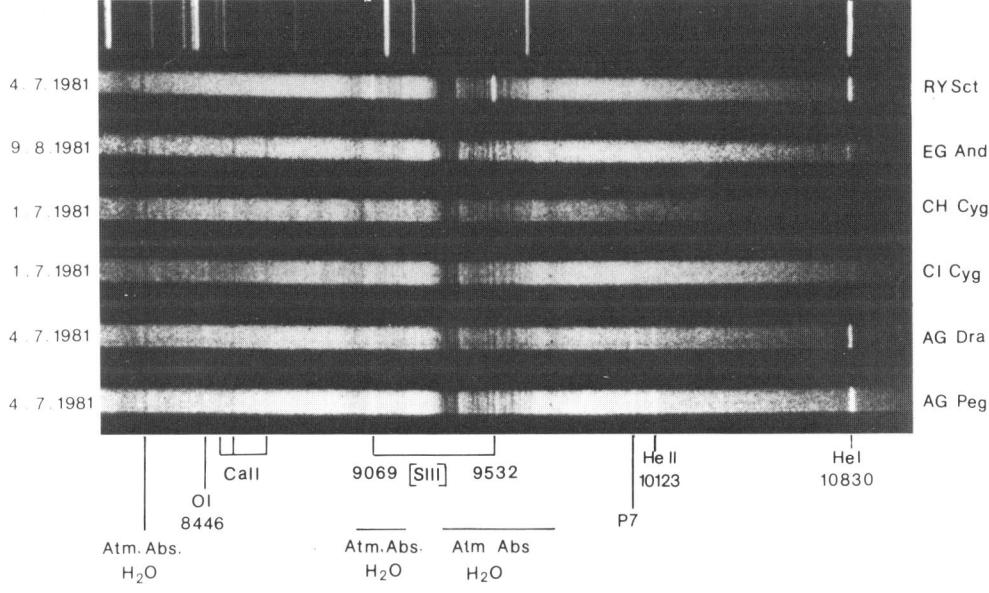

Fig.1

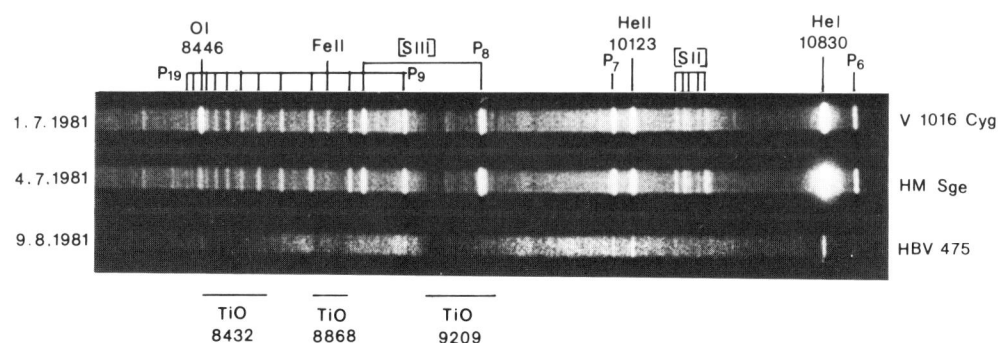

Fig.2

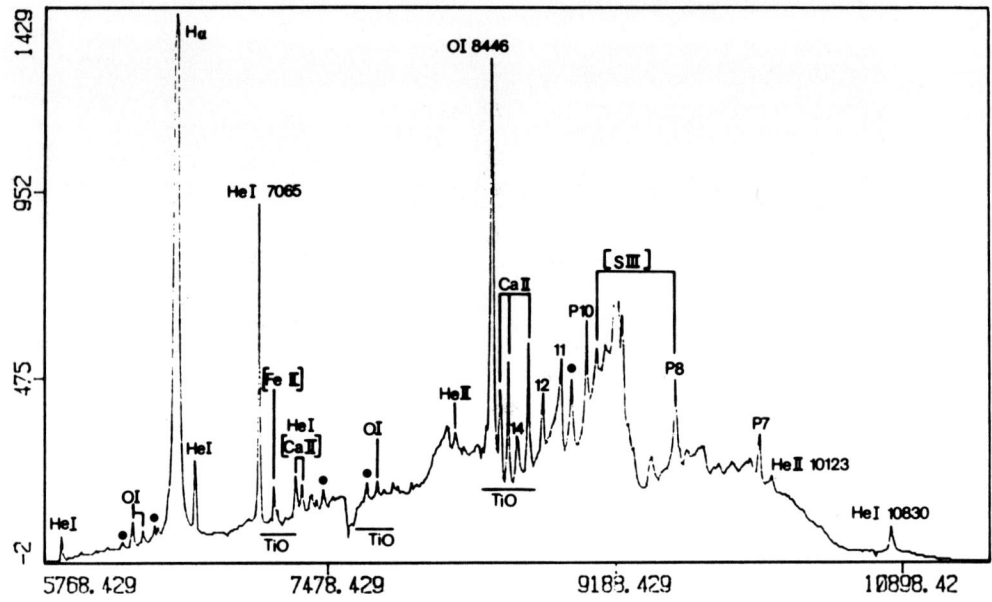

Figure 3. The near infared spectrum of RX Pup on 30 November 1980.

DISCUSSION ON NEAR INFRARED OBSERVATIONS

Hack: I was surprised to see that no emission of OI λ 8446 has been observed in CH Cyg, since the λ 1300 lines are present and very strong in emission, and both are expected to be the results of the Lyβ excitation mechanism.

Andrillat: OI 8446 is not visible in our spectrum. The continuum is strong can mask weak emissions. In any case if this is the reason, OI λ 8446 should be very weak.

PHOTOGRAPHIC INFRARED SPECTRA OF SYMBIOTIC STARS*

Y. Andrillat and L. Houziaux
Observatoire de Haute Provence and
Institut d'Astrophysique de
l'Université de Liège

Relatively few spectra of symbiotic stars have been recently published in the photographic infrared. We have observed six objects during the period 1962-1977 with a grating spectrograph attached to the newtonian focus of the 120-cm telescope at Observatoire de Haute Provence. The reciprocal dispersion is 230 A.mm^{-1} and the region 5800 to 8800 A has been covered using hypersensitized IN plates. The minimum equivalent width for an emission line to be seen is about 0.5 A. The spectra are displayed on plates I and II. We now briefly review the main spectral characteristics.

Z And. Only Hα and O I 8446 A appear as bright lines. The latter is however absent in November 1975. TiO bands appear at λ 6852, λ 7054-7126 and λ 7589. The presence of a band at λ 8432 is doubtful. Let us note that on a spectrum in the near ultraviolet taken on August 8, 1978 a strong Balmer continuum emission is seen.

AG Peg. If we adopt a 830-day period with a minimum at JD 2 440 928, the phases of our observations are (according to the order adopted on plate I) 0.106, 0.106, 0.460, 0.590, 0.878, 0.021, 0.304 and 0.312. Emission lines at Hα, He I λλ 6678, 7065 O I 8446 are seen at all phases except near minimum when only Hα is bright. Ca II lines at λλ 8598, 8662 are conspicuous absorptions as well as the TiO bands at λ 6852, λ 7054-7126 (strong) and 7589 A. The absence of TiO λ 8432 together with the strengh of the Ca II lines lead to a spectral type M3 III. On our near ultraviolet spectrum taken on August 11, 1978, the Balmer jump is not seen either in emission or in absorption.

*Les observations ont été effectuées à l'Observatoire de Haute Provence (C.N.R.S.)

BF Cyg. Shows strong emissions due to Hα and O I λ 8446. The latter considerably weakened in July 1976. Absorptions due to Ca II are absent. TiO bands are strong at λλ 7054, 7126 and at λ 7589. The presence of the band at λ 8432 indicates a spectral type later than M4 for the red component.

AX Per. Hα is a constant emission feature but He I λλ 6678, 7065 and O I λ 8446 are seen in emission only on the November 1976 spectra. Ca II absorptions are very weak but the TiO bands at λλ 6852, 7054-7126 and 7589 are strong. Again the absorption band at λ 8432 leads to a spectral type later than M4.

T Cr B. The period 1962-1977 is covered with 8 spectra. Hα is weak or absent. The Ca II absorptions are conspicuous as well as the TiO bands.

C I Cyg. Hα, He I λ 7065 and O I λ 8446 appear in emission. The Ca II lines are seen in 1972 and 1973. All TiO bands in the wavelength range show conspicuous absorptions. The spectral type is M4 or later.

TX C Vn. Five spectra are available over the period 1962-1974. Hα is weak in emission. Ca II lines appear distinctly in absorption, while O I is a shallow absorption feature, except in February 1966 where it is strong. The absence of molecular TiO bands, except for a weak band at 7054-7126 Å, leads to a spectral type earlier than for the other stars, probably M1.

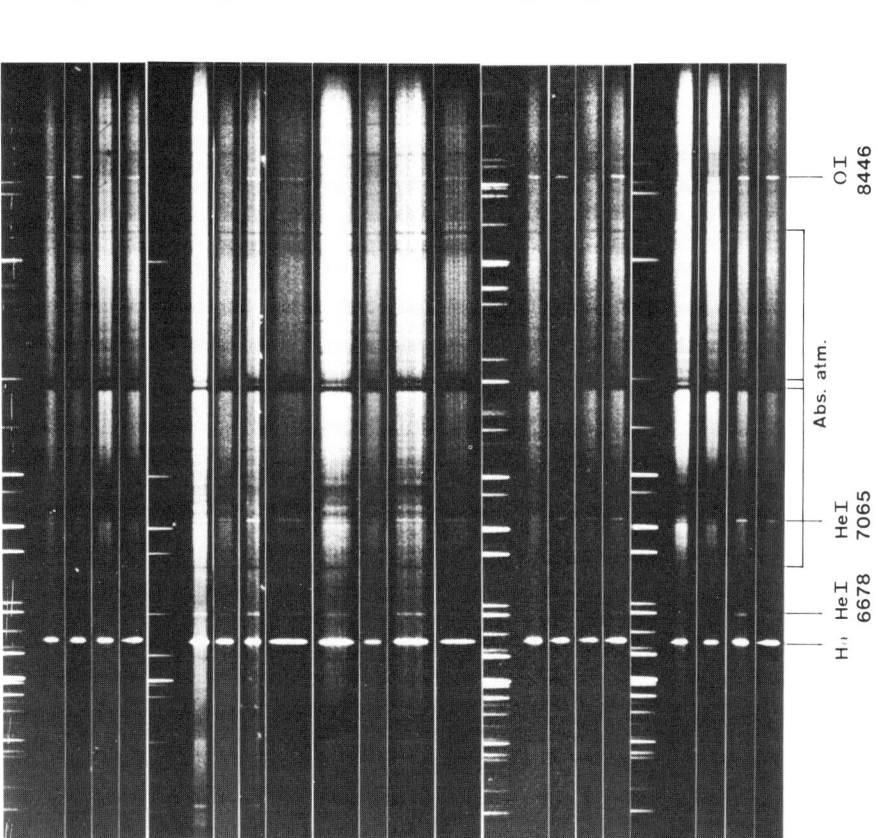

Plate I

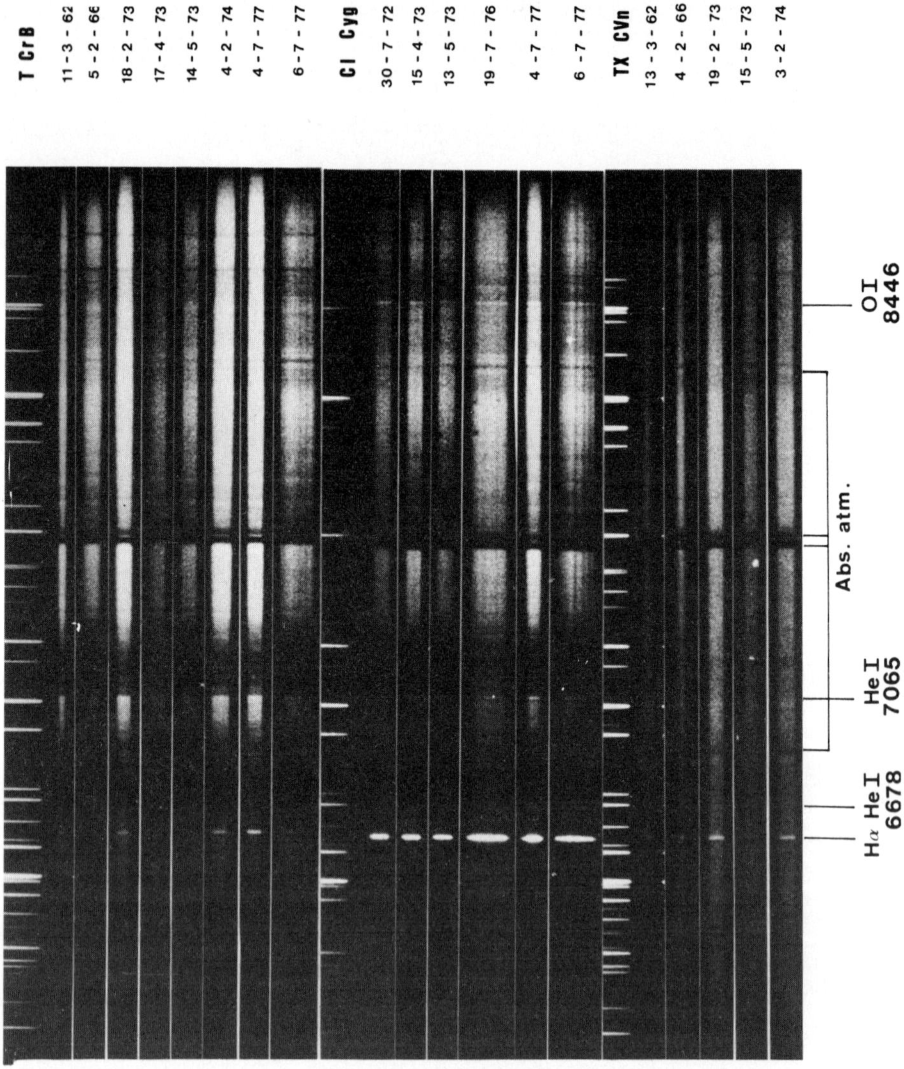
T CrB
11 - 3 - 62
5 - 2 - 66
18 - 2 - 73
17 - 4 - 73
14 - 5 - 73
4 - 2 - 74
4 - 7 - 77
6 - 7 - 77
CI Cyg
30 - 7 - 72
15 - 4 - 73
13 - 5 - 73
19 - 7 - 76
4 - 7 - 77
6 - 7 - 77
TX CVn
13 - 3 - 62
4 - 2 - 66
19 - 2 - 73
15 - 5 - 73
3 - 2 - 74
OI
8446
Abs. atm.
He I
7065
Hα He I
6678

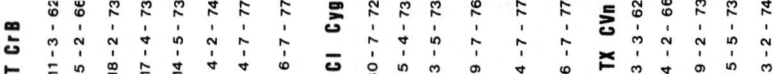

Plate II

PROPERTIES OF SYMBIOTIC STARS
FROM STUDIES IN THE OPTICAL REGION

Franco Ciatti

Asiago Astrophysical Observatory

1. Definition of Symbiotic Stars.
The traditional definition of Symbiotic Stars (SS) is that of objects which display a combination spectrum (e.g.Merrill,1950) that is emission lines requiring high-excitation conditions, superposed to the continuum and absorption features of a low-temperature star, most commonly an M-type giant. About one hundred of SS are known and listed today. It is anyway apparent that the classification criteria are rather rough, and since the excitation varies from the simple Me to SS with coronal emission, it is not well defined where a clear division should be made. As a result, the available lists include a very heterogeneous set of objects, probably different phases in stellar evolution. Moreover it has been remarked that SS show a rather confusing variety in their spectroscopic and photometric behaviour. Different intensities of both absorption and emission lines are reported from star to star. These facts indicate a high degree of individuality among SS, which is partly cause and/or effect of the adopted definitions.
In order to clarify this point, we recall the conditions proposed by Boyarchuk (1969): (a) absorption features of a late spectral type (TiO bands, metallic lines of low-excitation like CaI,CaII,FeII,etc.) must be seen; (b) emission lines of HeII,|OIII|, or highly ionised atoms must also be present (width for not more than 100 km/s); in addition he took into consideration that (c) the brightness can vary with amplitude up to 3 mag and period several years.
Some of these conditions appear not appropriate to Allen (1978), who has adopted the following different criteria: (a) the object must appear stellar; (b) emission from ions of IP $\gtrsim$ 55 eV (e.g.HeII) must be present, at some time, while evidence for a spectrum G or later must also exist (from optical or IR data); (c) when clear evidence of a late spectral type is lacking, IP must at some time exceed 100 eV (e.g.|FeVII|).Variability is not considered necessary. These criteria allow to include, besides the SS according to the definition of Boyarchuk, also the recurrent novae (T CrB, RS Oph) and the "very slow novae" (RR Tel, V1016 Cyg) or type II SS (Paczynsky and Rudak,1980). It is suggested that the differences among these types are only in the timescale between outbursts, correlated with the range of variability, and that such nova-like outbursts

may be a feature of all the objects which fulfill his criteria.
It has been also proposed to widen the definition of SS, as to include the so called "yellow SS", whose spectrum indicate a yellow-red dwarf-subdwarf star, instead of a red giant-supergiant (U Gem and Z Cam stars, also explosive variables). One can also consider similar to SS the VV Cep stars, showing the presence of an M supergiant and an OB star, on or above the main sequence (population I stars): they appear more luminous objects than classical SS, with which the evolutionary relation is not clear.

2. Photometric and spectroscopic evolution.

Typical SS present characteristic light variations, of more or less irregular shape and usually rather small amplitude ($\leq$ 1 mag) on timescales of months or years, which are described as a complex of small simultaneous flares. Occasionally at larger time intervals they also undergo larger variations or outbursts (3-5 mag), where the increase in brightness is steeper than the decline. The explosive character of these variations allows to define SS as "nova-like stars" (Of course T CrB and RS Oph have true recurrent nova outbursts). It is to be said that different stars show different light curves (regular, irregular, Mira-type), and in addition that the variability characteristics may change for the same stars from time to time (quasi-periodic trends from 200 to 900^d ; periods with small variations; different levels of minimum; etc.). During the brightness variations the stars are much redder at minimum with increasing (B-V) color, while the UV excess also increases. In the two-color diagram the data for SS are rather far from the main sequence trend, which is explained with more than one radiation source.

Typical spectroscopic variations are well correlated with the light curve, although they have not always been followed throughout full cycles, nor in all stars. We have in general, but mostly following the well studied properties of Z And, which is a prototype of the classical SS (e.g. Swings and Struve, 1941) :

i- Near minimum brightness, emission lines of low (e.g. permitted and forbidden FeII) and high excitation (e.g. HeII, |OIII|, in some cases up to |FeVII|) are present with large intensity. Together with the nebular spectrum, also an M-type giant (or supergiant) absorption spectrum is prominent with strong TiO bands.

ii- When the star brightens, the TiO bands become less conspicuous. Nebular and high-excitation permitted emissions progressively weaken, while the intensity of other emission lines may present fluctuations. An early-type shell spectrum develops, its continuum dominating the photographic region and covering the M spectrum.

iii- Around maximum light, the high-excitation and nebular emission spectrum disappears: only H, HeI, CaII, |OIII| remain. The emission/absorption ratio decreases toward the higher members of the Balmer series (emission from H_α to H_ϵ, then in absorption); the Balmer continuum is weakly present in absorption. The early-type shell spectrum is now well developed, with metallic lines (emission + absorption) of TiII, MgII, FeII, |FeII|, etc. In most cases also emissions of H and HeI have violet-displaced absorption components. At this time the spectrum is thus similar to that of the P Cyg-type stars.

It is often reported that the late-type absorption features completely disappear. Actually the most important TiO bands in the red-near IR are still visible during outburst, even when P Cyg features are revealed (Mammano et al.,1974).
iv- During the light decline we have the opposite trend. Shell characteristics of P Cyg components and metallic lines disappear, while a blue continuum becomes more evident. Higher Balmer members and their continuum come back to emission. Permitted and forbidden emissions develop again, the degree of ionization progressively increasing as luminosity declines (e.g.H,HeI,|OIII|,|NeIII|,|NeV|,|FeVII|). The ratio nebular/auroral lines increases. At the same time the late-type features like TiO bands are gradually enhanced over the continuum.
The general spectral behaviour suggests that some kind of matter ejection in a nova-like manner is taking place, and that this matter forms an envelope. It originates the shell features, and later on in the expansion leads to the appearance of observed emission lines from permitted and forbidden transitions. This is somewhat similar to a nova outburst, with obvious differences in the ejection velocity and mass, but in which a similar mechanism may be operating. At maximum, faint |FeII| and absence of other forbidden lines give indication of a rather dense envelope, or at small distance from the exciting source. Not high dilution effects tell us that the P Cyg shell should have $r \sim 5R_*$, that is of the order of 10^6 km. This is a much smaller dimension than expected from the expansion velocity, since we do not have a single expanding layer, as it occurs in the normal P Cyg-type stars where continuous ejection is observed.

3. The three-component model for SS.

The energy distribution of SS and color variations show that the continuum is steeper and Balmer jump increases when the star becomes fainter. Boyarchuk (1968) has thus supposed three sources of energy in SS, finding a satisfactory agreement between observed and theoretical distribution.
(1) An M-type giant (in general say G-M) which plays an important role in the visual-red spectrum, like a normal star of its class. The typical deep absorption features are reduced, but well seen in the red-near IR region where there are no other significant energy contributions. Bands of TiO,VO, and other luminosity criteria whenever possible, allow a classification as class III. We do not find S-type stars, but at least in UV Aur a C-type object is clearly present. Typical parameters are $T \sim 4000°$, $R \sim 10^2 R_o$, $M_v \sim -0.5$, masses probably of the order of 3-7 M_o.
(2) A hot and small component from which the strong blue continuum must come. It has typical values of $T \sim 10^5°$, $R \sim 0.5 R_o$, $M_v=0-2$, mass possibly of the order of 1-2 M_o. If one supposes that the late-type components of SS are normal giants, as favoured by the spectroscopic data, it follows that their hot components are located below the main sequence. In this region also central stars of PN, hot components of U Gem stars, novae, are also located, suggesting physical and evolutionary connections.
(3) Both stars are surrounded by a nebulosity, with $T_e \sim 17000°$, $n_e \gtrsim 10^6$-10^7 cm^{-3}, $R \sim 10^4 R_o$, mass of the order of $10^{-3} M_o$. Forbidden and permitted (at least in part) lines, usually found also in PN, are emitted from this

region and excited by the hot nucleus.
Forbidden lines of intermediate IP, e.g. OII 7325,SIII,ArIII, commonly among the strongest features in gaseous nebulae and nova ejecta, are absent or very weak in most SS, although lines of lower and higher IP (like |FeII| and HeII or sometimes |FeVII|)are usual in SS of comparable ionization stage. This absence is not due to an effect of density (which can be estimated from |SII|,|OII| doublets) but of ionization. The absence of these ionization stages is explained with a smaller dimension of the nebula, where ArIII and SIII are further ionized. SIII requires 35 eV, very close to the IP of OII, and OIII is indeed represented by strong lines.
Some permitted emissions are usually associated with the nucleus: CIII, NIII (showing the P Cyg effect), weak OIII, and (probably to some extent) H,HeI. This nuclear spectrum would indicate a temperature of $T_*=5\ 10^4 - 10^5$°, in agreement with the excitation stage in the nebula. Although the 4640 feature is typical of WR stars, it is not implied that the nucleus were an object of this type, where the emission is much broader. It is also possible that the "nuclear" emissions are produced in the innermost regions of the nebular shell.
On the other hand the low IP emissions of FeII,|FeII|, are observed simultaneously with |OIII| or even |FeVII|, a behaviour not recorded in the nebular spectrum of PN or novae. This presence of different IPs leads to suppose that they are not emitted in the nebula, but are possibly located in the extended atmosphere of the M giant or in a part of the envelope very close to it. In the dense atmosphere of the cool star single ionized atoms and TiO can be shielded, and further ionization or dissociation is prevented.
With the three-component model of SS one can determine the relative contribution of radiation from each source at any wavelength and time. It appears that the cool component does not undergo remarkable brightness variations, neither does the nebula, as confirmed by the spectral characteristics not changed since before the outburst. On the contrary the variations of the hot component are larger, so they determine the stellar variability on the whole. The temperature of the hot component should increase simultaneously with its magnitude; which is in agreement with the observed spectral and color changes of SS. From the computations in this model, the visual magnitude and temperature of the hot component result to vary in such a way that L_{bol} does not significantly change. In conclusion the outburst is most likely explained with the presence of a transient optically thick envelope surrounding the hot star. The individuality of SS is possibly due to a range in densities and velocities during this ejection phase. The observations of TiO bands together with the P Cyg-type spectrum support the interpretation that the shell is expanding around the hot star; since it is responsible for the high excitation, we understand why no HeII and faint HeI are observed during the P Cyg phase. We may think that also mass ejection takes place from the late-type object, especially if in the LPV stage, and it contributes significantly to the formation of the nebula. Also the ejection around the hot source can be itself a result of mass outflow from the cool companion. The nebular spectrum is similar before and after the outburst, suggesting that the distance of the nebula from the exciting component is large in

comparison with the radius of P Cyg-type shell. The weak nebular spectrum at maximum light can be understood in terms of partial volume of ejection, not completely shielding the exciting radiation, or of little amount of ejected mass (dilution before recombination).

4. Evidence for the binary nature.
It has been now assumed that SS are interacting binary systems, consisting of a red giant which provides the low-temperature spectrum and the required gas, and a hot star being the source of high-excitation lines which are difficult to be explained with the observed color temperature. Both are surrounded by a nebula of relatively small dimension and high density. All the basic observational data can be explained in this double-star hypothesis, which is the oldest and appears till now the simplest one. Of course it is not needed it is valid for all the objects classified SS, since they do not form an homogeneous class. Evidence for this interpretation is found as follows :
(a) The behaviour of the different components is well correlated with the spectral and photometric variations. The location of spectral features, indicating different physical conditions, is explained in the easiest way.
(b) Radial velocity curves with long periods (of several 10^2 days, mostly in agreement with light variations) are observed as in spectroscopic binaries. Lines of the red giant show an opposite trend to that found for high-excitation emissions associated with the hot source. The periodic changes are not accompanied by changes in the M spectrum. Among these cases we have e.g. AG Peg (820^d), BF Cyg (750^d), RW Hya(370^d), Z And (725^d), AX Per(880^d), CI Cyg(885^d), possibly R Aqr(9740^d). AG Peg (Hutchings et al., 1975) is rather strange having an $1M_0$ hot star with mass transfer toward the cool primary (not filling its Roche lobe), contrary to what expected. Although it is not clear, it might represent material lost after outburst. AG Peg was characterized by a protracted outburst, lasting much more than in other classical SS. If one include the recurrent novae among SS, T CrB is another known binary system with P=228^d, where the M giant looses mass onto the $2.6M_0$ companion, accreting through a disc.
Besides the objects with established periodicity, there is another group in which at least variations of radial velocity are recorded, and which deserve careful investigation. The orbital periods cannot anyway be easily detected when large orbits are present. Of course one must be careful in considering which lines may reflect an orbital motion, for instance most emission lines should arise from the surrounding region, and gas motion would add variable components to the data.
(c) A particular example is given by AR Pav, the first eclipsing variable among SS (see Thackeray,1959). During the orbital period of 605^d (probably indicated also by velocity variations) we observe that low-excitation lines like FeII,|FeII|, and TiO bands are better seen during the eclipse, while HeII disappears, H and HeI are weakened, and nebular emissions persist with small attenuation. We interpret these results with the occultation of a hot nucleus (primary star) exciting a dense nebula with stratification like in PN, by a cool giant filling its lobe.
While asimmetric lines led to the suspicion that also in Z And we may have a partial occultation, we have now a second case in CI Cyg with repea

ted eclipses observed in 1975 by Belyakina (1979) and during summer 1980 by many others. The period is P=885^d, and is accompanied by correlated spectroscopic variations.
(d) A rapid flickering (0.04 mag in 5^m) in blue light, together with larger and slower flares (0.1 mag in 15-20^m), has been reported for CH Cyg (Slovak and Africano,1978). This type of photometric activity, typical of the binary dwarf-novae, is a strong evidence for a binary system in which mass transfer between components is taking place. It is supported by the strong blue continuum detected at all phases over a typical M6III spectrum. The authors suppose a widely separated binary where the cool component (a SR giant?) is blowing a stellar wind, then interacting with the hot component. One must however note that CH Cyg (with low-excitation spectrum) is not a typical member of the SS group, where such flickering activity is not usually present (Walker,1977).
(e) The binary nature for SS appears to be confirmed by observations in the ultraviolet and infrared regions. Evidence is respectively reported for continuum of and lines excited by the very hot component, and continuum,features,in some cases typical variability of the cool giant.
(f) We may finally note how the proposed association of late giant and hot underluminous stars is displayed by other objects. Ordinary and dwarf novae where anyway the cool primary is a dwrf star, are binary systems. Because of the similarity between SS and some PN in spectral properties, it is to recall that also binary nuclei for PN are increasingly detected.

5. Spectroscopic properties and anomalies.

In describing the spectra of SS, and trying to deduce physical information from them, it is necessary to take into account that effects of stratification af layers and asymmetry are clearly present. Thus in some important lines, contributions from more than one region are probably involved, and derived parameters cannot be reliable ones. Otherwise different parts of the system, with particular velocity,temperature and density conditions, can be responsible for different lines. This seems to explain the difference in profile and velocity observed for the |OIII| lines at 4363 and 5007-4959 Å: the first one is usually narrow and variable, the other are broader and less affected by velocity variations.
They respectively indicate geometrical and kinematical properties of the inner zone where 4363 is emitted, and of the lower density extended region from which $N_{1,2}$ mostly come.For the same reasons displacements of different lines in radial velocity curves are recorded. There can also be contributions from a gaseous stream between components, affecting the regularity of an orbital motion: phase shifts are indeed observed from forbidden and permitted lines in the two line systems.
Attention must be paid to the interesting behaviour of some lines (see e.g. Swings,1970). In the HeI spectrum, a peculiarity is known for the enhanced singlet/triplet line ratio, which is moreover sometimes variable indicating time variations in the physical conditions. Singlet and triplet emissions also differ in showing different velocity curves. The OI 8446 triplet is usually strong in emission in SS, contrary to what observed in PN or novae: it is commonly explained with the mechanism of Ly$_\beta$ fluorescence, which is favoured in the case of expansion of the emitting gas. Two strongly correlated (same region, same atom) features at

6830-7090 Å have been recently discussed. They have no certain identification, are rather broad (20-30 Å), often variable; certainly they arise in a different region from that of forbidden lines. They are associated with only SS of high-excitation stage (IP$\geq$ 100 eV) where λ 6830 can be one of the strongest emissions. Allen (1980a) suggests they may be velocity broadened since formed in a rotating cloud like disc in dwarf-novae and X-ray binaries, which could be generated around the hot component by mass transfer from the cool giant. Also the feature at 6830 Å is not absorbed by TiO bands.

6. Single-star interpretations .
Owing to some difficluties among which the absence of radial velocity variations in many SS, single-star models have been also proposed. One of the alternative pictures described SS as single cool stars with very extended and dense coronae. High excitation should be in it produced from shock waves, or magnetic activity from the surface, but this remains the main difficulty. Usual absence of coronal lines, and the blue continuum detected at quiescence in several stars, are not understood in this type of model.

Otherwise it is proposed that the SS may represent a transition between the red giant and blue degenerate phases, thus obtaining two sources of radiation in the evolution of a single object. They would consist of a very hot, small, condensed core (remnant) surrounded by an extended atmosphere, possibly consisting of different layers or partial envelopes. This latter would be the M-type atmosphere now dissipating into space, and in whose outer parts the cool absorption spectrum can be formed, shielded in dense clumps, whereas the hot emissions would originate in the inner part. At the same time the nucleus is contracting and remains separated from the extended atmosphere.

A more recent version of this type has been proposed by Kwok to explain the origin of planetaries from objects like V1016 Cyg and HM Sge (e.g. Kwok et al.,1978; Kwok and Purton,1979). Widths and intensity evolution are predicted for different emission lines, and these tests for the model should be checked by appropriate observations. The transformation of the red giant through continuous mass loss should indeed lead to systems of lines typical of low-density from the red-giant slow remnant wind, of permitted lines due to the high density from the high-velocity new stellar wind of the exposed nucleus, and of forbidden lines of moderately high density from the shell which is being formed at the interface of the two winds by their interaction. In this evolution the first system should weaken with time. Some reported observational data may agree with this picture. We anyway think that it is still hard to explain in this way the origin of the cool continuum, its periodic variations, and the molecular features within which the emission spectrum is not obviously absorbed as foreseen in this case. It rather seems to indicate that an M star is present in the system.

7. The "very slow novae" and BQ|| stars.
A sequence of morphological properties, connecting typical SS (no ArIII and SIII, no IR excess, no radio emission) to compact PN like NGC 7027 (strong ArIII and SIII, IR excess, radio emission) can be found through objects like V1016 Cyg (strong radio emission, large IR excess, similar

ArIII and SIII doublets),HM Sge,V1329 Cyg. They form a small group of
stars which underwent a single large (∼5 mag) outburst, and then remain
at maximum or fade in a very long time, from which the name of "very slow
novae" (see e.g.Allen,1980b; Mammano and Ciatti,1975). Before the out
burst they presented colors and spectrum with low-excitation emission
typical for LPV. The emission spectrum, already present during the rise
in brigthness contrary to that found in ordinary novae, has successively
increased its ionization stage toward very high levels (up to |FeVI,VII|
and λ 6830),and is very rich in forbidden lines (Ciatti et al.,1979, and
references therein). Symbiotic characteristics are found in the visible-
near IR continuum and features of a very cool source. In particular at
least in V1016 Cyg there are Mira-type pulsations which appear to confirm
its binary nature, although the single-star model has been also applied
to this star.
These objects are correlated with the slow novae RR Tel,RT Ser, although
in these cases the spectrum at maximum indicated a true nova outburst (
see again Ciatti et al.,1979). In RR Tel the LPV pulsating before maxi
mum is still detected with the same periodicity (Feast, quoted by Allen,
1980b) as it may be the case for V1016 Cyg. A number of similarities
between these two objects have been remarked.
They are also suggested to represent early stages in the formation of PN,
at least some kinds of them, since the large envelope already results in
a very similar spectrum, and there is evidence for further expansion. In
V1016 Cyg the nebula expands with 35 km/s, while material is ejected by
the central star at 105 km/s.
Such "very slow novae" match the properties of the so called type II SS,
or "symbiotics with dust" in which strongly correlated radio emission
and IR excess from dust shell are recorded. Their cool components are
mostly of very late spectral type, and present variations typical of Mi
ra stars. One can explain all these differences with the classical SS,
by considering that emission and dust are strongly favoured by low-den
sity, and this means more extended envelopes, in whose outer parts this
condition can be reached. Mira stars which have considerable mass loss
by stellar wind well account for the formation of extended nebulae (la
ter spectral type - stronger mass loss - more extended nebulae - lower
density - forbidden lines and dust).The Mira characteristics are here
enhanced so that the question has been put on whether the dust is cause
or effect in symbiotic systems containing a Mira (Feast et al.,1977).
The photometric properties of this class make possible that other SS of
high excitation and unknown variability had underwent, on longer time
scale, a similar large outburst. May be there is a graduality in light
and spectral variations, similar to that among novae of different clas
ses, which would suggest a similar mechanism. It has been proposed that
the same accretion of matter lost by the giant may trigger in the hot
component H-shell flashes with increasing UV flux into the nebula, which
mimics the unveiling of degenerate core in the single-star model. Inter
val and duration of flashes would depend on the accretion rate, thus
providing a possible explanation for several cases of variability. A si
milar model applies to the novae, and RR Tel might represent a transition
case in a sequence dwarf novae-classical novae-SS.
Also the soft X-ray emission, absent in other SS of any spectral excita

tion, is found only for the "very slow novae". The intensity declines after the optical outburst, so that it can be related to the occurrence of thermonuclear surface burning.

These particular symbiotic objects are connected with the subclass of BQ|| stars which are also characterized by IR excess and sometimes by radio emission (Ciatti et al.,1974). Most of them are of lower spectral excitation than classical SS indicating that their hot components could be of lower temperature. There are no evidences of optical outbursts, while the spectrum shows typical lines of relatively high density (10^6-10^7 cm^{-3}) in the nebula, and support their binary nature with the presence of bands of late-type stars. They possibly represent a similar evolutionary phase in a different mass interval than the other SS.

8. Comparison of SS with other classes.

The spectroscopic properties of SS at different times have been compared with those of other classes of objects, like Wolf-Rayet (much denser and smaller envelopes), Of stars (lower excitation and small, dense shells), P Cyg-type stars, and Novae (larger ejection velocity and mass, transient nebulosity, binary nature). As already discussed, most remarkable is the similarity of the nebular spectrum of SS like Z And with that of PN of high excitation: forbidden lines are well represented, stratification effects are found, a great range in excitation conditions (e.g.NGC 7027) occurs, spectral variability is observed in both classes. For these reasons some SS were first classified as PN and the same is still possible true for other objects, where at least the cool spectrum is relatively less conspicuous. The main differences are in the intensity of λ 4363, weakness of |OII| 3727, in the Fe spectrum, in the presence of usually strong OI 8446, and absence of intermediate excitation lines (|ArIII|, |SIII|) in typical SS. From all these properties we infer that their nebulae have higher density (of the order of 10^{6-7} against 10^{3-4} cm^{-3}). A much closer similarity is found with compact planetaries like IC 4997 which is sometimes considered intermediate object between nebulae and stars.

Evolutionary connections among SS and PN are thus envisaged, with SS possible progenitors having smaller and more compact ejected envelopes, but the picture is not yet clear. It appears anyway that expansion velocities are comparable, as well as the location in the HR diagram and their kinematical and galactic distribution. Also in the binary star hypothesis for SS, we again recall the increasing number of binariety in well studied nuclei of PN too.

On the other hand Sahade (1965,1975) has most considered the relations of SS with other types of eruptive variables: U Gem and Z Cam, classical Novae, ultra-short-period binaries. In this discussion, the symbiotic phenomenon requires a binary system, with the presence of a hot subdwarf component together with a red giant (or supergiant) which is filling its Roche lobe. These conditions still hold for the other groups of variable stars undergoing outbursts related to mass accretion, except that the cool component is not a giant. In particular objects like T CrB are interpreted as transition cases between Novae and SS. Different orbital and variability periods imply however different evolutions. A discriminating parameter could be the evolutionary phase of the subdwarf compo

nent, larger outbursts corresponding to older and hotter stars.
In order to define the evolutionary status and perspectives for SS, we
need more accurate data and analysis. Unfortunately we do not have many
mass determinations for their components, and the evolutionary traks in
the region they cover in the HR diagram are not well known. We may add
some information on their distribution inside the Galaxy (Boyarchuk,1974)
although the relatively small number may be not appropriate for statistics. There is no particular distinction with the properties of PN, in
galactic latitude and longitude, and thus they have been assigned to the
old disk population. One can also obtain that the total number of SS
would not significantly exceed 10^3, that is they are rather rare objects
in the Galaxy. We can also see that the abundances are not well determined for these stars. Boyarchuk reported that 8 SS do not differ in chemical composition from the solar atmosphere, but different results have
been presented elsewhere. One has to be careful in studying line intensities which likely come from a range of different layers in the system:
better calibrated high-dispersion spectra are needed for this purpose.
Without going into details, we may finally quote the works on the origin
and evolution of SS presented by Tutukov and Yungel'son (1976), Paczynsky and Rudak (1980), all in the frame of binary model with mass exchange, as well as that of Bath (1977) where the properties of the outbursts
and their similarity to those of novae are explained.

References
Allen,D.A. 1978, IAU Colloquium N°46, Hamilton NZ
Allen,D.A. 1980a, Monthly Notices Roy.Astron.Soc. 190,75
Allen,D.A. 1980b, Monthly Notices Roy Astron.Soc. 192,521
Bath,G.T. 1977, Monthly Notices Roy.Astron.Soc. 178,203
Belyakina,T.S. 1979, Isvestia Crimskoi Astroph.Obs. 49,133
Boyarchuk,A.A. 1969, Non-periodic phenomena in variable stars,Budapest,
                page 395
Boyarchuk,A.A. 1974, IAU Symposium N°67, Moscow, page 377
Ciatti,F.,D'Odorico,S.,Mammano,A. 1974, Astron.Astrophys. 34,181
Ciatti,F.,Mammano,A.,Vittone,A. 1979, Astron.Astrophys. 79,247
Feast,M.W.,Robertson,B.S.C.,Catchpole,R.M. 1977, Monthly Notices Roy.
                Astron.Soc. 179,499
Hutchings,J.B.,Cowley,A.P.,Redman,R.O. 1975, Astrophys.J. 201,404
Kwok,S.,Purton,C.R.,FitzGerald,P.M. 1978, Astrophys.J. 219,L 125
Kwok,S.,Purton,C.R. 1979, Astrophys.J. 229,187
Mammano,A.,Ciatti,F. 1975, Astron.Astrophys. 39,405
Mammano,A.,Rosino,L.,Yildizdogdu,S. 1974, IAU Symposium N°67, page 401
Merrill,P.W. 1950, Astrophys.J. 111,484
Paczynsky,B.,Rudak,B. 1980, Astron.Astrophys. 82,349
Sahade,J. 1965, IAU Colloquium on Variable Stars,Bamberg, page 140
Sahade,J. 1975, 20th Liège Colloque d'Astrophysique, page 303
Slovak,M.H.,Africano,J. 1978, Monthly Notices Roy.Astron.Soc. 185,591
Swings,P. 1970, Spectroscopic Astrophysics, Univ.of Calif.Press,page 189
Swings,P.,Struve,O. 1941, Astrophys.J. 93,356
Thackeray,A.D. 1959, Monthly Notices Roy.Astron.Soc. 119,629
Tutukov,A.V.,Yungel'son,L.R. 1976, Astrofizika 12,521
Walker,A.R. 1977, Monthly Notices Roy.Astron.Soc. 179,587

SPECTRA OF INDIVIDUAL SYMBIOTIC STARS

Nancy A. Oliversen and Christopher M. Anderson
Washburn Observatory, University of Wisconsin-Madison

ABSTRACT

In connection with the discussion of the individual symbiotic stars, we present low resolution, flux calibrated spectra and high resolution H α profiles for each of 12 objects.

I. INTRODUCTION

Since 1978 September, we have conducted a regular program of spectrophotometric observations of a sample of 12 symbiotic stars. The observations are obtained at low resolution (c. 8 A) with a Boller and Chivens cassigrain spectrograph and at high resolution with the Washburn Observatory echelle spectrograph (Schroeder and Anderson 1971) both of which are used on the Observatory's 0.9-meter telescope. In both cases an intensified Reticon detector (McNall and Nordsieck 1976) is used. The observing techniques and reduction procedures are described in detail in Anderson, Oliversen and Nordsieck (1980) and Oliversen (1981). In addition to the standard flat field, dark/sky subtractions and extinction correction the low resolution spectra are calibrated against flux standards while the high resolution line profiles are wavelength calibrated on every object. The low resolution spectra thus give an accurate relative flux distribution over the wavelength range 3900--5900 A. The relative fluxes differ from the true fluxes by a factor of order unity which is the ratio of slit and seeing losses on the object to the average thereof during the standard star observations. The fidelity of the wavelength scale of the line profiles has been checked against IAU velocity standards and appears to be about 3 km s^{-1}.

II. DISCUSSION

Representative spectra of each of the objects are given in Figures 1 through 12. In table 1 we summarize in a semiquantitative

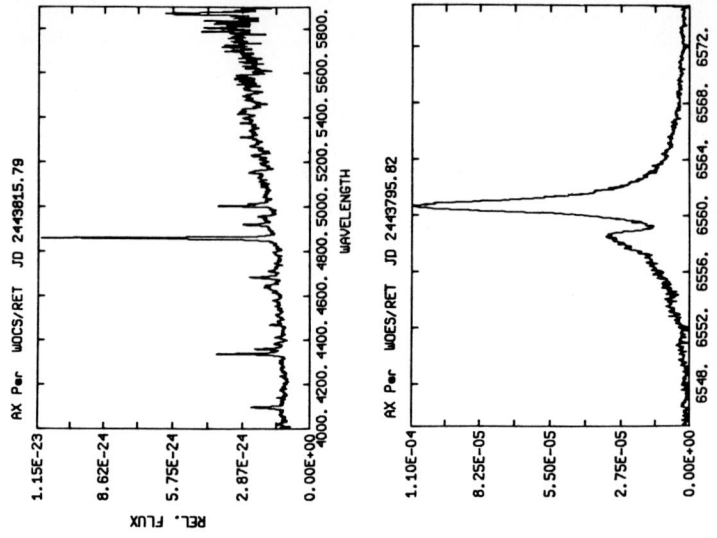

Figure 2

Figure 1

SPECTRA OF INDIVIDUAL SYMBIOTIC STARS

73

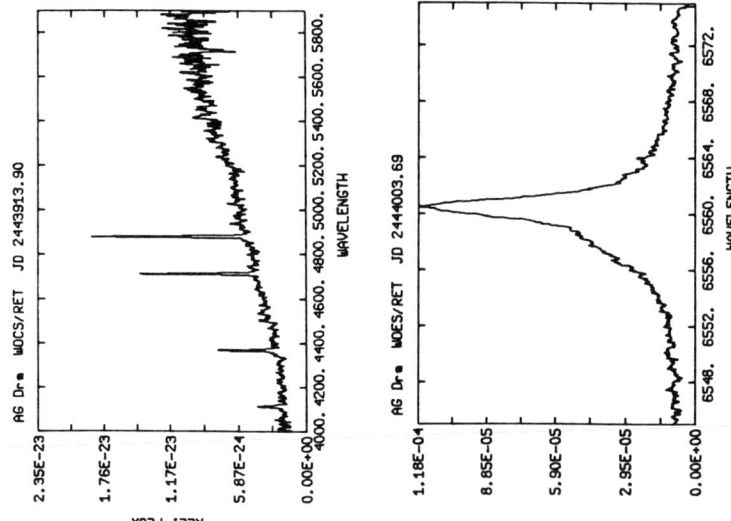

Figure 4

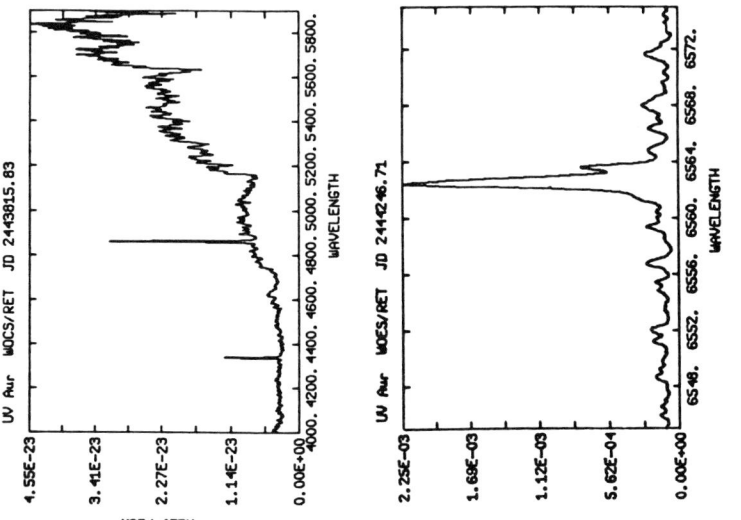

Figure 3

TABLE 1
SUMMARY OF OPTICAL SPECTROPHOTOMETRIC DATA ON SYMBIOTIC STARS

STAR NAME	K/M STAR OBSERVED	Hα PROFILE	H	INTENSITIES wrt Hβ (note 1)		
				λ 5007 [O III]	λ 4363 [O III]	λ 4686 [He II]
HD 4174	Y		.3:	--	--	--
AX Per	Y--N		.3	.3--.1	.3--1.5	.1--.5
UV Aur	Y		.4	--	--	--
AG Dra	Y		.3	--	--	.4--.7
MWC 603	Y,v		.3	.08	.2	.1
CH Cyg	Y		.2	--	--	--
BF Cyg	N		.3:	2.1	.3	.05:
CI Cyg	Y,v		.4	.4	.7--1.5	.3--.5
V 1016 Cyg	N		.3	2.2	.6--.9	.5
HBV 475	N		.3	.5	--	.4--1.2
AG Peg	Y		.4	.5	.2--0	.5 vb
Z And	Y		.3	.05	--	.4--1.2
HM Sge	N		.3:	2:	π	.5:
RR Tel	Y:		.3	1.2	.9	.6
He 2 177	N		.2	.7	.5	.5
RY Sct	N			--	--	--
FR Sct	Y			--	--	--
AR Pav	Y,v		.3	.7	.5	.3
RS Oph	Y,v		.2	--	--	--
R Aqr	Y		.3	1.8	.6	--
HK Sco	Y		.2	--	--	.4
AS 295 B	N		.2	.4	--	.03

	λ 4922	λ 5876	SPECIES PRESENT			
STAR NAME	He I s	He I t	[N II]	[Fe II]	[Fe II]	[Fe VII]
HD 4174	--	--	N	--	--	--
AX Per	.1	.5	N	N	N--Y	Y--N
UV Aur	--	--	N	--	--	--
AG Dra	.05:	.5	N	N	N	N
MWC 603	.1:	.3	N	N	Y	N
CH Cyg	.05:	.3:	N	Y	Y	N
BF Cyg	.05	.3	Y:	N	Y	N
CI Cyg	.05	.3	N	Y	Y	N
V 1016 Cyg	.05	.2	Y	Y	Y	Y
HBV 475	.08	.4	N	Y:	Y	Y
AG Peg	.07	.3:	N	N	Y	N
Z And	.07	.3:	N	N	N	Y
HM Sge	.01:	.5	Y	N:	N:	π
RR Tel	.05	.1	Y	Y	Y	Y
He 2 177	.1	.4	Y	N	Y	Y
RY Sct	--	--	Y	N	N	N
FR Sct	--	--	Y	Y	N	N
AR Pav	.1	.4	N	N	Y	N
RS Oph	--	1.0	N	N	N	N
R Aqr	.05	.2	Y	N	N	N
HK Sco	--	.5	N	N	Y	N
AS 295 B	.01	.2	Y	N	N	N

Note 1. [O III] $\lambda 4363$ Intensity is w.r.t. Hγ

way the general characteristics of the spectra. Included in this tabulation is information on ten southern objects observed in 1979 September-October at CTIO.

HD 4174 (EG And) Fig. 1

At low resolution the spectrum of the late-type component is most obvious while the emission lines can be weakly detected. The Hα line is usually weakly in emission with a FWHM of 100 km s^{-1}. In 1980 June, at a phase of 0.6 according to the ephemeris of Smith (1980) the profile was deeply reversed with the absorption core going well below the continuum of the cool star.

AX Per Fig. 2

AX Per has one of the most radically variable spectra in the sample. It has shown at some times a virtually featureless continuum with only relatively low excitation emission lines and hydrogen lines with deep reversals on the blue (Hα FWHM 180 km s^{-1}) side while at other times the molecular band heads of TiO etc. dominate the continuum, emission lines up to Fe VII are present and the Hα profile has only a slight blueward asymmetry (FWHM 80 km s^{-1}). The former condition is illustrated in figure 2.

UV Aur Fig. 3

The carbon star continuum dropping off rapidly into the blue is the most obvious characteristic of UV Aur at low resolution. Only the hydrogen lines are seen in emission and these have at times completely disappeared. When the emission lines are present, the Hα profile is unusually narrow (FWHM 40 km s^{-1}) main peak with a small subsidiary peak roughly 50 km s^{-1} to the red of the main peak.

AG Dra Fig. 4

In our low resolution spectra the continuum of AG Dra is featureless, slopping gradually downward to the blue. The emission lines of hydrogen, He I and He II are easily detected. The Hα profile has a strong sharp central peak (FWHM 95 km s^{-1}) and broad wings (375 km s^{-1} at I/I_c = 2) with the blue wing substantially stronger than the red. This latter condition is termed " the blue asymmetry ". As noted by Smith and Bopp (1981) around phase 0.5 ±.25 the blue wing appears to develop a slight reversal. Smith and Bopp attribute this to the variation in strength of one of the components of a two component system. Our spectra, which detect the stellar continnum and remain linear and unsaturated over the full intensity range of the profile, indicate that that an absorption feature develops in the existing profile.

MWC 603 Fig. 5

Through out our survey the continuum spectrum of MWC 603 has show moderately obvious molecular band heads. The emission lines of H, He I, He II and [O III] are seen but very little if any variations have been noted. The Hα profile shows a slight blue asymmetry and FWHM of 40 km s^{-1}. Occasionally small shoulders or other features appear on the blue wing of the emission.

CH Cyg Fig. 6

The continuum of CH Cyg varies from nearly featureless to totally dominated by molecular features. Hydrogen and neutral helium emission lines are always present. Singly ionized iron has been detected. The Hα profile is usually strong with a very deep central reversal and the blue peak 10 to 20 percent stronger than the red. However episodes have been encountered during which the relative strengths of the two peaks reversed on time scales of a few days. This phenomenon has been discussed by Anderson, Oliversen and Nordsieck (1980).

BF Cyg Fig. 7

At our low resolution, the continuum of BF Cyg is featureless although on KPNO echelle plates the absorption line of the late component are weakly visble. Hydrogen and neutral helium are seen as is [O III] and the relative strengths of these lines undergo substantial variations. The Hα profiles shows the blue asymmetry, FWHM of 120 km s^{-1} and some variations in slight features on the blue wing.

CI Cyg Fig. 8

The molecular features in the continuum of CI Cyg are always moderately obvious and the emission lines of hydrogen, He I, He II and [O III] are seen and their relative intensities vary in a manner correlated to some degree with the eclipses. The blue asymmetry of the Hα always has had some degree of reversal which becomes markedly stronger during the eclipses. Furthermore, the heliocentric velocity of the center of the reversal is least negative (-16 km s$_1^{-1}$) near the phase of the eclipse and reaches a minimum (-30 km s^{-1}) one half period later. This might indicate that this feature originates in a flow from the cool component to the small, hot object at a rate of about 7 km s^{-1}.

V 1016 Cyg Fig. 9

The continuum of V 1016 Cyg is but weakly detected and appears featureless. The emission line spectrum is the richest in the sample with both premitted and forbidden lines from neutral species all the way up to six times ionized iron. The relative strengths of the lines show only minor variations. The Hα profile is nearly

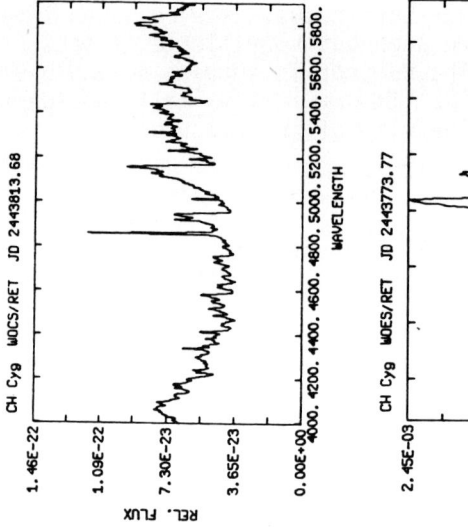
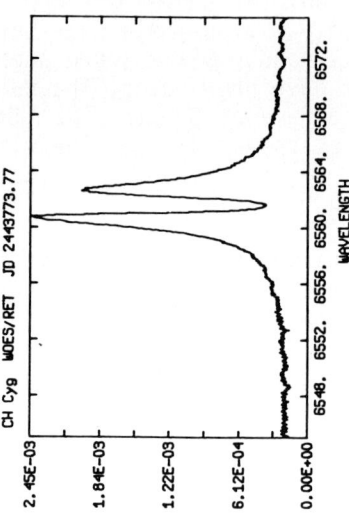

Figure 6

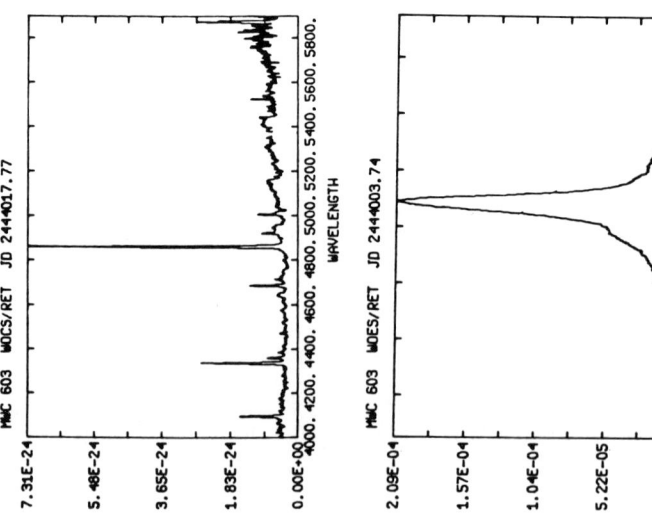

Figure 5

SPECTRA OF INDIVIDUAL SYMBIOTIC STARS

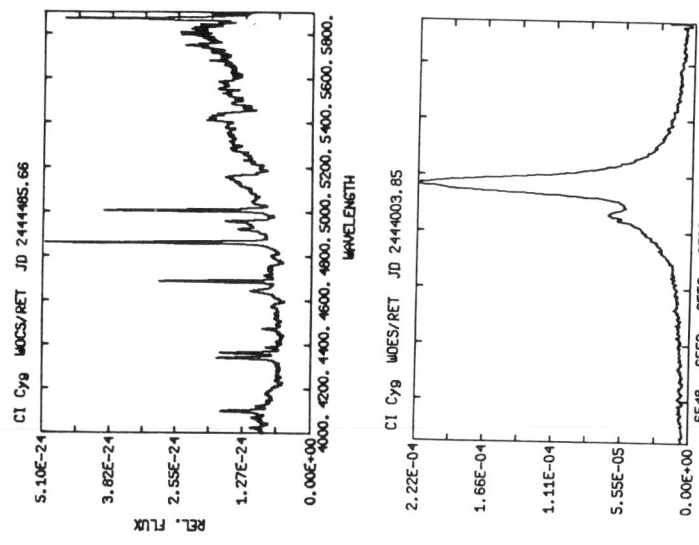

Figure 8

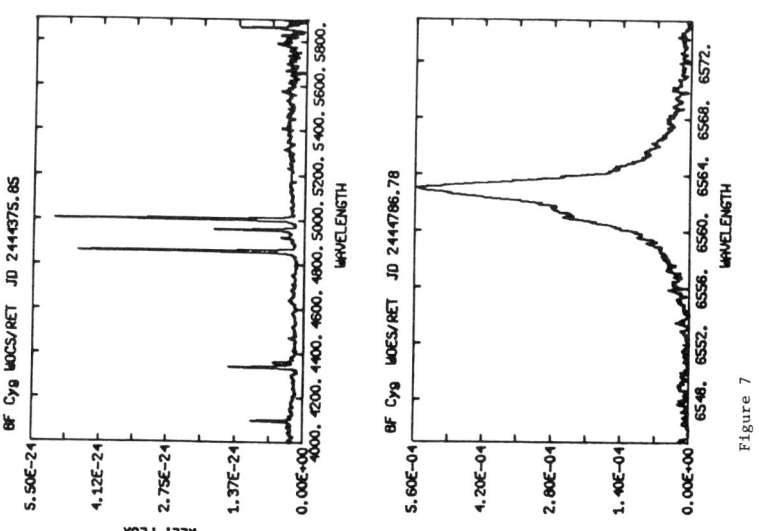

Figure 7

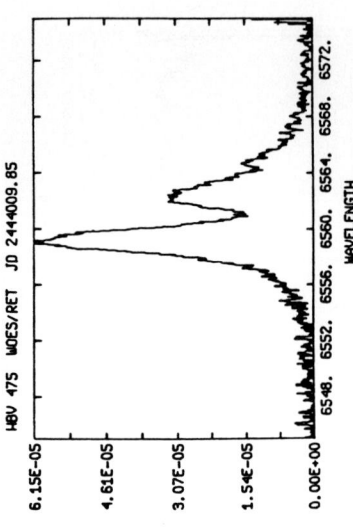
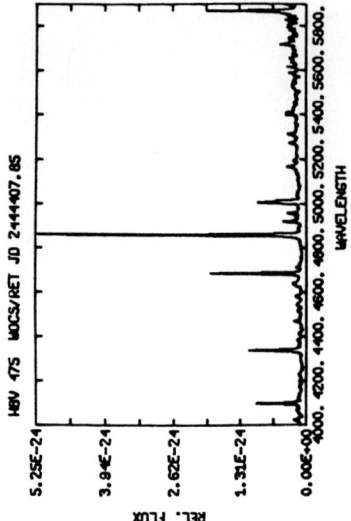

Figure 10

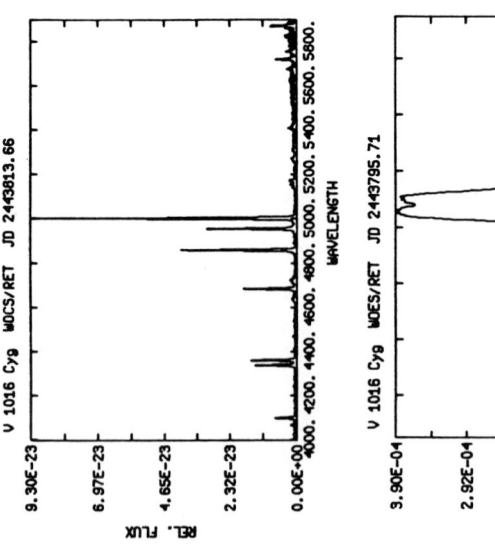
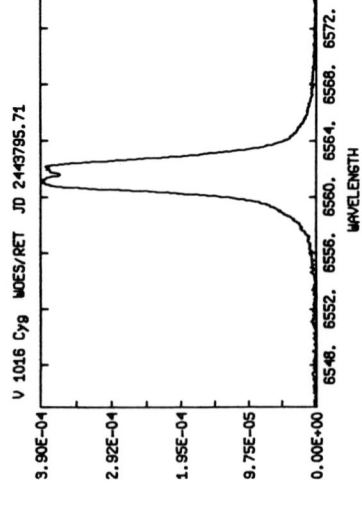

Figure 9

SPECTRA OF INDIVIDUAL SYMBIOTIC STARS

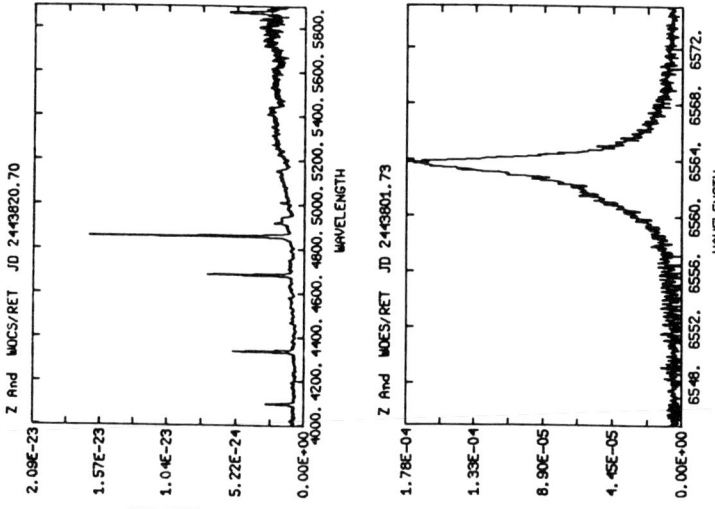

Figure 12

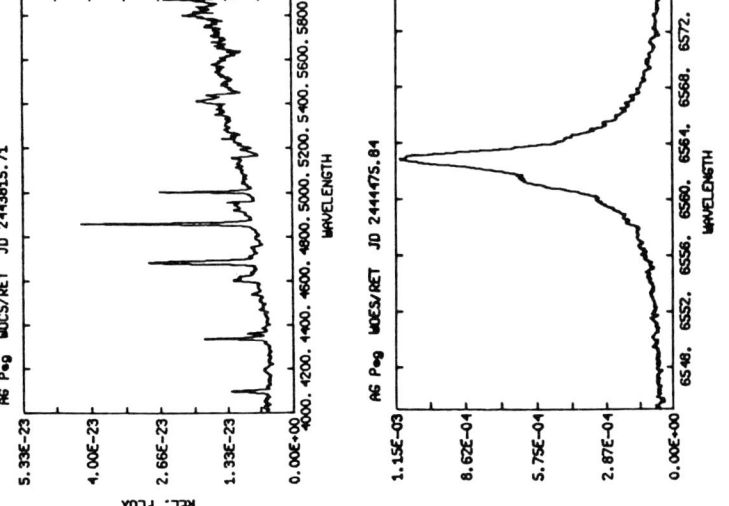

Figure 11

symmetric with a FWHM of 115 km s^{-1}. There is a slight reversal in the peak of the line in which some minor variations may have been detected.

HBV 475 (V 1329 Cyg) Fig. 10

At low resolution HBV 475 has most of the same ionic species as does V 1016 Cyg although with different strengths and its continuum is undetected. The Hα profile has undergone profound variations from the 500 km s^{-1} wide multiple component appearance shown in the figure to a nearly symmetric 170 km s^{-1} FWHM shape.

AG Peg Fig. 11

At our low resolution the continuum of AG Peg shows the evidence of late component only moderately well. The emission spectrum is the classic one with hydrogen, He I , He II , [O III] and Fe II. Even at low resoilution the extraordinary width of He II λ 4686 is obvious. The Hα profile shows the typical blue asymmetry with a reversal appearing in the blue wing occasionally. Several episodes or events during which the profile became remarkably disturbed have been observed.

Z And Fig. 12

Through the first two years of our program the molecular feature in the continuum grew progressively stronger, but have recently again declined in strength. When the molecular spectrum was most obvious, He II λ 4686 was substantially stronger than Hβ. The[O III] lines are only weakly visable in our spectra. The Hα profile has been seen in two very different shapes, one the standard blue asymmetry with a FWHM of 80 km s^{-1} the other a much broader line (FWHM c. 270 km s^{-1}) which has a reversed, somewhat narower peak.

III. CONCLUSION

We trust that this atlas of some of the visual spectroscopic characteristics of symbiotic stars will promote interest in and discussion of these most interesting objects.

REFERENCES

Anderson, C. M. , Oliversen, N. A. and Nordsieck, K. H. 1980. Ap.J. 242, 188.
McNall, J. F. and Nordsieck, K. H. 1976, Proc. I.A.U. Coll. No.40, 26-1.
Oliversen, N. A. 1981, Ph D. Thesis, Univ. of Wisconsin.
Schroeder, D. J. and Anderson, C. M. 1971 Publ.A.S.P.83, 438 .
Smith, S. E. 1980, Ap.J., 237, 831.
Smith, S. E. and Bopp, B. W. 1981, M.N.R.A.S., 195, 733.

VISUAL SYMBIOTIC SPECTRA OBTAINED WITH THE HAUTE PROVENCE
MULTIPHOT DETECTOR

G. Muratorio
Observatoire de Marseille, France

M. Friedjung
Institut d'Astrophysique, Paris, France

Spectra of symbiotic stars have been obtained with the Haute Provence
Multiplhot detector system. This has 512 pixels, so one can study either
a large spectral region at low resolution, or a smaller region with a
higher resolution.

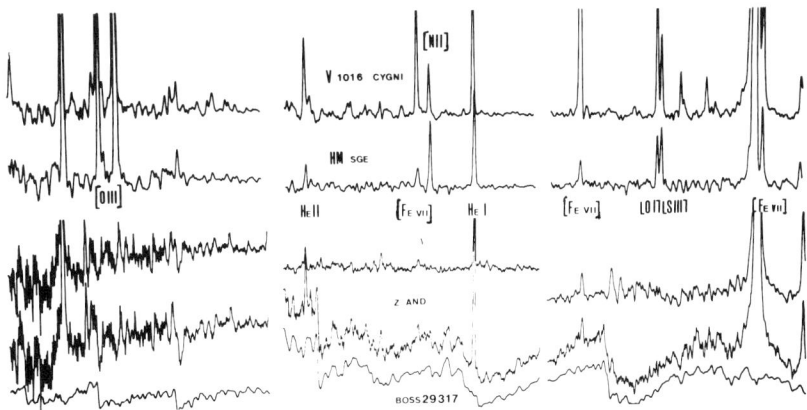

The figure shows the low resolution spectra of V1016 Cyg, HM Sge and
Z And in summer 1980 in three spectral regions: ~4800-5400 A, 5400-6000 A,
and 6000-6700 A. The spectrum of Z And (fourth row) is compared with that
of the M3III star Boss 29317 (fifth row), with a comparable strength of
the TiO bands. By subtraction we obtained the spectrum in the third row.

DISCUSSION ON OPTICAL OBSERVATIONS

<u>Nussbaumer</u>: May I draw the attention of the visual observers to the value of observations covering the [FeVII] lines. To determine electron temperatures it would be particularly valuable to obtain relative intensities for two lines where one should originate from ^1D ($\lambda 6087$ or 5721) and the other from ^1G (e.g. λ 3587). For a full list of the visual [FeVII] lines, see for example Nussbaumer and Osterbrock, 1970, Ap. J. <u>161</u>, 811. The value of comparing these line intensities with the [FeVII] $\lambda 2015$ line is mentioned in my talk on UV lines.

<u>Kafatos</u>: There is a puzzling fact about the optical line profiles. Why do the optical lines generally show considerable structure (large velocities, multiple components), whereas the UV lines show very little - if any - structure. One would expect intuitively the situation to be exactly opposite to what one sees, because the UV lines presumably come from regions closer to the hot component.

<u>Oliversen</u>: Good question: it seems likely that the H_α emission comes from a more extended region, probably consisting of more than one region. The blue reversal could be due to absorption in the systems by accreting material.

<u>Viotti</u>: Did you find in the symbiotic stars large variations in the H_α line profile, and what is the typical time scale of these variations?

<u>Oliversen</u>: Yes. Often the H_α profile will change from the single 'blue asymmetrical' profile to a profile with a reversal (or emission component) on the blue wing. Typical time scales for this are of the order of months to a few years.
V1016 Cyg has been constant in its H_α profile.

<u>Slovak</u>: We have made observations at H_α separated by a few minutes to hours and have noticed no short time variations.

<u>Swings</u>: Measurements of the line widths of H_α, HeI and [NII] in a series of peculiar emission line objects (Swings and Andrillat, Astr. Astrophys. in press), tend to support the interacting winds theory for the formation of planetary nebulae (see e.g. Kwok, these proceedings).

UV LINE EMISSION OF SYMBIOTIC STARS

H. Nussbaumer
Institute of Astronomy, ETH-Zentrum,
CH-8092 Zürich, Switzerland

ABSTRACT

General characteristics of emission line spectra from symbiotic stars are outlined. Data from some special line ratios in the 1000 Å - 3000 Å range, and others connecting the visual and the far UV lines are presented, and their application to symbiotic stars is discussed. Integrated fractional abundances for ions easily observed in the far UV are given to facilitate abundance determinations for nebular conditions. It is found that the physical conditions of the regions emitting the emission line spectra differ considerably among different symbiotic stars.

1. INTRODUCTIONARY REMARKS

In this review I concentrate on the diagnostic possibilities offered by studying the IUE (International Ultraviolet Explorer) line spectra, the continuum will be treated by Slovak (1982). The nebular spectrum as seen in the visual already tells us, that symbiotic objects contain emission regions where elements are typically one to three times ionised, with some higher ionisation stages for heavier elements. The well known [N II], [O III], [Fe VII] and other nebular lines appear with intensity ratios showing that electron temperatures and densities of $T_e > 10^4$ K and $N_e > 10^5$ cm^{-3} prevail. At 1100 Å $< \lambda <$ 3200 Å we not only expect to observe the strong resonance and intercombination lines from the ions already observed in the visual, but also those ions which do not possess low lying metastable levels, such as the very important missing links like C^{2+}, C^{3+}, N^{2+}, N^{3+}, O^{3+}.

2. THE UV SPECTRA (1100 Å - 3300 Å)

2.1. General appearance

The UV spectra of symbiotic stars are characterised by emission lines from neutral to several times ionised atoms. The easily observed emission lines are much stronger than the underlying continuum; absorption features can occasionally be detected. In some symbiotic objects high ionisation spectra such as N V, Ne V, Mg V (and perhaps Fe VII) are observed, whereas in others only once or twice ionised atoms are detected. As examples I show in Figure 1 the spectra of Z And (Altamore et al. 1981) and RW Hya (Kafatos et al. 1980). From these figures and other descriptions (e.g. Penston et al. 1981, Nussbaumer and Schild 1981) we find the typical nebular lines: recombination lines of He I, He II, the collisionally excited allowed, forbidden, and semiforbidden lines originating from the energetically lowest terms, as well as the O III Bowen lines excited by resonant absorption of He II Ly α. The identified Fe II multiplets form a contrast by not being restricted to resonance multiplets like UV 1, 2, 3, they cover the whole Fe II bound state energy domain. CH Cyg shows a different picture; its spectrum alternates between one with absorption lines of high ionisation stages (C III, C IV, N V etc.) and emission lines of singly ionised and neutral atoms (Hack 1979).

2.2. Preliminary conclusions about electron temperatures and densities

At a first glance these spectra could be mistaken for those of planetary nebulae. However, inspection of the C III λλ1907, 1909 doublet high resolution spectra (or descriptions) published by Keyes and Plavec (1980) (AG Peg), Kafatos et al. (1980) (RW Hya), Penston et al. (1981) (RR Tel), Nussbaumer and Schild (1981) (V 1016 Cyg), Altamore et al. (1981) (Z And), shows that in all these objects the λ1909 component is much stronger than the usually absent λ1907 line. According to Figure 5 of Nussbaumer and Schild (1979) we can therefore safely assume that the region emitting C III has electron densities $N_e > 10^6$ cm^{-3}. Altamore et al. (1981) have studied the N III] multiplet at λ1749 emitted by Z And. With the relative emissivities calculated by Nussbaumer and Storey (1979) they find $N_e \approx 2 \cdot 10^{10}$ cm^{-3}. They find the same result when studying the O IV] λ1401 multiplet for which the relative emissivities were calculated by Flower and Nussbaumer (1975). Penston et al. (1981) find for RR Tel $N_e \approx 5 \cdot 10^6$ cm^{-3}, this from the already mentioned C III λ1908 and from N IV] λ1485 for which the relative emissivities are given in Figure 2 of Nussbaumer and Schild (1981). For V 1016 Cyg Nussbaumer and Schild (1981) find $N_e \approx 3 \cdot 10^6$ cm^{-3} mainly from the N IV] λ1485 multiplet.

From symbiotic UV spectra thus far investigated we obtain $10^6 < N_e[\text{cm}^{-3}] < 10^{11}$. This is considerably higher than the typical value of $N_e \lesssim 10^4$ cm^{-3} found in planetary nebulae, though for the young

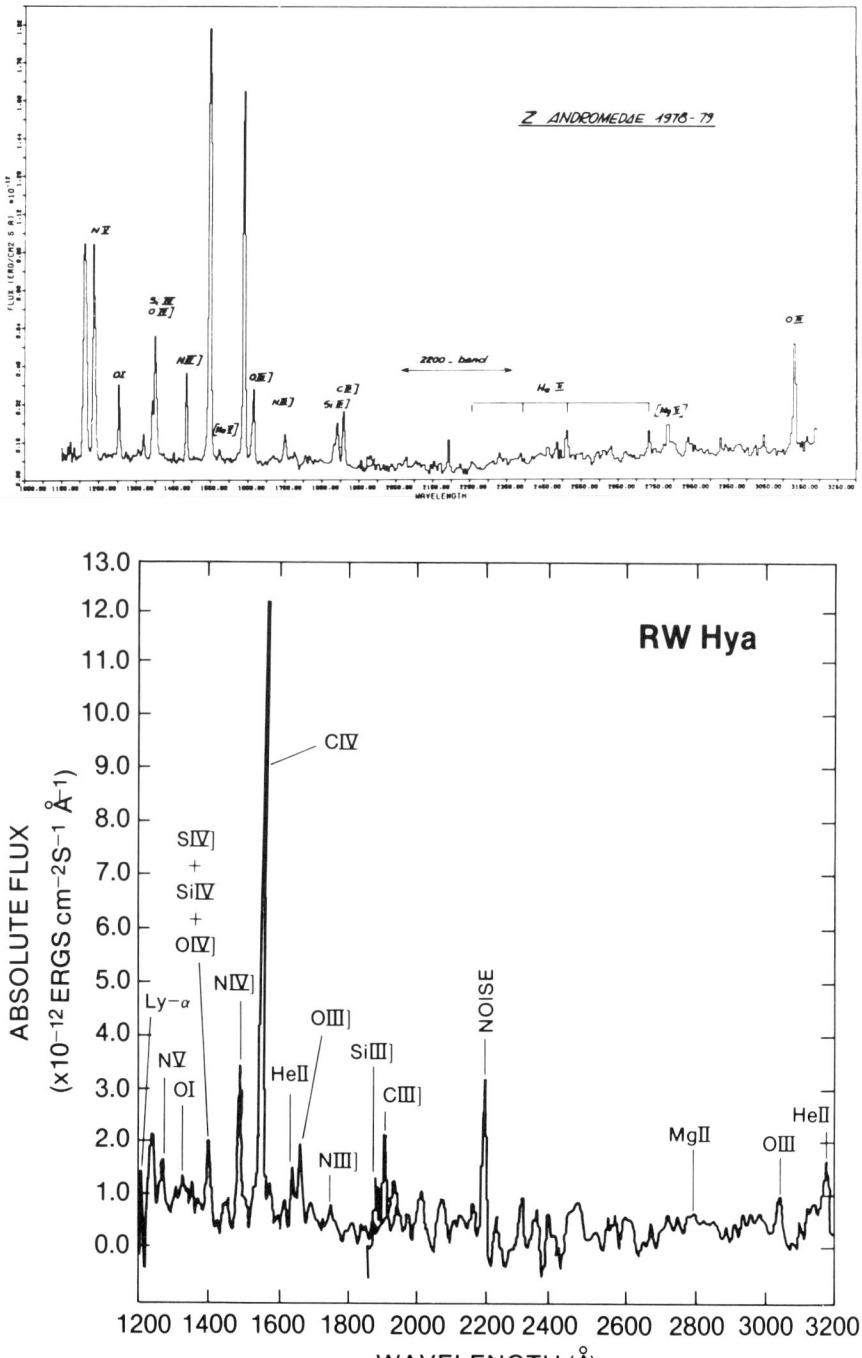

Figure 1. IUE low resolution spectra

planetary nebula IC 4997 a density of $5 \cdot 10^5$ cm^{-3} was derived by Flower et al. (1979); as a reminder: the innermost part of the solar corona has 10^8 cm^{-3} ≲ N_e ≲ 10^9 cm^{-3}.

What about the electron temperatures at which these spectra are emitted. For Z And Altamore et al. (1981) derive 60000 K ≲ T_e ≲ 80000 K. To find this result they have assumed C III] and N III] to be emitted from the same geometrical region; as a first approximation this is probably justified. They are on shakier grounds with the additional assumption that the abundance ratio N/C has the solar value (see next section). For V 1016 Cyg Nussbaumer and Schild (1981) find 8000 K < T_e < 25000 K; this temperature results from energy balance calculations and is thus very model dependent. For RR Tel Penston et al. (1981) claim excitation temperatures of 12000 K to 19000 K. This claim is backed by comparing recombination lines with resonance lines, like C III λ2296 with C IV λ1550 (for the theory see Storey (1981)). However, the temperatures thus found are only limits. They also call on the Si II lines λλ1264, 1533, 1817; from their relative intensities they deduce T_e ∿ 12000 K. But Jordan (1969) who calculated the relevant emissivity ratios found large discrepancies between observation and calculation which she blamed on the quality of the atomic data; from my own experience with Si II (Nussbaumer 1977) I cannot but share her suspicion. From the O III λλ2321, 1660, 1666 they find T_e = 19000 K by assuming N_e ∿ 10^5 cm^{-3}, but with their previously determined N_e ∿ $5 \cdot 10^6$ cm^{-3} the curves of Nussbaumer and Storey (1981) give T_e ∿ 40000 K for the observed ratios in RR Tel.

For RW Hya Kafatos et al. (1980) opt for T_e = 12500 K. They chose this temperature because they find that abundances of C, N, O, Si, S determined with a model of this temperature fit the cosmic abundances best; although they do not force a cosmic abundance on their model, indeed they find some significant abundance differences.

Sahade (1980) considers that the simultaneous presence of Mg II, N V and intermediate ionisation stages indicates a large range in excitation temperatures. Sahade may be right that some symbiotic spectra originate in chromospheres, coronae, or their transition regions, but a N V itself is no proof of T_e ∿ 10^5 K; in the V 1016 Cyg model of Nussbaumer and Schild (1981) the N V doublet is formed at T_e ∿ 25000 K.

2.3. Abundances

For symbiotic stars the IUE wavelength range provides very advantageous conditions for determining relative abundances, in particular of C, N, O and He. For C, N, O the collisionally excited ground state transitions of the different ionisation stages are available. From He we observe recombination lines like He II λ1640 and the n → 3 series (λλ3203, 2733, 2511, 2385 etc.) and He I recombination lines like some of the 2s ^{3}S - np ^{3}P lines. To determine abundances relative to H, we have to rely on lines of the Balmer series.

Nussbaumer (1980) has tabulated relative abundances of planetary nebulae as determined with IUE observations. For O/H where the solar value is $8.3 \cdot 10^{-4}$ the planetary nebula ratios vary from $1.1 \cdot 10^{-4}$ to approximately solar; for C/N where the solar ratio is 4.7, planetary nebula ratios vary from 0.5 to 6. This should warn us against assuming that symbiotic stars have solar abundances, when planetary nebula show such large variations. A review on abundance determinations and related problems, with a large number of references for gaseous nebulae has been given by Peimbert (1980). The C, N, O, He abundance is of central interest when trying to form an evolutionary picture of our objects; so are Si and Mg when investigating dust formation. Of these elements the most elusive is Mg. It is very likely to be present mainly as Mg^{2+}, but there are no Mg III lines in either the visual or the IUE wavelength range. The situation is as bad with Mg IV and only Mg V 2p^4 ^{3}P - ^{1}D $\lambda\lambda 2783, 2929$ establish contact with Mg. Mg II $\lambda\lambda 2796, 2803$ are of course well known and observed. However, as Mg^+ can also exist in the H^o region of our objects, self absorption in that doublet may be substantial; interstellar absorption proper, due to its small line width, could be more easily disentangled.

An abundance determination based on the $\lambda < 3000$ Å spectra is that of Nussbaumer and Schild (1981) for V 1016 Cyg. I list their results (model 2) to compare them with solar abundances (logarithmic abundances relative to 12 for hydrogen); the solar ratios are from Lambert (1978) for C, N, O, and from Withbroe (1971) for the other elements.

	He	C	N	O	Ne	Mg	Si
V 1016 Cyg	11.30	8.28	8.08	8.43	8.00	7.85	7.11
Sun		8.67	8.00	8.92	7.50	7.54	7.55

We have thus at least one symbiotic object with abundance ratios different from the sun, and it would be astonishing if it remained the only one.

Can we determine relative abundances in a simple way? The total energy from the m times ionised atom X in the transition i → j is

$$L_{ij}(X^m) = \int_V N(X_i^m) A_{ij} h\nu_{ij} dV \qquad (1)$$

$N(X_i^m)$ is the particle density of the m times ionised element X in the state i. In the case we treat, the integral will cover the H$^+$ region. We shall assume constant T_e and N_e throughout the emitting region. With the expansion

$$N(X_i^m) = \frac{N(X_i^m)}{N(X^m)} \cdot \frac{N(X^m)}{N(X)} \cdot N(X) \qquad (2)$$

and defining

$$P(X^m) = \int_V \frac{N(X^m)}{N(X)} dV, \qquad (3)$$

the expression for L_{ij} takes the form

$$L_{ij}(X^m) = \frac{N(X_i^m)}{N(X^m)} \cdot N(X) \cdot P(X^m) \cdot A_{ij} h\nu_{ij} \qquad (4)$$

where

$$N(X) = \sum_n N(X^n), \text{ and } N(X^n) = \sum_i N(X_i^n). \qquad (5)$$

As we are mainly interested in lines which are excited from the ground term, we shall work with the most simple atomic model: (a) the atom consists of 2 states, (b) the lines are optically thin, (c) the upper state is collisionally excited from the ground state and radiatively de-excited (no collisional deexcitation); (b) may be problematic for allowed lines and (c) for intercombination and forbidden lines. In an equilibrium situation the relation

$$N(X_1^m) q_{1i} N_e = N(X_i^m) A_{i1} \qquad (6)$$

is valid. The collisional excitation rate coefficient is defined as

$$q_{mn} = \frac{8.63 \cdot 10^{-6} T_{mn}}{g_m \sqrt{T_e}} \exp(-\Delta E_{mn}/k T_e) \quad [cm^3 s^{-1}] \qquad (7)$$

g_m is the statistical weight of the state m, and T_{mn} is the averaged collision strength. The emitted luminosity ratio of two lines of the elements X and Y is

$$\frac{L_{ij}(X^m)}{L_{k\ell}(Y^n)} = \frac{N(X)}{N(Y)} \frac{N(X_i^m)}{N(X^m)} \frac{N(Y^n)}{N(Y_k^n)} \cdot \frac{P(X^m)}{P(Y^n)} \frac{A_{ij}}{A_{ij}} \cdot \frac{\lambda_{k\ell}}{A_{k\ell}}$$

$$= \frac{N(X)}{N(Y)} \frac{P(X^m)}{P(Y^n)} \frac{T_{ij} g_\ell \lambda_{k\ell}}{T_{k\ell} g_j \lambda_{ij}} \cdot e^{-(E_i - E_k)/k T_e} \qquad (8)$$

The observed flux ratio is

$$\frac{F^{obs}_{ij}(X^m)}{F^{obs}_{k\ell}(Y^n)} = \frac{L_{ij}(X^m)}{L_{k\ell}(Y^n)} \cdot 10^{-c(f(\lambda_{ij})-f(\lambda_{k\ell}))} \quad (9)$$

where the extinction curve $f(\lambda)$ could for example be obtained from Seaton (1979) and the extinction coefficient c has to be determined separately. Disregarding the error sources in our primary assumption as well as those in F^{obs}, c, f, T, the remaining unknown quantities are the ratio $P(X^m)/P(Y^n)$ and the temperature T_e. Accepting the dangerous assumption of constant T_e, we may ask how strongly the uncertainty in the absolute value of T_e effects the resulting $N(X)/N(Y)$. Take as an example a $N(C)/N(O)$ determination with the lines O III λ1666 and C III λ1909. The upper states of the two transitions differ by 7614 cm^{-1} = 1.49 10^{-12} erg. The function $\exp(-(E_i-E_k)/kT)$ thus has values of 0.34, 0.49, 0.58, 0.76, 0.90 for $T_e/10^4$ K = 1, 1.5, 2, 4, 10. With a reasonable notion of T_e the error in $N(C)/N(O)$ should be $\lesssim$ 30%.

I have calculated $P(X^m)$ for several elements for the following conditions. A central star with a blackbody radiation of temperature T* and radius R* ionises a spherically symmetrical nebula with constant N_e and T_e, which begins at a radial distance r, and is semi-infinite. $P(X^m)$ is calculated for the region where H is ionised; in that region most of the elements of Table 1 will be at least singly ionised. When obtaining the results of Table 1 the dielectronic recombination coefficients of Aldrovandi and Péquignot (1973) were employed. But for $T_e \lesssim$ 20000 K autoionising states lying just above the first ionisation edge can significantly influence these dielectronic recombinations, as has been shown by Storey (1981). Thus, shifts in the ionisation balance have to be expected as more accurate recombination coefficients become available.

From Table 1 we see that for the conditions mentioned, the twice and three times ionised elements constitute mostly more than half of the total abundance. Thus extending expressions (8) to the two most important of the observable ionisation stages should already permit good estimates. Table 1 cannot replace a proper model calculation, but a feeling for the effects of variations in some of the relevant physical parameters can be acquired; it can serve to estimate relative abundances and to correct for unobserved ionisation stages.

Table 1 Integrated relative fractional abundances $P(X^m)/\Sigma_i P(X^i)$ for singly (m=1) and more highly ionised atoms

		$T_e = 15000$ K $N_e = 10^6$ cm^{-3}					$T_e = 40000$ K $N_e = 10^6$ cm^{-3}				
	T*	m=1	2	3	4	5	m=1	2	3	4	5
He	60000	0.96	0.02				0.93				
	100000	0.95	0.04				0.96	0.04			
	150000	0.83	0.17				0.84	0.16			
	200000	0.69	0.31				0.71	0.29			
C	60000	0.13	0.86	0.07			0.47	0.49			
	100000	0.01	0.66	0.32			0.11	0.79	0.09		
	150000		0.32	0.56	0.12		0.06	0.58	0.24	0.11	
	200000		0.22	0.52	0.26		0.05	0.44	0.27	0.24	
N	60000	0.44	0.54	0.07			0.51	0.47			
	100000	0.07	0.52	0.40			0.11	0.66	0.23		
	150000	0.04	0.24	0.59	0.09	0.04	0.06	0.40	0.48	0.03	0.03
	200000	0.03	0.16	0.51	0.08	0.22	0.04	0.29	0.45	0.03	0.19
O	60000	0.50	0.49				0.52	0.47			
	100000	0.07	0.89	0.03			0.09	0.88	0.03		
	150000	0.03	0.78	0.05	0.11	0.02	0.05	0.80	0.06	0.09	
	200000	0.02	0.65	0.03	0.10	0.16	0.04	0.68	0.04	0.16	0.06
Ne	60000	0.53	0.47				0.41	0.58			
	100000	0.05	0.91	0.04			0.05	0.92	0.03		
	150000	0.02	0.79	0.05	0.11	0.02	0.02	0.81	0.04	0.10	0.02
	200000	0.01	0.65	0.27	0.11	0.18	0.02	0.68	0.02	0.10	0.16
Mg	60000	0.08	0.48	neutral Mg: 0.44			0.04	0.54	neutral Mg: 0.43		
	100000	0.04	0.85	0.03			0.02	0.85	0.03		
	150000	0.02	0.74	0.07	0.11	0.01	0.01	0.76	0.05	0.11	0.02
	200000	0.02	0.60	0.03	0.12	0.20	0.01	0.62	0.02	0.09	0.21
Si	60000	0.20	0.76	0.04			0.51	0.42			
	100000	0.08	0.43	0.35	0.13		0.41	0.47	0.05	0.04	
	150000	0.05	0.16	0.28	0.50		0.26	0.32	0.08	0.31	0.01
	200000	0.04	0.10	0.20	0.50	0.16	0.19	0.23	0.06	0.32	0.18

3. CIRCUMSTELLAR OR NEBULAR EMISSION

Altamore et al. (1981) consider that in Z And the emission lines are formed in a solar type transition region around the M 6.5 star. In such a transition region the ionisation of the elements is mainly due to collisions by free electrons, and the ionisation temperature T_{ion} is approximately equal to the electron temperature T_e; where T_{ion} is defined as the electron temperature T_e, at which a given ion reaches its maximum fractional abundances when subjected to collisional ionisation. For V 1016 Cyg Nussbaumer and Schild (1981) propose a planetary nebula type model. In that case ionisation is due to the radiation field of the central star and for stellar temperatures we are probably concerned with,

($T^* < 200000$ K) one finds $T_e < T_{ion}$. C IV may serve as a practical example. According to Nussbaumer and Storey (1975) C^{3+} has $T_{ion} \approx 90000$ K but in the V 1016 Cyg model C IV is mainly emitted at $T_e \approx 15000$ K. (Examples of curves of fractional ionisation for several elements may be found in Jordan (1969)). Michalitsianos et al. (1980) consider the UV line emission from R Aqu to originate from a $N_e \approx 10^6 - 10^7$ cm^{-3}, $T_e \approx 15000$ K nebula of a few times 10^{14} cm, placed around a M7 star and ionised by a white dwarf placed at the edge of the nebula. Hack (1979) after including the IUE spectra in her study of CH Cyg is still undecided on the model to opt for. Analysing the low resolution spectra of the four symbiotic stars YY Her, SY Mus, CL Sco, BX Mon Michalitsianos et al. (1981) tend towards a double star model with an accretion disc. Are all these models realised to some degree, or are some of the models just fantasy? Progress could be more easily achieved if we had better notions about T_e, in particular if we could decide whether $T_e \approx T_{ion}$ or $T_e \neq T_{ion}$. For the well observable lines in the IUE domain there are not many pairs of lines fulfilling the requirements for T_e determination. The O III lines at $\lambda\lambda 2322, 1667$ are fine for $N_e \lesssim 10^4$ cm^{-3}, for higher N_e that ratio is strongly density dependent, and up to now all the UV emitting regions of symbiotic stars have been found with $N_e > 10^3$ cm^{-3}. The case is similar for [Ne IV] for which Penston et al. (1981) give the following vacuum wavelengths: $\lambda(^4S^o_{3/2} - {}^2D^o_{3/2}) = 2424.97, \lambda(^4S^o_{3/2} - {}^2D^o_{3/2}) = 2422.43, \lambda(^4S^o_{3/2} - {}^2P^o) = 1601.5$. I have calculated the intensity ratio for a range of N_e and T_e embracing those found in symbiotic stars, with collision strengths and transition probabilities from Giles (1980) and Zeippen (1981). The emissivity ratios $\epsilon(\lambda 1601)\epsilon/(\lambda 2424)$ of the total multiplets $^4S^o - {}^2P^o$ and $^4S^o - {}^2D^o$ are given in Fig. 2. The ratio is a good T_e indicator for $N_e < 10^4$ cm^{-3} and for $N_e > 10^{10}$ cm^{-3}. Thus for Z And where Altamore et al. (1981) derive $N_e \cong 2 \cdot 10^{10}$ cm^{-3} the [Ne IV] ratio is insensitive to density variations and should allow an accurate T_e determination; the practical problems will be the simultaneous detection of both multiplets. Penston et al. (1981) observe in RR Tel $F(\lambda 1601)/(F(\lambda 2422) + F(\lambda 2424)) = 2.6$. As the reddening in the two multiplets is comparable we neglect this effect for the present qualitative discussion. This ratio is compatible with $10^6 < N_e$ [cm^{-3}] $< 10^7$ for any T_e.

In the visual spectra of some symbiotic stars [Fe VII] is detected. Some work on the visual lines of [Fe VII] was done by Nussbaumer and Osterbrock (1970), however they omitted the transitions from $3d^2\ {}^1S$. From Ekberg (1981) we now know the energy of this term, the strongest transition from ^1S occurs at 2015 Å. Penston et al. (1981) list for RR Tel an unidentified feature at $\lambda 2015.33$. This transition when compared to the other forbidden [Fe VII] lines, is exceedingly sensitive to variations in T_e in the range $T_e \lesssim 60000$ K, as is shown in Fig. 3 (Fig. 3 is based on a preliminary configuration calculation of Nussbaumer and Storey (1982)). But we need calibrated fluxes in the visual domain.

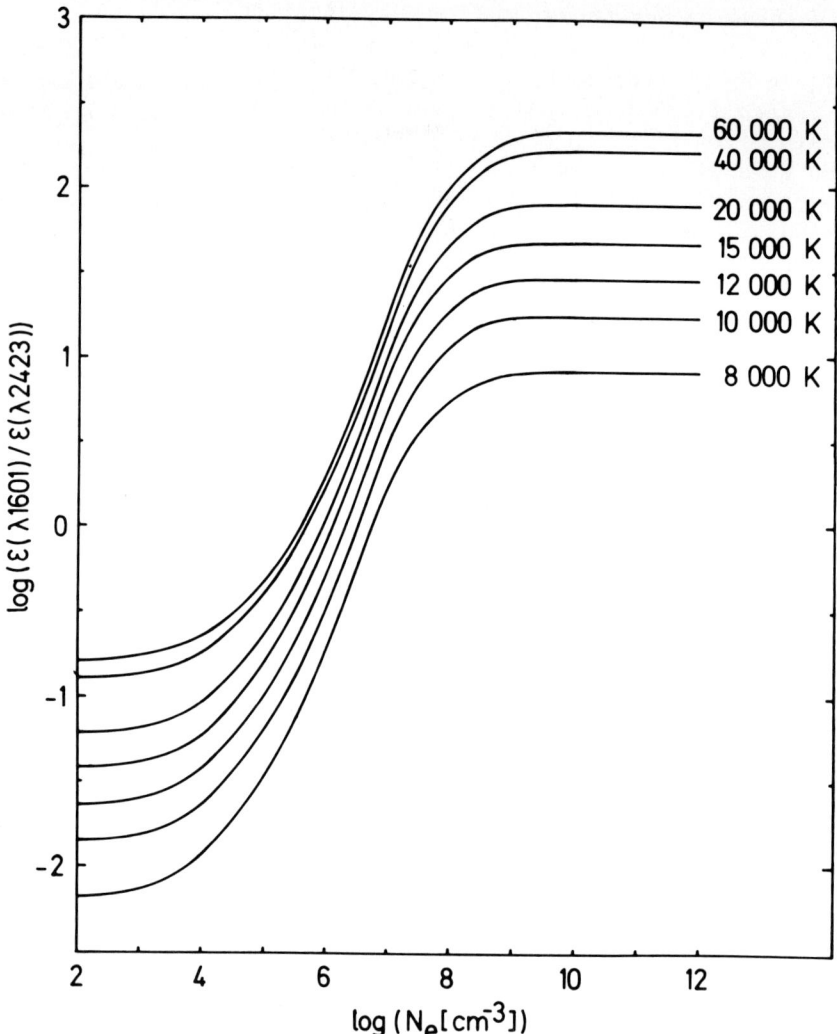

Figure 2. [Ne IV]: Ratios of emissivities in the multiplets $\epsilon(^4S^o - {}^2P^o)/\epsilon(^4S^o - {}^2D^o) = \epsilon(\lambda 1601)/\epsilon(\lambda 2423)$. The doublet at $\lambda 2423$ consists of $\lambda\lambda 2422, 2423$; the $\lambda 1601$ components should be separated by approximately 0.2 Å. ϵ was calculated in erg/s per ion.

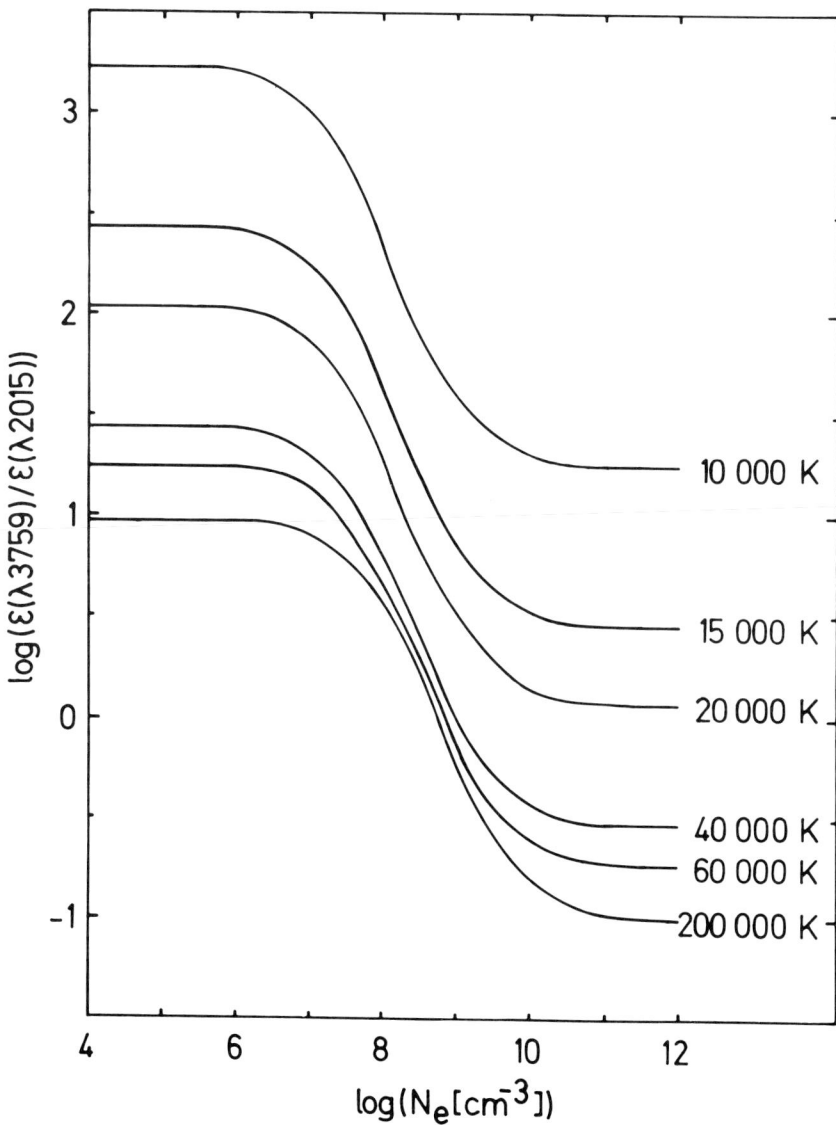

Figure 3. [Fe VII]: Ratios of emissivities in the lines
$(^3F_4 - {}^1G_4)/\varepsilon({}^1D_2 - {}^1S_o) = \varepsilon(\lambda 3759)/\varepsilon(\lambda 2015)$

In Z And [Fe VII] λ3758 is one of the most prominent emission lines (Figure 4 of Altamore et al. 1981). If Altamore et al. are right with their estimate of $N_e \gtrsim 2 \cdot 10^{10}$ cm^{-3}, $T_e \gtrsim 4 \cdot 10^4$ K, then the unreddened flux in λ2015 should be approximately three times as strong as in λ3758; with $E_{B-V} \approx 0.3$ this ratio will be reduced by approximately a factor 3.

The C III lines deserve our attention as well. Kafatos et al. (1980) observe λλ1176, 1247, 1908 in RW Hya, but they only measure the flux in λ1908. For RR Tel λλ1247, 1908, 2296 were identified, Penston et al. (1981) give fluxes for λλ1908, 2296. If the observed ratio of F(λ2296)/F(λ1909) $\approx$ 0.01 is uniquely due to collisional excitation then RR Tel emits the C III spectrum at $T_e \gtrsim 40000$ K (see Fig. 4 of Nussbaumer and Schild 1979). At this T_e, and $N_e \approx 10^6$ cm^{-3} the $^3P^o - ^1S$ transition at λ1247 would not be observable as the calculated ratio is F(λ1247)/F(λ1908) < 0.001. However, as mentioned earlier, both λ1247 and λ2296 can be formed by recombination in the C IV region (Storey 1981); working with these lines therefore needs careful investigation in each case. Qualitative information is also provided by the absence of a line. Thus Altamore et al. (1981) suspect $T_e \approx 40000$ K, $N_e > 10^{10}$ cm^{-3} for Z And. For these conditions an unreddened ratio of F(λ2296)/F(λ1909) $\gtrsim$ 0.1 is expected; the scientific advantage of setting observational limits to that ratio is thus obvious. If the λ1176 multiplet was not placed in the low sensitivity range of IUE, it could help to verify our models as well.

At this stage I want to appeal to those taking spectra in the visual range. Qualitative description of temporal changes in the symbiotic objects are now clearly insufficient, we need calibrated spectra that give us fluxes in [energy/(time surface)]. Thus the spectrum of HM Sge as published by Blair et al. (1981) can be directly compared with IUE spectra from the same object; coverage down to the ultraviolet cutoff would of course add even more value.

We must admit that our knowledge about the nature of symbiotic objects is at present on the level of inspired guesswork. Whether some type of stellar corona has to be accomodated is not a futile speculation since X-ray radiation has been detected in symbiotic stars. If an emitting region with $T_e \approx 10^6$ K exists we might also expect coronal lines to appear in the IUE range. A list of coronal lines as observed in the sun between 1000 Å and 3000 Å has been published by Sandlin et al. (1977) and Sandlin and Tousey (1979); you might want to check your spectra against these lists. R Aqu may serve as an example. Zirin (1976) reports a detection of the Fe XIII 3s^{2}3p^2 $^3P_o - ^3P_1$, $^3P_1 - ^3P_2$ transitions at λλ10747, 10798. If these transitions should still be seen, then we could also expect the $^3P_1 - ^1D_2$ λ2579 line which has been seen by Sandlin et al. (1977) in the solar spectrum.

4. LINE PROFILES

Apart from the ionisation structure, the radiation and electron temperatures, the particle density and the elemental abundances, we are also interested in the geometrical and dynamical configurations of our pets. Here the shapes of the emission lines hold a clue. In Figure 4 high resolution IUE spectra of C IV $\lambda\lambda 1548, 1551$ are shown. It is usually a strong doublet, and features which are real and not just instrumental noise should appear on both components.

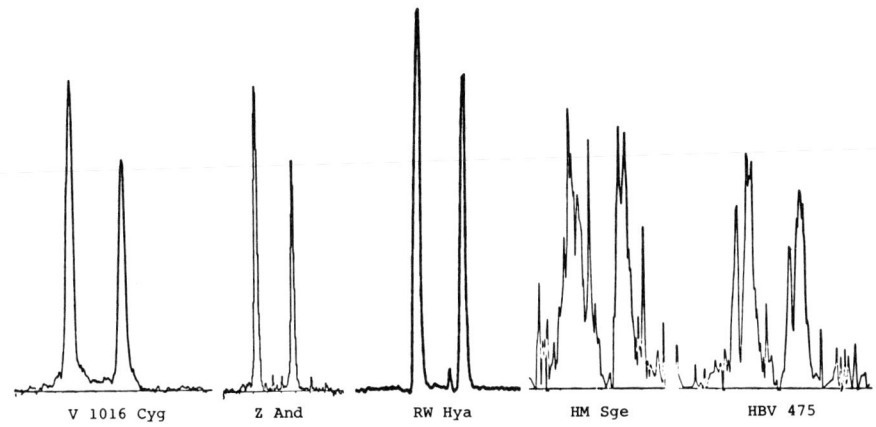

Figure 4. Line shapes of the C IV doublet at $\lambda\lambda 1548, 1551$

In Z And, V 1016 Cyg, and RW Hya the profiles are approximately symmetrical and a spherically symmetrical model can certainly be advocated. The profiles are much more complicated in HM Sge and even more so in HBV 475. Are the profiles of HM Sge and HBV 475 composed of several emission components, or are absorption lines eating into broader emission lines? In addition to the absorption feature at the rest wavelengths seen in HBV 475, I want to draw attention to an emission red shifted by ~ 1 Å in both HM Sge and HBV 475. Is this a sign of a particle jet leaving those objects, away from us at $v \geq 190$ km/s, or does it originate from mass transfer in a binary system, or do we observe double star systems with C IV emission from both? Well, it is probably none of these, but an instrumental effect, at least it has been seen in some more symbiotic objects (though not in others) at exactly the same position, Z And shows it but neither V 1016 Cyg nor RW Hya. We cannot altogether rule our a physical explanation; P.J. Storey and myself are looking into the possibility of satellite lines of the type $2s(^2S)n\ell - 2p(^2P^o)n\ell$ being responsible for these lines.

A problem that becomes apparent when studying high resolution IUE spectra concerns the flux ratios $F(^2S_{1/2} - ^2P^o_{3/2})/F(^2S_{1/2} - ^2P^o_{1/2})$ within the C IV multiplet. Instead of the expected ratio of 2, the observation give 1.7 for V 1016 Cyg, 1.3 Z And, and 1.2 for RW Hya. Nussbaumer and Schild (1981) mention that interstellar absorption could offer an explanation. However, if it should turn out that other classes of emission objects, for example planetary nebulae, do not show that effect, then the interstellar explanation would probably have to be dropped.

5. CONCLUSION

In our quest for understanding symbiotic stars, emission lines are likely to play a key part. They can be interpreted in a straight-forward way to find typical values for N_e and T_e, abundance determinations, however, are already more demanding. When calculating T* from recombination lines we usually assume that the ionising star emits as a black body. Results from such preliminary investigations may already tell us whether the emission lines are emitted in a gas with predominantly radiative or collisional ionisation. Emission line profiles may hold clues to relations between the stellar and nebular parts of the objects. These preliminaries will be necessary before refined models can be elaborated.

I suspect that we shall find widely differing conditions in different symbiotic stars. Thus RR Tel and V 1016 Cyg have both similar densities of $N_e \sim 10^6$ cm^{-3}, but whereas V 1016 Cyg emits the bulk of its emission lines at $T_e \sim 15000$ K the evidence from O III and C III lines is rather in favour of $T_e \sim 40000$ K for RR Tel; in Z And where the N III] and O IV] lines require $N_e \sim 10^{10}$ cm^{-3}, the strong C III] $\lambda 1909$ combined with the absence of C III $\lambda\lambda 1176$, 2296 indicate $T_e < 30000$ K. I do not want to preempt later discussions but I feel that placing an astronomical object in the class of symbiotic stars is simply a declaration of ignorance. I therefore suggest that we slightly modify the definition of a symbiotic object by including a declaration of ignorance in the following way: A symbiotic star is an astronomical object with the following properties:

a) its spectrum shows the typical nebular emission lines combined with the spectrum of a cool star
b) on the time scale of years its luminosity shows variations which may attain several mag
c) it cannot be clearly classified as something else.

Point (c) does not form part of the standard definition (e.g., Boyarchuk 1975). I propose its addition with the intention of turning "symbiotic stars" into a transitory class. Our task then consists of solving the enigmas attached to each object and unveiling its true nature. The advent of IUE (International Ultraviolet Explorer) has opened new roads in this endeavour.

REFERENCES

Aldrovandi, S.M.V., Péquignot, D.: 1973, Astron. Astrophys. 25, 137.
Altamore, A., Baratta, G.B., Cassatella, A., Friedjung, M., Giangrande,A., Ricciardi, O., Viotti, R.: 1981, Astrophys. J. 245, 630.
Blair, W.P., Stencel, R.E., Shaviv, G., Feibelman, W.A.: 1981, Astron. Astrophys. 99, 73.
Boyarchuk, A.A.: 1975, IAU Symp. No. 67, Reidel Dordrecht.
Ekberg, J.O.: 1981, Physica Scripta 23, 7.
Flower, D.R., Nussbaumer, H.: 1975, Astron. Astrophys. 45, 145.
Flower, D.R., Nussbaumer, H., Schild, H.: 1979, Astron. Astrophys. 72, L1.
Friedjung, M.: 1982, this volume.
Giles, K.: 1981, Mon. Not. R. astr. Soc. 195, 63P.
Hack, M.: 1979, Nature 279, 305.
Jordan, C.: 1969, Astrophys. J. 156, 49.
Jordan, C.: 1969, Mon. Not. R. astr. Soc. 142, 501.
Kafatos, M., Michalitsianos, A.G., Hobbs, R.W.: 1980, Astrophys. J. 240, 114
Keyes, C.D., Plavec, M.J.: 1980, NASA-GSFC Symp. "The Universe in Ultraviolet Wavelengths"
Lambert, D.L.: 1978, Mon. Not. R. astr. Soc. 182, 249.
Michalitsianos, A.G., Kafatos, M., Hobbs, R.W.: 1980, Astrophys. J. 237, 506.
Michalitsianos, A.G., Kafatos, M., Feibelmann, W.A., Hobbs, R.W.: 1981, Astrophys. J. (submitted)
Nussbaumer, H., Osterbrock, D.E.: 1970, Astrophys. J. 161, 811.
Nussbaumer, H., Storey, P.J.: 1975, Astron. Astrophys. 44, 321.
Nussbaumer, H.: 1977, Astron. Astrophys. 58, 291.
Nussbaumer, H., Storey, P.J.: 1979, Astron. Astrophys. 71, L5.
Nussbaumer, H., Schild, H.: 1979, Astron. Astrophys. 75, L17.
Nussbaumer, H.: 1980, Proc. Second European IUE Conference, Tübingen, p.xℓiii.
Nussbaumer, H., Schild, H.: 1981, Astron. Astrophys. 101, 118.
Nussbaumer, H., Storey, P.J.: 1981, Astron. Astrophys. 99, 177.
Nussbaumer, H., Storey, P.J.: 1982, in preparation.
Peimbert, M.: 1980, NASA-GSFC Symp. "The Universe in Ultraviolet Wavelengths"
Penston, M.V., Benvenuti, P., Cassatella, A., Heck, A., Selvelli, P., Beeckmans,F., Macchetto, F., Ponz, D., Jordan, C., Cramer, N., Rufener, F., Manfroid, J.: 1981, preprint.

Sahade, J.: 1980, NASA-GSFC Symp. "The Universe in Ultraviolet Wavelengths: The First Two Years of IUE".
Sandlin, G.D., Brueckner, G.E., Tousey, R.: 1977, Astrophys. J. 214, 898.
Sandlin, G.D., Tousey, R.: 1979, Astrophys. J. 227, L107.
Seaton, M.J.: 1979, Mon. Not. R. astr. Soc. 187, 73P.
Slovak, M.H.: 1982, this volume.
Storey, P.J.: 1981, Mon. Not. R. astr. Soc. 195, 27P.
Swings, P., Struve, O.: 1941, Astrophys. J. 93, 356.
Withbroe, G.L.: 1971, The Menzel Symposium, NBS Spec. Publ. 353, p. 127.
Zeippen, C.J.: 1981, Mon. Not. R. astr. Soc. (in press).
Zirin, H.: 1976, Nature 259, 466.

DISCUSSION ON UV EMISSION LINES

Friedjung: The CIII temperature gives the temperature of the CIII region, but not necessarily of regions of higher ionization stages. Our model for Z And was probably oversimplified; lower ionization stages are probably photoionized by a warm continuum. The question is how the high ionization stages are formed; are they photoionized or formed by collisions in something like the solar transition region?

Nussbaumer: Friedjung has touched on an important point. CIII informs on regions where CIII is formed, and the physical conditions there might be different from the CIV region. One must even be cautious with different lines from the same ion. Thus in the 1908 and the 1176 or 2296 multiplets of CIII the essential contributions to the total strength may come from different regions.

Viotti: The permitted (1176...)over intercombination (1909 A) emission line ratio is quite sensitive to the presence of a (even largely) diluted hot radiation field because of the radiative excitation of the permitted lines which dominates over collisional excitation.

Nussbaumer: I agree that radiative excitation in allowed multiplets like CIII λ 1176 has to be taken into account in the kind of model you (Altamore et al. 1981) suggest for Z And; on the other hand it can be neglected in the model that Schild and I propose for V1016 Cyg.

Kwok: In an extended object like V1016 Cyg where the size exceeds one arcsec, there are certainly variations in temperature and density within the nebula. How sensitive is the line interpretation on the density/temperature structure?

Nussbaumer: You are right in principle, one has to allow for density and temperature structure. But if you intend to say that this is more important for an extended object than for a small one, then I disagree. As counter examples (which may both be relevant to symbiotic stars) I can cite the solar transition region where T_e and N_e change by a factor 100 over 1000 km, and planetary nebulae where they hardly change over vastly extended volumes.

Kafatos: I would like to emphasize that for intercombination lines — which are optically thin — one should use the absolute intensities as well as relative intensities in the multiplets. For example, for the CIII lines it makes a lot of difference as far as the emission measure is concerned, if the temperature is $\lesssim 30000K$ or if it is $\gtrsim 40000K-100000K$. When one does this for objects like RW Hya, V1016 Cyg and Z And, one finds that indeed the temperatures are low ($\lesssim 20000K$) characteristic of photo-ionization.

Nussbaumer: The absolute fluxes certainly contain important information. In our objects the intercombination lines should be optically thin. Thus if we have a reasonable knowledge of N_e and T_e we can immediately gain some rough knowledge on the size of the emitting region. The doubts you have about the model proposed for Z And might be answered by the authors of Altamore et al. paper.

Viotti: The discussion about the electron temperature will be continued by Friedjung in his review on models.

Houziaux: What is the dilution factor used in the ionization computations, and what is the importance of dielectronic recombination on intercombination lines at the electron densities of symbiotic stars?

Nussbaumer: For the calculation of Table 1 I took the semi-infinite shell model as used by Nussbaumer and Schild for V1016 Cyg. The exact value of the inner boundary is not crucial in the present context. Dielectronic recombination is unlikely to be crucial as excitation mechanism for the well observed intercombination lines. It may however be the principal mechanism for other lines; examples are given by Storey (1981, Mon. Not. 195, 27P).

Keyes: Do you include charge-exchange in your calculations?

Nussbaumer: They are included as far as they have been published.

Keyes: Are you aware of any atomic data (transition probabilities, collision strengths) for the short-wavelength component of the SiIII intercombination multiplet $\lambda 1884$. This feature is present in Z And (Altamore et al. 1981), and atomic data would enable a direct determination of the electron density to be made.

Nussbaumer: There is a thesis by Nicolas at NRL where these atomic data are given. Burke has done a new calculation for the collision strength of the total $^1S - {}^3P^o$ multiplet.

ULTRAVIOLET PROPERTIES OF THE SYMBIOTIC STARS

Mark H. Slovak and David L. Lambert
University of Texas at Austin, U.S.A.

1. Introductory Remarks

Prior to the launch of the IUE satellite in early 1978, the only symbiotic star previously detected in the ultraviolet by earlier UV satellites, such as the OAO-2, TD-1 and ANS experiments, was AG Pegasi = HD 207757 (Gallager et al. 1979). These broad-band observations indicated that the symbiotics as a class may show a significant ultraviolet flux and thus they became natural candidates for a survey with the IUE satellite. The following is an interim report on a survey of the symbiotics, both at low and, for AG Pegasi and CH Cygni, at high resolution.

Our IUE program began in June, 1979 and is continuing through 1981. The main thrust of the program has been to attempt to verify that the symbiotics are, for the most part, binary systems and form a natural extension of the short period cataclysmic variables. We hoped to accomplish this goal by directly observing the hot component in the UV, where it is uncontaminated by the flux of its giant companion.

Our early hope that the symbiotics would appear as UV "bright" objects has been realized. As a class, they display a diverse and complex spectrum in the UV, exhibiting both a rich emission-line spectrum, and on the deeper exposures, a definite continuum. We have obtained multiple low resolution spectra on a sample of twelve symbiotics, including the classical prototype Z And, BF Cyg, AX Per, and the eclipsing systems CI Cyg and AR Pav. High resolution spectra were taken on AG Peg around its orbital cycle to study both line profiles and radial velocity variations. Some early results of our survey were presented by Lambert et al. (1980); Sahade and Brandi (1980) discussed a similar survey, dividing the symbio stars into two broad groups based on the appearance of the UV emission-line spectra.

The following sections present a general discussion of the UV spectra of the symbiotics, including both the emission lines and the continua. As it is somewhat premature to draw general conclusions based on a small sample, the emphasis is biased towards a discussion of individual stars. AG Pegasi is used as an illustrative, albeit atypical, example.

2. The survey Selection

The observational sample for which we have extensive results consists of the brighter, and perhaps the more interesting, symbiotics. The sample forms a relatively homogeneous group, having been selected by the following criteria:

1. The stars were (originally) in a <u>quiescent</u> state, outside of eruption. However, during the course of the survey CH Cyg, AX Per and most spectacularly AG Dra have undergone eruptions;

2. the stars are classified as S type symbiotics (Allen 1978) showing relatively normal stellar-like infrared colors. Multiple infrared observations have established the lack of any significant variability;

3. the stars included the known eclipsing symbiotics, such as CI Cyg and AR Pav, and the only well established spectroscopic binary, AG Peg.

By observing the S type symbiotics, the interpretive aspect becomes more tractable since the variability of the late type component does not additionally confuse the question, as for the D type symbiotics. As importantly, the lack of any significant variability permitted the UV observations to be combined with ground based optical and infrared data with a degree of confidence.

3. Ultraviolet Spectra of the Symbiotics

Ultraviolet spectra have been obtained for the above sample, using the SWP and LWR cameras aboard the IUE satellite in both the low and high resolution modes. Particular attention has been paid to the known binary systems, either eclipsing or spectroscopic, to assure coverage around the orbital cycle and both outside and during eclipse.

a. Low resolution spectra

The low resolution ($\lambda = 6$ A) data show a rich and varied emission-line spectra for all of the stars in the sample with the exception of CH Cyg. The SWP bandpass includes the highest number of lines, dominated by NV (1239, 1243 A), CII (1335 A), SiIV 1394, 1403 A), NIV] (1486 A), CIV (1548, 1551 A), HeII (1640 A), OIII] (1666 A), NIII] (1749 A), SiIII] (1892 A), and CIII] (1909 A). The LWR bandpass also includes the CIII]

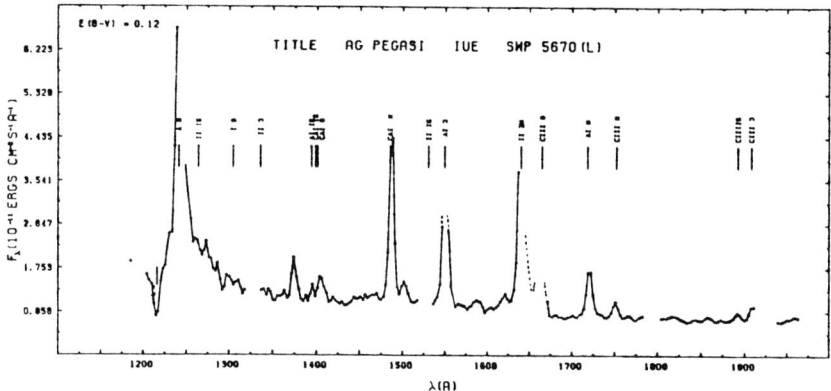

Figure 1. Low resolution SWP spectrum of AG Pegasi.

feature, in addition to various HeI lines and the MgII doublet (2796, 2802 A). Thus, the lines present span a wide range in excitation, and represent both resonance and intercombination transitions.

Sequential SWP and LWR spectra are shown for AG Peg in Figures 1 and 2, respectively. Gaps in the data arise from reseaux features falling on the spectral order; dashed lines indicate saturated portions. The data have been de-reddened using a value of $E(B-V) = 0.12$, determined from the existence of the 2200 A feature in the LWR spectrum. While atypical because of the line breadth as compared to most symbiotics, the spectra show the impressive diversity of emission features.

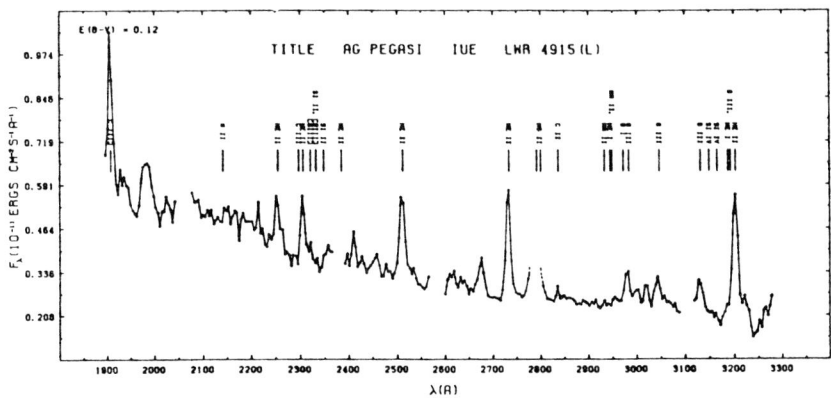

Figure 2. Low resolution LWR spectrum of AG Pegasi.

While the line ratios differ from star to star, several general comments can be made. CIV invariably appears as the strongest feature, followed closely by HeII. In AG Peg, the nitrogen lines (NV and NIV]) are also exceptionally strong. The intensity ratios are orbitally modulated

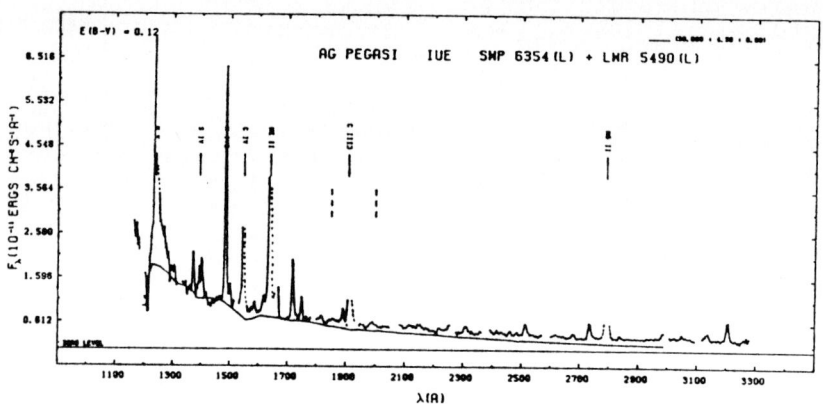

Figure 3. Combined low resolution spectra for AG Pegasi.

an effect which is most clearly seen in the eclipsing systems. In AG Peg, the MgII lines weaken and strengthen around the orbit, in a similar fashion to the behaviour of the MII lines in the optical (Hutchings, Cowley and Redman 1975). The HeII/NIII] and CIV/NIII] ratios in AR Pav decrease remarkably upon ingress to eclipse, reminiscent of the HeII/[OIII] ratio in the optical (Thackeray and Hutchings 1974).

By deliberately saturating the emission lines, the underlying continuum can be detected. Two deeply exposed spectra of AG Peg have been combined in Figure 3 to show the ultraviolet continuum. In the SWP bandpass the continuum is well represented by a Kurucz (1979) stellar model atmosphere with $T_e = 30,000$ K and $\log g = 4.50$. However, longward of 1900 A, the observations deviate significantly from the model, showing an excess flux presumably arising from Balmer continuum emission.

The UV continua of Z And and BF Cyg are also well represented by stellar model atmospheres, giving confidence to the interpretation of these stars as binary systems. However, in the eclipsing systems CI Cyg and AR Pav, the hot component is obscured by the large optical depth in the accretion "shell" which surrounds the stars. The continuum appears rather flat (f_λ = constant).

b. High resolution spectra

For two of the brightest symbiotics (AG Peg and CH Cyg) we have obtained a series of high resolution spectra, primarily in the SWP bandpass. Representative line profiles for AG Peg are shown in Figure 4.

A remarkable diversity of the line profiles is evident, again somewhat unrepresentative of the typical symbiotic. The breadth of the lines gives a Wolf-Rayet appearence to the features, indicative of a strong stel

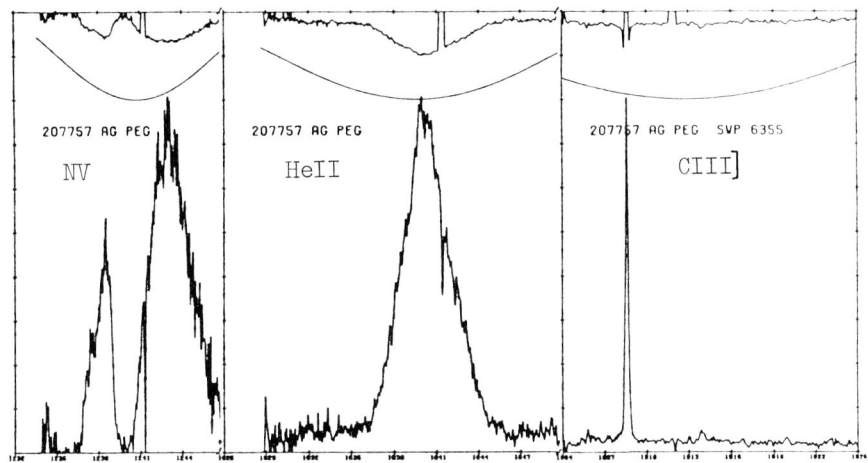

Figure 4. High resolution IUE profiles for AG Pegasi.

lar wind. The UV resonance doublet shows a classical P Cygni profile, whereas the HeII line is quite symmetric nimicking the HeII (4686 A) feature. The narrow CIII] is more characteristic of a true nebular emission line. The stratification of the emission line region is thus clearly evident.

CH Cyg shows a strong OI line at 1300 A in the SWP high resolution spectra, in addition to a castellated MgII feature in the LWR data. Each component of the MgII doublet shows the same structure, and Wing and Carpenter (1980) argue that the two absorption components are due to interstellar absorption and circumstellar absorption, respectively. Excluding the interstellar contribution to the profile, the MgII profile resembles the H_α emission line (Anderson, Oliversen, and Nordsiek 1980).

4. Combined Ultraviolet and Ground-based Data

While awaiting the absolute flux calibration for the IUE high resolution spectra, our analysis has concentrated on the continuum distribution derived from the low resolution data. By combining the IUE results with ground-based optical and infrared fluxes, we have derived absolute energy distribution for our sample, covering the range from 1200 A to 34,000 A. We have calculated mean fluxes across various bandpassas selected to avoid the emission lines, and thus the flux distributions truly represent the continuum.

Low resolution UV spectra have been combined with optical data of similar resolution, obtained with a Cassegrain Digicon Spettrograph (CDS) at the McDonald Observatory, and are shown for AG Pegasi in Figure 5. A partial identification of line features is given longward of 2200 A; notable

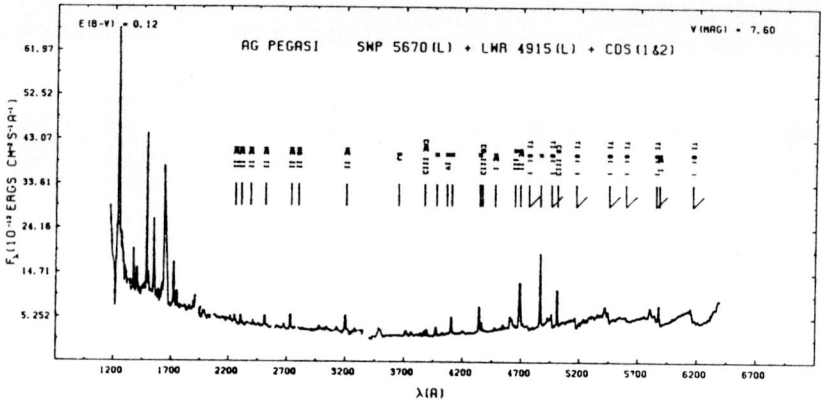

Figure 5. Combined IUE and optical spectra for AG Pegasi.

are the HeII emission lines juxtaposed across the continuum containing TiO bandheads, arising from the α and γ series.

Using broad-band Johnson JKL fluxes, which appear uncontaminated by emission, the absolute energy distribution displayed in Figure 6 was derived for AG Peg. In this figure, the Johnson U and B fluxes are shown, clearly demonstrating the contamination of the emission lines and indicating the danger of estimating the properties of the hot component from the (U-B) - (B-V) color diagram. The bandpasses of the filters are indicated by the horizontal bars central on individual data points.

The data in Figure 6 are shown on a magnitude scale, where the fluxes have been converted to the same energy scale as the Johnson V magnitudes. The energy distribution is compared to a composite stellar distribution, comprised of an O9"V" star (V = 11.0) and an M3III star (V = 8.40). The composite curve is derived using the intrinsic Johnson UBVRIJKL colors in addition to UV colors determined from the OAO-2, TD-1, and ANS satellites (Parsons 1981). The intrinsic distributions are scaled to the observations upon specifying the V magnitude of the individual components and a value for the reddening. Good agreement with the observations is obtained except in the region where the nebula most strongly contributes (see figure 3), confirming the analysis of Gallagher et al. (1979).

5. Concluding Observations

The IUE observations of the symbiotic stars have revealed ultraviolet properties which rival diversity of the optical features. Nonetheless, the UV data have for the first time permitted the hot component to be studied relatively uncontaminated by the giant companion, which dominates the optical regime. The UV observations provide convincing evidence that

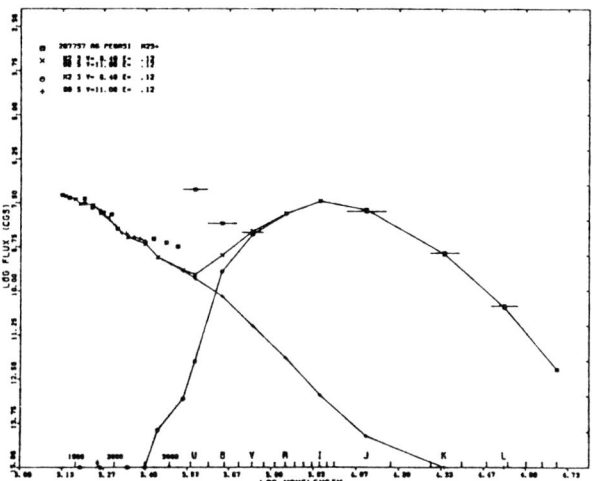

Figure 6. Absolute energy distribution for AG Pegasi.

indeed many of the symbiotics have hot stellar companions, imbedded in the enshrouding nebula or accretion "shell" fromed from the wind arising off of one or possibly both of the components. Another possibly impulsive, as opposed to steady-state, contributor to the nebula could come during the eruptive periods. The large range in excitation and density of the symbiotic nebulae can be understood by a combination of such processes.

Much remains to be done in the ultraviolet, as the initial observations have only really served to point the direction future research should take. Of immediate interest is the study of an eruptive symbiotic, such as AG Dra, to define the active, as opposed to quiescent, properties. Hopefully, such observations would lead to an understanding of the source of the symbiotic eruptions, currently believed to be a mass-transfer "burst" (Bath 1981). The final goal which should motivate any program, observational or theoretical, is to place the symbiotics in a stellar evolutionary sequence so that their progenitors, as well as their progeny, may be clearly identified.

The authors would like to thank the staff of the IUE Observatory for their extensive and competent assistance. This research was supported in part by the NASA Grant NSG 5379.

REFERENCES

Allen, D.A.: 1979, in F.M. Bateson, J. Smak, and I.H. Urch (eds.), "Changing Trends in Variable Star Research", IAU Colloquium No.46, University of Waikoto, Hamilton, New Zealand, p.125.

Anderson, C.M., Oliversen, N.A., and Nordsieck, K.H.: 1980, Astrophys. J. 242, 188.

Bath, G.T.: 1981, in R.E. Stencel (ed.), "North American Workshop on Symbiotic Satrs", JILA, Boulder, p.20 (abstract).

Gallagher, J.S., Holm, A.V., Anderson, C.M., and Webbink, R.F.: 1979, Astrophys. J. 229, 994.

Hutchings, A.D., Cowley, A.P., and Redman, R.O.: 1975, Astrophys. J. 201, 404.

Kurucz, R.L.: 1979, Astrophys. J. Suppl. 40, 1.

Lambert, D.L., Slovak, M.H., Shields, G.A., and Ferland, G.J.: 1980, in "The Universe at Ultraviolet Wavelengths", NASA, Greenbelt, p.461.

Parsons, S.B.: 1981, submitted to Astrophys. J.

Sahade, J., and Brandi, E.: 1980, in"The Universe at Ultraviolet Wavelengths", NASA, Greenbelt, p.451.

Thackeray, A.D., and Hutchings, J.B.: 1974, Mon. Not. Roy. astr. Soc. 167, 319.

Wing, R.F., and Carpenter, K.G.: 1980, in "The Universe at Ultraviolet Wavelengths", NASA, Greenbelt, p.341.

DISCUSSION ON UV PROPERTIES

Keyes: The Zanstra temperatures derived from HeII 1640 for several objects, notably AG Peg, AG Dra, and RW Hya, are in the range 80000K – 120000K, which are considerably hotter than the Kurucz models used in your synthesis. In view of the fact that the IUE spectral range is on the Rayleigh-Jeans tail for such temperatures, have you attempted any fitting with black-body or model distributions hotter than 50000K?

Slovak: At this stage in the analysis, I have only used stellar models which are currently available. The Kurucz models, of course, only go up to T_{eff}= 50000K. My comparison is more of an indication that a simplistic comparison. Using a composite distribution of two stars alone is not sufficent to explain those symbiotics where the nebular contribution is significant. Good fits are derived for Z And and BF Cyg, but I cannot claim these are unique.

Kafatos: We have attempted similar fits for a few symbiotics (e.g. YY Her, SY Mus, etc.) and we find that they are not fitted well by a

Kurucz model. The UV lines are very important because they show that the ionizing photons come from a $\sim$100000K hot subdwarf or a $\sim$100000K boundary layer. The continuum above $\sim$2000 A is most probably nebular and therefore it is highly suspect to just fit the far UV continuum (between 1200 and $\sim$ 1800 A).

<u>Slovak</u>: UV lines are indeed better indicators of UV radiation. What we may be seeing in the UV continuum is reprocessed radiation e.g. through an accretion disk.

<u>Viotti</u>: Dr. Cassatella at VILSPA did a lot of careful fitting of the UV continua of symbiotic stars. In the case of Z And (and of other stars, see e.g. the figure on page 9 of this book) the slope of the continuum in the far UV is close to Rayleigh-Jeans one, and has been fitted with a 100000K black-body. The difference between the observed continuum and the black-body one is easily fitted by a 15000K nebular (= ff + bf + 2 photons) continuum.

<u>Slovak</u>: The continuum fitting using non-stellar contributions in addition to the Rayleigh-Jeans tail of a hot star with T_{eff} 100000K, does indeed provide a better fit to the observed continua. The two photon contribution, however, is bound to be small at the densities considered in typical symbiotic nebulae.

<u>Houziaux</u>: I would like to ask Viotti: (i) one may wonder it is legitimate to just add or subtract continuum fluxes; (ii) how such temperatures (100000K) compare with Zanstra temperatures (from HeII lines e.g.).

<u>Viotti</u>: (i) Yes, as far as we assume that they originate in two separate regions. (ii) In AG Dra the hot continuum and HeII temperatures are nearly the same and may suggest a common origin.

<u>Kafatos</u>: I would like to go back to a point raised by Dr. Plavec about the importance of symbiotics where the hot source of the UV continuum is directly seen. In at least RW Hya, SY Mus and AG Peg this continuum comes from the Rayleigh-Jeans tail of hot subdwarfs with $T_{eff} \gtrsim 40000K$.

<u>Plavec</u>: A temperature of $\sim$40000K is still low to explain the HeII emission.

<u>Kafatos</u>: That is correct, but since this is a lower limit one does much better than the Kurucz models in explaining both the far UV continuum and the ionizing radiation needed to explain the UV lines.

<u>Michalitsianos</u>: A number of objects observed with IUE appear to outwardly resemble an early type star in terms of effective temperature and luminosity. As such a compact object through accretion may superficially resemble an early main sequence star, but if compared to model

atmospheres do not fit theoretical expectation. What UV and optical tools exist by which we can differentiate an early main sequence star from an object with an accretion shell?

Slovak: High resolution IUE (and other ultraviolet spacecrafts, i. e. Space Telescope) observations of various line profiles would prove invaluable. Detailed studies of line profile (as opposed to continua) would give such fundamental parameters as log g, Doppler broadening, rotational broadening, etc. permitteing a distinction between stellar models and accretion disks.

Nussbaumer: In V1016 Cyg we have reproduced the observed IUE ultraviolet continuum with a combination of a 160000K black-body star, and a recombination and two-photon continuum from a $T_e \sim$ 15000K gas, rather similar to the drawing shown by Viotti for Z And. The hot star continuum dominates for $\lambda <$ 1600 A.

Slovak: I have no IUE observations of V1016 Cyg myself, so I cannot comment on the ultraviolet continuum. Your results are encouraging for a binary interpretation of this system; the contribution of the hot gas is clearly an important component.

Kwok: How can we be sure of the relative contributions to the infrared colors from (i) the photosphere, (ii) ff and bf continua, and (iii) dust emission?

Slovak: I attempted to simplify the interpretive aspect by showing distributions for Allen's S type symbiotics. These stars are characterized by non-variable, stellar-like infrared continua. The infrared flux appears to arise entirely from the photosphere of the late type component with possibly some small contribution from the ff + bf emission in the nebula.

Keyes: If indeed the cool component's IR photometry is reasonably constant as is the case for several S type symbiotics in Slovak's sample, then we probably do not see much of a contribution of ff radiation radiation in the IR. This is because our observations with IUE and the Lick image tube scanner from 1200 to 8000 A show that the Balmer continuum level (3200-3600A) plus Balmer emission features can vary by as much as 0.3 magnitudes during periods of relative quiescence for objects such as AG Peg, while the available IR photometry for that star does not show corresponding large variations.

Slovak: Yes, the ff plus fb continuum falls steeply in the Johnson JHKL bandpasses.

Cassatella: I do not expect that it will be possible in the case of all symbiotic stars to see the presence of the hot stellar component in the UV. This is for example the case for o Ceti (M giant plus probably a white dwarf) whose UV spectrum only shows Balmer free-free and free-bound emission.

Slovak: The IUE observations indeed show that in many of the symbiotics, the hot stellar component is hidden by the nebular material, which reprocesses the radiation field of the hot component ($T_{eff} \sim 100000K$ using the HeII 1640 A line to derive a Zanstra temperature) and the UV continua appear cooler ($T_{eff} \sim 10-30000K$) and non stellar.

Viotti: Dr. Ricciardi of our Institute has compared the intensities of the UV emission lines of the symbiotic stars observed at high resolution. Figure 1 shows the line intensities in V1016 Cyg, Z And, and AG Dra as compared to those in RR Tel. This kind of comparison may allow to make hypothesis on the structure of the emitting envelopes and to select among the possible models. Evident in the figure is that three stars, V1016 Cyg, Z And and RR Tel, have quite similar UV spectra, in spite of the fact that different models have been proposed by Nussbaumer and Schild (1981), Altamore et al. (1981) and Penston et al. (1981) for the three stars. On the contrary, the figure seems to suggest that the physical conditions of the emitting regions should be almost the same. In the case of AG Dra it is clear that the HeII recombination lines are about one order of magnitude stronger with respect to the other lines than those in RR Tel, probably as the result of a larger density of ionizing radiation.

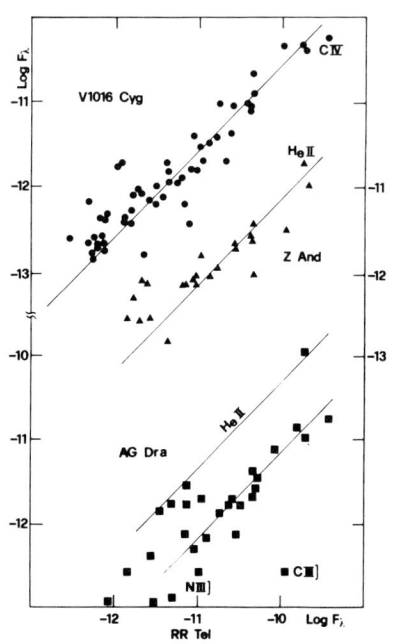

Figure 1.

X-RAY OBSERVATIONS OF SYMBIOTIC STARS

David A. Allen
Anglo-Australian Observatory

ABSTRACT

Observations are reported of 19 symbiotic stars made with the imaging proportional counter of the Einstein Observatory. Three of the objects (HM Sge, V 1016 Cyg and RR Tel) were detected as soft X-ray sources. All three have shown slow-nova eruptions in the past 40 years. The data are interpreted as support of a model for slow novae involving thermonuclear events on white dwarfs which accrete from M giant companions. Symbiotic stars in their steady state, not being detected X-ray sources, are presumed to be powered by the accretion process alone.

The observations are summarized in table 1.

The full paper is published in Monthly Notices R. Astr. Soc. (1981), volume 197, 739-743.

Table 1. X-ray and other data on symbiotic stars

NAME	F_x	N_e	IP	CS Dust	F_{2cm}	Sp.	Opt.Var
AX Per	<9	low	var	N	-	M	R
AS 201	<6	high	25	Y	<17	G	N
He2-38	<6	low	80	Y	<6	M	N
He2-106	<11	low	100	Y	40	M	N
BD-21 3873	<7	extreme	70	N	-	G	N
He2-127	<11	low	100	Y	<12	M	N
Hen 1092	<4	medium	80	N	<18	K	N
HD 330036	<7	low	40	Y	-	G	N
He2-171	<12	low	120	Y	<12	M	N
Hen 1242	<4	high	70	N	<22	M	N
V 455 Sco	<8	medium	100	N	<12	M	R
AE Ara	<5	high	50	N	<18	M	R
H1-36	*	low	90	Y	91	-	N
Y CrA	<6	high	90	N	<20	M	R
AS 295B	<12	high	250	-	-	M	R
HM Sge	83±6	medium	rising	Y	57	M	S
CI Cyg	<5	medium	100	N	-	M	R
V 1016 Cyg	75±5	low	100	Y	110	M	S
RR Tel	18±3	medium	120	Y	54	M	S

F_x = X-ray Flux 0.2 - 2 keV x 10^{-14} ergs cm^{-2} s^{-1}

N_e = Electron density class

IP = Ionization Potential in eV
CS Dust = Is Circumstellar Dust present?
F_{2cm} = Radio Flux at 2 cm in mJy

Sp. = Spectral type of the cool star
Opt.Var. : N = No history of Variability
            R = Rapid variations in the form of minor nova-like outbursts typically every few years and of 3-4 magnitudes amplitude.
            S = Slow-nova outbursts with time scales of many decades and amplitudes near 8 mag.
* H1-36 is discussed elsewhere

X-RAY DETECTION OF THE SYMBIOTIC STAR AG DRACONIS

Christopher M. Anderson, Joseph P. Cassinelli
Nancy A. Oliversen and Roy V. Myers
Washburn Observatory, University of Wisconsin-Madison
and
W. T. Sanders
Department of Physics, University of Wisconsin-Madison

ABSTRACT

The symbiotic star AG Draconis was observed by the HEAO-2 Imaging Proportional Counter (IPC) and found to be an unusually intense source of very soft X-rays.

I. INTRODUCTION

AG Draconis was observed with the IPC on 1980 April 11, for an effective exposure time of 8800s. During the exposure, 2345 net IPC source counts were detected. The pulse-height distribution is shown in Figure 1. The spectrum is very soft, consistent with all of the counts being caused by photons of energy less than 0.28 keV.

II. DISCUSSION

X-ray spectrum analysis is performed by folding a source model spectrum through the detector response function and comparing the result with the data via a χ^2 test. The source of the X-rays is assumed to be a volume of gas at a single temperature having an emission spectrum (continuum plus lines) as calculated by Raymond and Smith (1979). The source spectrum may be attenuated by a foreground column of material having the opacity per hydrogen nucleus of the interstellar medium given by Brown and Gould (1970). The source temperture, T, emission measure, EM = N_e^2 x Volume, and the foreground column density, N_H, are free parameters in the fitting process. In addition, the value of the IPC gain used by the detector response function has been allowed to vary ± 10% to accomodate uncertainties in the IPC calibration.

The results of the spectral fitting are shown in Figure 2. Below the solid line is the 90 % confidence region in the T versus N_H plane, determined using the procedure described by Lampton, Margon and Bowyer (1976). Also shown are contours of the best-fit emission measures. If we adopt the M_V of an ordinary K3 III star then the object's V = 9.4 indicates a distance of 700 pc. From the data of Heiles (1974) we then estimate a foreground interstellar hydrogen column density of 3×10^{20} cm^{-2} for this distance in this direction. A further limit on the likely values of the emission measure can be made following the technique described by Nordsieck, Cassinelli and Anderson (1981). The absence of various coronal lines in the visable spectrum at the 1 A equivalent width level limits allowable models to the region of the EM versus T plane below the various lines in Figure 3. The combination of these loci transforms to the long-dashed line in Figure 2. Together these numbers and reasonable estimates of the uncertainties therein lead to the following characteristics for the X-ray source in AG Dra :

$$T = 1.1 \times 10^6 \text{ K}$$
$$EM = 2.6 \times 10^{55} \text{ cm}^{-3}$$
$$N_H = 3 \times 10^{20} \text{ cm}^{-2}$$
$$L_x (.2-1.0 \text{ keV}) = 5 \times 10^{32} \text{ ergs s}^{-1}$$

Because of the uncertainty in the IPC gain and its relatively broad smearing function and because of the multiplicity of and uncertainty in the free parameters in the fitting procedure, all of these characteristics are uncertain by factors of the order of 2.

The implications of these observations are discussed in more detail by Anderson, Cassinelli and Sanders (1981) who find that if the source of the luminosity is to be associated with mass transfer from a red-giant to a white dwarf, an extraordinarily large mass loss rate is required. On the other hand, this luminosity would be produced if about 10% of the acoustic flux expected in a K3 III star were converted to X-rays in a chromosphere-corona region. Efficiencies of this order are not unusually high.

Recently we have investigated as an alternate input model spectrum a simple Planck function. With the same foreground column density adopted above we find that the best fit temperature is 2.5×10^5 K and the emitting area corresponds to a sphere of a thousasnd kilometers radius. Just at the lower limit of acceptability in the model fits is a temperature of 1.5×10^5 K and a radius of 1.4×10^4 km .

The X-ray observation was made during the rise to the most recent, unusually bright maximum. The object was to be reobserved, but the untimely demise of the HEAO-2 spececraft made this impossible.

X-RAY DETECTION OF THE SYMBIOTIC STAR AG DRACONIS

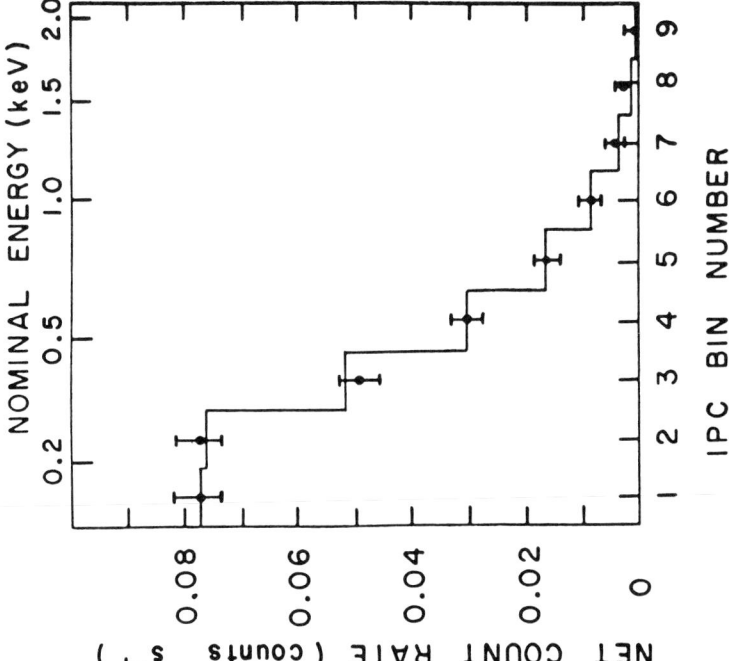

Figure 1. Pulse height distribution in IPC counts per second per energy bin versus IPC bin number. Error bars indicate 1 σ statistical uncertainty. The solid line is the best fit model with N_H fixed at 3×10^{20} cm^{-2}, and $T = 1.3 \times 10^6$ K, and EM = 7×10^{54} cm^{-3}.

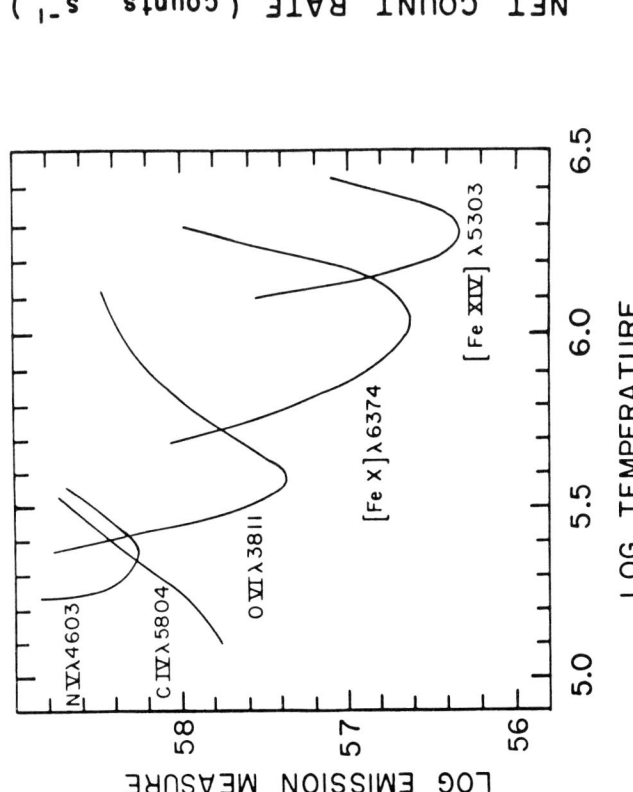

Figure 3. The EM, T plane for AG Dra. The absence, at the 1 Å equivalent width level, of each of the indicated coronal lines restricts the object to the region below the various lines.

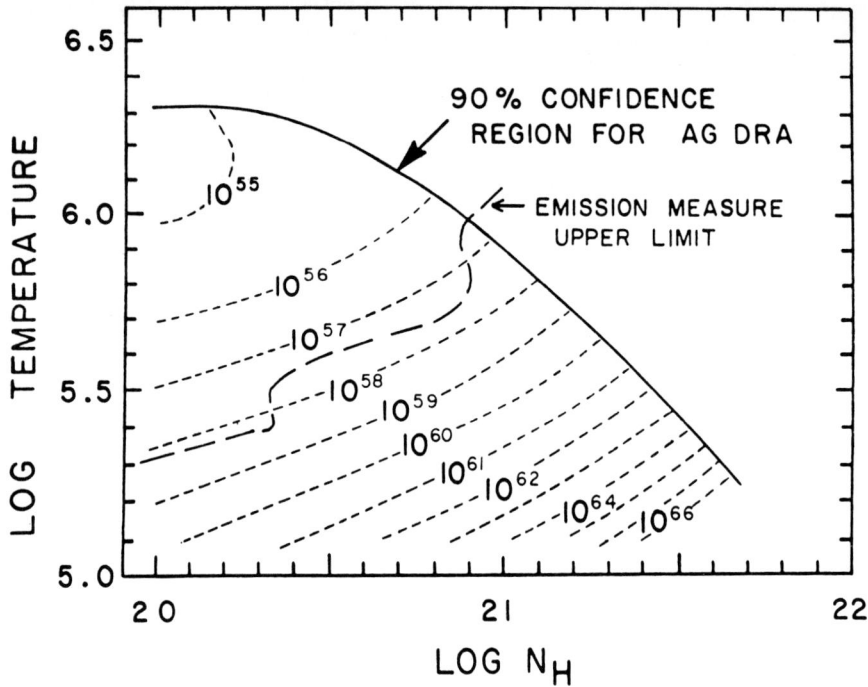

Figure 2. The region below and to the left of the solid line is the 90% confidence region for the AG Dra temperature and absorption column density. Inside the 90% confidence region are contours of the best fit emission measures. Allowance has been made for a ± 10% untertainty in the IPC gain.

REFERENCES

Anderson, C. M., Cassinelli, J. P. and Sanders, W. T. 1981, Ap.J. 247, L 127.
Brown, R. L. and Gould, J. R. 1970, Phys. Rev. D, 3d Ser., 1, 2252 .
Heiles, C. 1974 Astr. Ap. Supl., 20, 37.
Lampton, M, Margon, B., and Boyer, S 1976, Ap.J., 208, 177 .
Nordsieck, K. H., Cassinelli, J. P. and Anderson, C. M. 1981, Ap.J. 248, 678.
Raymond, J. C. and Smith B. W. 1979 , private communication.

DISCUSSION ON X-RAY OBSERVATIONS

Keyes: The HeII 1640 A Zanstra temperature of AG Dra before outburst interpreted with Hummer and Mihalas (1970) log g = 5.0 models, is about 120000K, and the stellar radius, implied by integration of the observed ultraviolet flux is very roughly 2.5×10^4 km.

Viotti: We have estimated the hydrogen column density from the Ly interstellar line in the high resolution IUE spectra of AG Dra obtained after the outburst, and showing a detectable continuum down to ~1200 A. We found log N(HI) ~ 20.2 in agreement with the X-ray determination and with the weakness of the 2200 A band.
As regards the other three X-ray sources, I note that they also are radio sources, and I would like to know what are the physical implications of this.

Kwok: In response to Dr. Viotti question, the X-ray emission from HM Sge, V1016 Cyg, and RR Tel may be the result of wind interactions in these systems. If part of the kinetic energy from a hot star wind is dissipated in the collision process, enough X-rays can be produced to be observable.

SESSION II

DISCUSSION ON INDIVIDUAL STARS

Chairmen: M.F. McCarthy and M. Hack

Introductory reports on:

Z ANDROMEDAE (R. Viotti, A. Giangrande, O. Ricciardi and A. Cassatella)

CH CYGNI (M. Hack and P.L. Selvelli)

CI CYGNI (A.G. Michalitsianos, M. Kafatos, R.E. Stencel, A.A. Boyarchuk)

V1016 CYGNI (A. Cassatella)

V1329 CYGNI (J. Grygar and D. Chochol)

AG DRACONIS (N.A. Oliversen and C.M. Anderson)

SY MUSCAE (A.G. Michalitsianos and M. Kafatos)

AR PAVONIS (M.H. Slovak)

AG PEGASI

RX PUPPIS (M. Kafatos and A.G. Michalitsianos)

HM SAGITTAE (S. Kwok)

RR TELESCOPII (P.A. Whitelock, R.M. Catchpole and M.W. Feast)

PU VULPECULAE (T.S. Belyakina, R.E. Gershberg, Y.S. Efimov, V.I. Krasno-babtesev, E.P. Pavlenko, P.P. Petrov, K.K. Chuvaev)

Symbiotic Stars: Hot Star with a Cool Envelope?

Z ANDROMEDAE: THE PROTOTYPE?

R. Viotti, A. Giangrande, O. Ricciardi
Istituto Astrofisica Spaziale (CNR), Frascati, Italy

A. Cassatella
Astronomy Division ESTEC, Villafranca Satellite Tracking
Station of ESA, Madrid, Spain

Z And is considered as the "prototype" of the symbiotic stars. Besides its symbiotic spectrum, the star is also known for its characteristic light curve (and for the related spectral variations). Since many theoretical speculations on Z And and similar objects have been based on the luminosity and spectral variations of this star, it is important to make a critical analysis of the observational data concerning it.

1. The light curve (see Mattei 1978) can be synthetically described as characterized by two (or more) phases = the active phase which starts with a $\sim$ rapid increase of the visual luminosity by about two magnitudes, followed by a kind of oscillation with decreasing amplitude (= transition phase), until a quiescent phase ($V \sim 11^m$) is reached.
The minima (or maxima) are separated on the average by about 632 days. However, the interval between two successive maxima is not constant but varies from 310 to 790 days (Mattei 1978), indicating that the luminosity variations are not periodic. At present Z And is in quiescent since many years, and a new outburst could well occur very soon.

2. The optical spectrum is extensively described in many review papers on symbiotic stars (see e.g. Merrill 1958, Sahade 1960, Swings 1970). At minimum luminosity and quiescence it has a typical symbiotic spectrum with a blue region rich in strong emission lines, having a wide range of ionization energy from FeII to [FeVII]. A weak blue continuum, with a large Balmer excess, is present under the emission lines, but longwards of about 5000 Å a red continuum emerges with several strong TiO absorption bands typical of an $\sim$ M6.5 stellar spectrum (Altamore et al. 1979). At the light maxima the highest ionization energy lines are much weaker with respect to the lower ionization emission lines, and to the blue continuum, which then largely washes out the M-type spectral features.

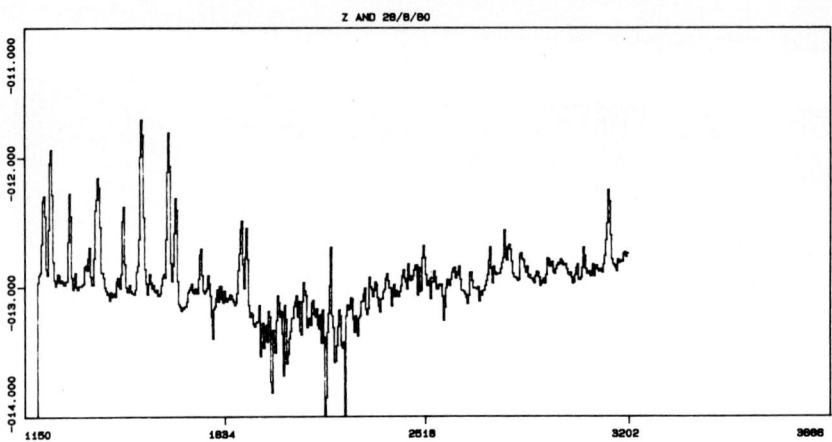

Figure 1. The low resolution ultraviolet spectrum of Z And in August 1980. Ordinates are log fluxes in erg cm^{-2}s^{-1}A^{-1}.

Hydrogen and FeII lines show a P Cygni profile, and give evidence for outflowing matter at velocities of about 180 km s^{-1} (Swings and Struve 1941).

3. <u>The ultraviolet spectrum</u>. Like in the other symbiotic stars a large number of emission lines can be identified in the IUE spectra belonging to many different ionization energies, from OI, MgII, FeII to HeII, NV, OV and MgV. At high resolution the emission lines appear narrow. The relative intensity of intercombination lines of CIII, NIII and OIV suggests an electron density of about 2×10^{10} cm^{-3} (Altamore et al. 1981).
At low resolution (Figure 1) a weak continuum is detectable with a depression near 2200 A suggesting an interstellar extinction of E(B-V) = 0.35. According to results we have just obtained, the continuum of the dereddened spectrum (Figure 2) can be fitted by the sum of two components: (a) a hot black body with T$\sim 10^5$°K, and (b) a nebular type continuum with $T_e \simeq 1.5 \times 10^4$°K, for which free-free, bound-free and two photon (negligible) emission have been considered. This result is quite in agreement to that of Boyarchuk (1968) who suggested for Z And a three component model — M giant, hot source, nebular source — based on the optical observations. The black body temperature of 10^5°K is rather uncertain because the slope of the continuum near 1300 A is close to the Rayleigh-Jeans distribution. This temperature is also critically dependent on the derived value of the interstellar extinction.

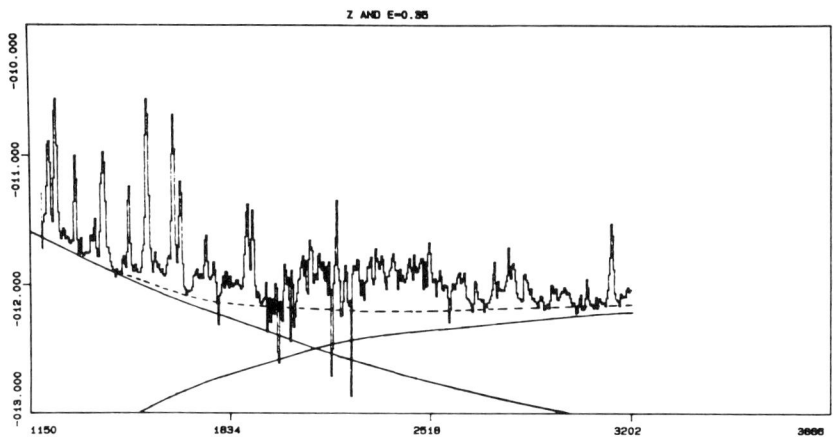

Figure 2. The low resolution ultraviolet spectrum of Z And corrected for an interstellar extinction of E(B-V)=0.35. The continuum (dashed line) is fitted by the sum of bb and nebular emission.

4. <u>The infrared spectrum</u>. Infrared photometry of Z And was recently obtained in September 1980 at the Wyoming Infrared Telescope by M. Ferrari-Toniolo and P. Persi (paper in preparation), and in July 1981 at Almeria by Eiroa et al. (this volume). The energy distribution between 1 and 10μ agrees well with a black body spectrum with a 3200°K temperature, without any clear evidence for a dust emission. We have also found that the "nebular" component of Figure 2, when extrapolated to the infrared, does not seem to contribute significantly to the 10μ flux.

Observation of low weight near 20μ made by Woolf (1973) suggests a rapid rise of the continuum beyond 10μ. It would be important to confirm or disprove this result because of its strong model implications.

No significant differences have been found among the different IR observations so far reported in the literature, which could support that the cool spectral component of Z And is not variable.

The total power emitted by the cool component is about 2.6×10^{-8} erg cm^{-2}s^{-1}, to be compared to about 2×10^{-9} emitted in the UV continuum, and ~1×10^{-9} in the UV emission lines. The 10^5°K hot continuum would contribute with about 2×10^{-8} erg cm^{-2}s^{-1} to the total energy output per sec from Z And. This value is of the same order of magnitude of the total radiation power of the cool component, and this could be an argument in favour of the binarity of Z And.

5. <u>The problem of Z Andromedae</u>. In order to know the nature of this star and in particular to understand the origin of the large light variations,

it is important to study the star when it is in a quiescent phase. In fact, during the outbursts the star(s) is (are) certainly surrounded by a massive gaseous envelope which should mask the inner regions, while during quiescence we can look deeper into the object, and it should be easier to derive the basic parameters of the star(s).

From the study of the ultraviolet spectrum of Z And Altamore et al; (1981) argued that the thickness in the line of sight, derived from the ratios of the NV and SiIV resonance lines, when combined with the measured emitting volume, suggest that the geometry of the line formation region is not compatible with formation in an extended nebula. Also the stellar radiation density in the NIII] region should be very low, otherwise the multiplet lines relative intensities would be that characteristic of Boltzman equilibrium contrary to the observations.
Finally the emission lines are too narrow to be formed in a wind coming from a hot star, or in a rotating disk (unless the inclination to the line of sight is very small). Altamore et al. concluded that all the arguments are consistent with the formation of the lines around the cool star, in a solar-type transition region. The hot continuum component, confirmed by the present observations, may come from an optically thick region, that is either a hot subdwarf, or a hot active region on the surface of the cool giant, while the nebular continuum should originate in an extended region surrounding Z And.

During the present quiescent phase the star is still variable, with small luminosity variations. Sahade et al. (1981) reported variations of the intensity of the ultraviolet emission lines. In addition we have also noted that the relative intensity of the SiIII] and CIII] lines changed during 1979-81. The long time-scale variation of the intensity of H_α found by Altamore et al. (1979) is confirmed, but without any evidence for a periodicity. Large changes in the H_α profile were reported by Oliversen and Anderson (this volume).

Clearly it would be important to know more about this activity of the star during quiescence, since it could be of help in understanding the physical processes which are responsible for the outbursts of the star.

Further studies are required mainly with respect to:
(a) Systematic near-infrared photometry to look for any variation of the M star, and accurate observations near 10-20 μ to search for a possible dust emission.
(b) Detailed study of the optical and ultraviolet spectrum using well exposed spectra in order to derive the emission line profiles and radial velocities with better accuracy, and to measure the continuum level and, possibly the (interstellar or atmospheric) absorption lines

at high resolution (in this context it would be important to observe a number of representative symbiotic stars with the Space Telescope).
(c) X-ray observation with the EXOSAT satellite, since Z And may well be a (probably soft) X-ray source detectable with the experiment.

REFERENCES

Altamore, A., Baratta, G.B., Viotti, R.: 1979, Inf. Bull. Var. Stars, 1636
Altamore, A., Baratta, G.B., Cassatella, A., Friedjung, M., Giangrande, A. Ricciardi, O., Viotti, R.: 1981, Astrophys. J. 245, 630.
Boyarchuk, A.A.: 1968, Soviet Astr. -A.J. 11, 818.
Mattei, J.A.: 1978, J. Roy. astr. Soc. Canada 72, 61.
Merrill, P.W.: 1958, in "Etoiles à Rayes d'Emission", Mem. Soc. Liège, p. 436.
Sahade, J.: in Stars and Stellar Systems, Vol.6, Stellar Atmospheres, T. L. Greenstein (ed.), University of Chicago Press, p. 498.
Swings, P.: 1970, in Spectroscopic Astrophysics, G.H. Herbig (ed.), University of California Press, Berkeley, p. 498.
Swings, P., Struve, O.: 1941, Astrophys. J. 93, 356.
Woolf, N.J.: 1973, Astrophys. J. 185, 229.
Sahade, J., Brandi, E., Fontenla, J.M.: 1981, Rev. Mex. Astr. Astrophys. in press.

DISCUSSION ON Z ANDROMEDAE

Plavec: On your diagram matching the continua, you use hydrogen continuum at $T_e=15000K$. However, on your NIII] diagram determining the electron density, you use $T_e=80000K$. Why? Is there any observational evidence for such a high electron temperature? Do you observe a continuous hydrogen contribution with this electron temperature? If not, and if N_e is not sensitive to T_e, why assume such a high temperature.

Viotti: In the Altamore et al. paper we found relative consistency between the emission measures for the intercombination and resonance lines in agreement with our model. The corresponding ff+bf contribution of the high temperature region to the UV continuum is lower than that observed suggesting a different source for that continuum. In other words the NIII] lines and the UV continuum are formed in different regions of Z And.

Kafatos: I would like to point out that when we used the OIV] or NIII] multiplet ratios in some stars we found consistent single-valued densities, but in some other stars one gets different values from different ratios of the same multiplet. I believe that this is a real effect, pointing to different density regions rather than an uncertainty in atomic parameters.

Keyes: The ratios of the NIII] 1754/1749 and 1748/1749 should be locked in the ratio of the A-values, as both 1754 and 1748 lines share a common upper level. This provides a useful check on the accuracies of the line intensity measurement. Are the discrepancies you find in densities derived from other line ratios significant compared to the measurement uncertainty implied by your 1754/1749 and 1748/1749 density values?

Viotti: Our results for Z And, RR Tel and AG Dra show that at least the collision strengths of the OIV] and NIII] lines are not well known, and this may partly explain the dispersion of the density values. In addition the intensity of the 1749 line seems weaker than expected, but it is difficult to decide whether this is a real effect or an instrumental one (reseau mark).

Nussbaumer: When employing the OIV] 1401 or the NIII] 1749 multiplets to determine electron densities, you have to allow for uncertainties in atomic data and observations. In practice this means that if the observed intensity ratios lie close to the high or low density limits, then you can no longer expect to obtain very accurate N_e. Thus Viotti just showed us that for RR Tel he obtains $N_e = 10^9 \mathrm{cm}^{-3}$; well it might be more prudent to conclude $N_e \leq 10^9 \mathrm{cm}^{-3}$.

A REVIEW OF THE PROPERTIES OF THE SYMBIOTIC STAR CH CYGNI

M. Hack, P.L. Selvelli
Astronomical Observatory of Trieste, Italy

Introduction

The spectrum of CH Cygni usually has the appearance of a normal M6 III star, but at intervals of several years, it has phases of activity during which it shows the characteristics of a symbiotic star, that is, a blue continuum extending shortward of the Balmer discontinuity, and strong emission lines of He I, Fe II, [Fe II], [S II], [O I], [O III]; the Balmer lines and the H and K lines of Ca II present emission wings. Outbursts were observed: 1) in September 1963; the star returned to its normal phase by August 1965 (Deutsch, 1964); 2) in June 1967 (Deutsch, 1967); the spectrum was observed by Faraggiana and Hack (1971) until December 1970, when the activity phase was over; 3) another outburst was observed in August 1977 (Morris, 1977; Fehrenbach, 1977), and at the time of our last spectroscopic observations (June 1981) it was still going on. During this last period of activity several observations were made with IUE in both the low and high resolution modes, from 1175 A to 3100 A.
The first observed outburst lasted less than two years, the second one a slightly more than three years; the present one is at the end of its fourth year.
The most obvious hypothesis for explaining symbiotic stars is that they are binary systems formed of a late-type giant and a hot subdwarf which occasionally has outbursts. The other possibility is that the cool star may be surrounded by an extended corona, heated by shock waves during episodic phases of eruptive mass loss. This corona is supposed to be responsible for the blue continuum and for the emission lines. A third hypothesis, that a hot star is surrounded by a cool envelope, is ruled out at least in the case of CH Cyg, by our spectroscopic observations of 1967-70 and of 1977-81, because there is clear evidence that the blue continuum is filling up the M6 photospheric lines, deleting them almost

completely, especially in the blue-violet region of the spectrum. Several symbiotic stars present orbital motions; however, CH Cygni does not show any evidence of periodic radial velocity variations. Instead, our observations of the 1967-70 episode supported the hypothesis that shock waves heat the envelope surrouding the M6 giant. However, photometric observations by Slovak and Africano (1978) indicate that CH Cygni presents a rapid flickering, characteristic of cataclysmic variables, most of which are close binary systems. We will summarize the results of the observations of the 1967-70 activity phase in the photographic spectral range (3300-4900), and of the activity phase which started in 1977 in the spectral range 6700-3400 A and 3100-1175 A, and we shall compare the phenomena observed during the two outbursts.

The 1967-70 outburst

The M6 spectrum is veiled by a continuum which partially fills the absorption lines and increases in intensity toward the violet. This continuum was absent in August 1965 and in September 1970; it appeared at the moment of the explosion, in June 1967, and reached a maximum in August 1968. The color temperature was about 10000 K. No measurable Balmer discontinuity was observable. Emission lines of He I, Fe II, [Fe II], [S II] were present. In July 1968 λ 5007 [O III] appeared. The Balmer lines had red and violet emission wings, R/V $>$ 1, while the H and K lines presented a P Cygni contour. At some epochs H and K presented two absorption cores. Absorption chromospheric lines, which were completely missing in July 1967, appeared in the ultraviolet continuum in July 1968. Only lines of metallic ions were present. In May 1970 the absorption lines of low excitation at $\lambda <$ 3800 give radial velocities systematically more negative by about 20 km/s than the other lines having the same low level. The resonance lines behave like the ultraviolet lines. The H and K absorption cores have presented two well-separated components since July 1968; the sharpest of the two in July and August 1968 has a radial velocity of -170 km/s, which in 1969-70 reaches a value of -120 km/s, while the broad component decreases from -80 km/s in July-August 1968 to -10 in 1969-70. The photospheric lines (i.e. lines from excited levels of neutral metals) have an almost constant velocity, included between -60 and -65 km/s, which we assume to be the velocity of the center of mass of the star. The forbidden and permitted emission lines of Fe II have generally velocity, included between -50 and -60, but systematically less negative by 5 to 10 km/s than the photospheric velocity. Radial velocities for a large number of Balmer lines could be measured only on the spectrograms taken in July and August 1968, i.e. at the epoch of maximum intensity of the blue continuum. There is a

Balmer progression, the radial velocity of Hβ, Hγ, Hδ and H7 being about −60 km/s in July 8-11 and −70 on August 14, while the other members of the series give −45 ± −50 in July and −55 ± −60 in August, i.e., values equal to or less negative than the photospheric ones. The picture emerging from these data is that just after the outburst (spectrogram of July 15, 1967) the layers where Hβ and H and K are formed are expanding relatively to the photosphere. One year later one layer has the same velocity as the photosphere; another, identified with the sharp component of the Ca II resonance lines, is expanding outward at −170 km/s; then the latter continues to expand at −120 km/s and the former falls toward the photosphere.

The outburst starting in 1977

The most striking difference between the outburst of 1967 and that of 1977 is the greater intensity of the blue continuum during the latter. The spectra taken in July and August 1968, at the epoch of maximum intensity of the blue continuum, and those taken in September 1977, when the continuum observed during the present outburst was at minimum intensity, appear comparable, but on February 1978 the photospheric absorption lines are almost completely filled up by the continuum extending longward of Hβ. Another striking difference between the two outbursts is found in the Balmer line emission wings. The upper members of the series had two emission wings in 1967-70, while from 1977 onward they present P Cygni inverse contours. The H and K lines of Ca II, on the contrary, have similar contours at the two epochs. Hα, Hβ, Hγ, Hδ, have two emission wings with V/R $>$ 1. Hence during this last outburst we have the peculiar occurrence of the simultaneous presence of upper layers, where the H and K lines are formed, which are expanding relatively to the photosphere, and of lower layers, where the higher members of the Balmer series are formed, which are moving toward the photosphere. Also the absorption lines of Sc II, Ti II, V II, Mn II, Sr II have an inverse P Cygni contour and about the same velocity as the Balmer lines with n $\gtrsim$ 9. The emissions of Fe II, on the contrary, have a radial velocity of about −75 km/s, systematically more negative than that of the forbidden lines, −65 km/s (Faraggiana, 1980).

Since February 1978 no strong variations in the spectrum have been observed. One of the most evident variations is that of the H and K cores. Two components are clearly present in September 1977 and February and June 1978; the shortward component, at about −140 km/s, is sharp, the longward one at about −110 km/s is broader. In September 1978 the two components are blended; in July 1979, February 1980, and September 1980 the two components are blended, the longward one being the faintest;

in January and March 1981 the emission wing is much fainter than at previous epochs. In June 1981 the two components are clearly separated again, the longward component being the sharper, just the contrary of what was observed in September 1977-June 1978. Several spectra of the red part of the spectrum have been taken since September 1977. Hα has strong emission wings with V/R $>$1. The forbidden lines of O I were absent in September 1977 and have been present as two strong sharp emissions since July 1979 . Instead, no O III forbidden lines have been observed during this outburst. The radial velocity of the Hα absorption core ranges between about -60 and -65 km/s.

The ultraviolet spectrum observed with IUE

Low resolution spectra were obtained in 1978, 79 and 80; high resolution spectra in the near UV were obtained in March 1979, and in the far UV in September 1980. Comparison of the observed flux with the Kurucz models indicates that no one of them agrees with the observations. Faint discontinuities are observable at λ1500 and λ1700, as if the strong discontinuities present in theoretical spectra for effective temperatures ranging between 8000 and 7500 K were partly filled by continuous emission extending up to 1300 Å.

On September 1, 1980 the flux reached its maximum value after the beginning of the present phase of activity (3×10^{-12} erg s^{-1} cm^{-2} Å^{-1} at 1400 Å).

The line spectrum shows rather different behaviors for the far UV and the near UV ranges. At short λ, absorptions of once-ionized metals (mainly FeII and NiII) dominate the spectrum, while only a few emissions (i.e. OI 1300, 1355 and 1641.2, NI 1745, FeII m=191, SiIII λ 1892 and CIII λ 1908) are present. At longer λ, on the contrary, the spectrum shows numerous emissions that, except for MgII 2800, MgI 2852, CII 2324 and AlII λ 2669.2 are all attributable to FeII.

It is evident that the resonance fluorescence mechanism, i.e. absorption in the far UV region followed by re-emission toward longer λ, is responsible for the FeII emissions. In this respect CH Cyg is very similar to the peculiar Be star HD45677, where the FeII lines show the same behaviours (Selvelli, Stalio 1980; Stalio, Selvelli 1980).

The OI λ1300 resonance triplet shows relative intensities that disobey the theoretical 5/3/1 ratio.This fact together with the presence of the semiforbidden 1641.3 emission, which shares the same upper level with the 1300 lines and has comparable intensity, can be explained in terms of "trapped" 1300 resonance radiation and multiple resonance scattering processes. The excitation mechanism for the strong OI triplet emission seems to be Lyβ fluorescence (Swings, 1955). Two other possible mechanism are recombination from O^{++} and continuum fluorescence. Both seem unlikely, the first one because no quintet lines (e.g. $\lambda\,6158$ and $\lambda\,3947$) are observed, and the second because the far UV continuum is rather weak and the $\lambda\,4368$ line is not present.

Apart from FeII and OI, the other emissions in the UV range are the resonance lines of MgI $\lambda\,2852$ and MgII $\lambda\,2800$ and the intersystem lines of AlII $\lambda\,2669.2$ and SiIII $\lambda\,1892.0$. The MgII resonance doublet presents two absorption cores, the sharp one, probably of interstellar origin, at rest velocity, and a broader one shortward shifted at about -110 km s^{-1}. The emission wings give V/R<1.

MgI, AlII and SiIII, together with MgII, AlIII and SiIV, form two isoelectronic sequences. Inside each sequences the same terms scheme applies, but the energies of the corresponding terms are different. It is commonly agreed that the population of the lowest terms of the above species is attributable to collision by electrons with $T_e \approx$ some e V. This explains why all the intersystem lines of the MgI, AlII, SiIII sequence (i.e. $\lambda\,4571$, $\lambda\,2669$ and $\lambda\,1892$ respectively) are observed; the low lying metastable term is easily populated by collisions. On the contrary, only the resonance line of MgI $\lambda\,2852$ is present because the upper term of the resonance lines of AlII ($\lambda\,1670$) and SiIII $\lambda\,1206$ needs higher electron energies to be populated. In the MgII, AlIII, SiIV sequence only the resonance doublet of MgII $\lambda\,2800$ is observed. Either the electron energy is not sufficient to populate the first excited level of AlIII and SiIV, or these ions are formed in a higher density region where collisional de-excitation takes place. But the presence of the SiIII $\lambda\,1892$ and CIII $\lambda\,1908$ emissions seems to favour the first hypothesis.

Conclusion

These observations indicate that: a) the velocity of the star, as given by the photospheric lines, ranges irregularly between -65 and -55 km/s; b) a rarefied outer layer, where the forbidden lines are formed, has about the same velocity as the star, plus or minus 5-10 km/s; the intensity of the forbidden lines of FeII relative to the permitted ones varies, from one outburst to another; in 1964 [Fe II]/Fe II>1, in 1967-70 [Fe II]/Fe II$\sim$1 and during the present episode [Fe II]/Fe II<1, thus indicating that the density of the envelope during this outburst is larger than during the previous ones; c) between the photosphere and the outer envelope where the forbidden lines are formed, there are several regions characterized by different motions. The general behavior seems to be the following: the upper layers (where the strong resonance lines of Ca II and Mg II are formed) are expanding outward, and sometimes dividing into two layers, one continuing to expand at a velocity greater than escape velocity, and the other falling back toward the star, while the lower layers (where the higher members of the Balmer series are formed) are falling toward the stellar surface. Intermediate layers, where the first members of the Balmer series and the strong permitted lines of Fe II and other abundant ions are formed, are sometimes observed to move outward, sometimes inward with a velocity differing no more than 20 km/s from that of the photosphere; d) the blue-ultraviolet continuum has a distribution which is not attributable to a hot star. Hence, if the presence of a hot companion can be discarded, the blue-ultraviolet continuum must be formed in the dense, optically thick parts of a layer above the photosphere, where also the sharp "chromospheric" absorptions of ionized metals are formed.

It is generally believed that the red giant stage is followed by the planetary nebula stage. It is possible that the outburst episodes observed in CH Cygni, with mass-loss estimated of the order of a few times 10^{-8} the solar mass per episode, indicate that CH Cygni is in the transition phase.

Note added in proof

Andrillat (1982) reports that in spectra taken in 1977 and in 1981, the OI λ 8446 emission (that feeds the 1300 triplet in the cascade following the excitation by Ly β) is missing. This seems to rule out the proposed Ly β fluorescent mechanism as responsible for the OI emissions. Comments on this matter together with more details on the UV observations will be reported in a paper that is to appear soon in Astronomy and Astrophysics.

References

Andrillat, Y.: 1982, This colloquium, p. 47.
Deutsch, A.J.: 1964, Ann. Rep. Mt. Wilson and Palomar Obs., 11
Deutsch, A.J.: 1967, IAU Circ. No. 2020
Faraggiana, R.: 1980, Astr. Astrophys. 84, 366
Faraggiana, R., Hack, M.: 1971, Astron. Astrophys. 15, 55
Fehrenbach, Ch.: 1977, IAU Circ. No. 3102
Morris, S.C.: 1977, IAU Circ. No. 3101
Selvelli, P.L., Stalio, R.: 1980, Second European IUE Conference Tübingen ESA SP-157, p.155
Slovak, M.H., Africano, J.: 1978, Mon. Not.R. astr. Soc. 185, 591
Stalio, R., Selvelli, P.L.: 1980, Proceedings of the Conference "The Universe At UV wavelengths" GSFC
Swings, P.: 1955, Aurorae and the Airglow, Eds.Eb.Armstrong & A. Dalgarno. Pergamon Press Ltd., Oxford, p.249

RECENT INCREASE OF THE ACTIVITY OF THE SYMBIOTIC STAR CH CYGNI

Drahomir Chochol, Ladislav Hric
Astronomical Institute of the Slovak Academy of Sciences
Tatranska Lomnica, Czechoslovakia

CH Cyg has exhibited a great deal of activity in the recent times. We obtained two spectra with dispersion of 9 A/mm with the 2 m telescope of the Ondrejov Observatory on the nights of July 4/5 and August 6/7, 1981. The differences between these spectra are conspicuous. The description of the spectral changes during this period follows:

The Balmer lines became broader, the R wing of the early Balmer shell lines being partly suppressed by absorption. The inverse P Cygni profiles became apparent from H_8 to H_{30}. The most conspicuous feature is the split

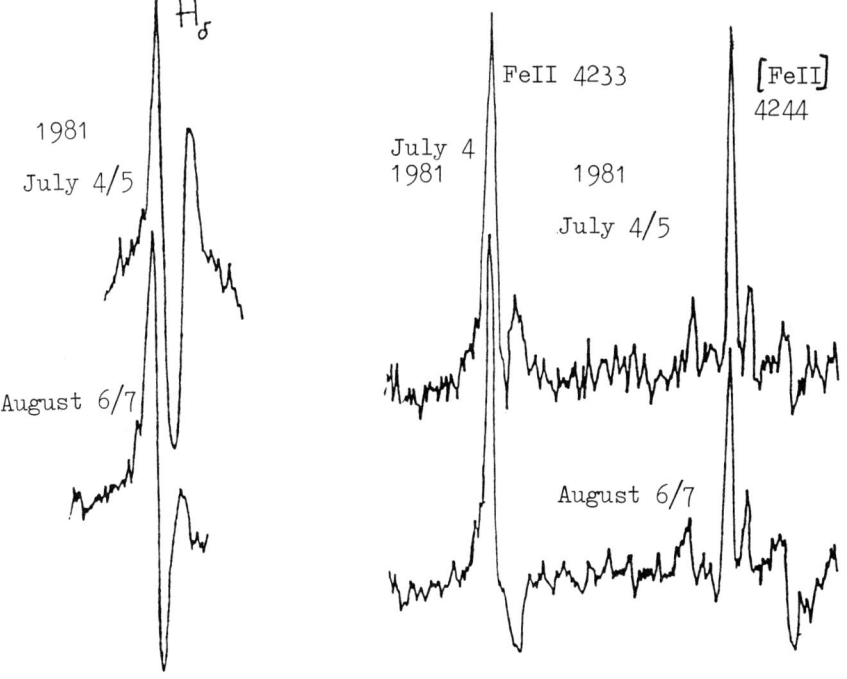

ting of the TiII absorptions into two components. CaII became broader with several absorption cores. The shell V≫R profile of the FeII lines changed to an inverse P Cygni profile, while no change of the [FeII] line profiles was observed. An increase of the blue continuum is apparent.

Our spectroscopic observations support a binary model in which a cool component fills its Roche lobe from time to time, thus inducing mass transfer towards the gainer which is a hot subdwarf. An accretion disk is formed around the hot companion which falls onto the star. During the outburst, which can be explained by a thermonuclear runaway on the surface of the gainer, the central bright hot region became larger. This is the cause of the stronger absorptions which arise in the accretion disk. According to the shapes of the line profiles, the following layers in the accretion disk around the hot companion can be distinguished, in the sequence away from the hot companion: H I, CaII, TiII, FeII. [FeII] lines are present in a common envelope. The shape of the shell lines with R<V shows that absorption saturates the R emission as the matter recedes from us when falling down into the surface of the gainer. The recent activity of the star is also confirmed by the photoelectric photometry from Skalnate Pleso Observatory, too. The V magnitude fluctuates around $\sim 6^m$. The brightness during recent activity is higher than than during the recent photometric history of the star. The observed flickering (Slovak and Africano 1978, Mon. Not. R. astr. Soc. 185, 591) is due to a hot spot on the accretion disk like in the dwarf novae.

POLARIMETRY OF CH CYGNI

V. Piirola
Observatory and Astrophysics Laboratory
University of Helsinki, Finland

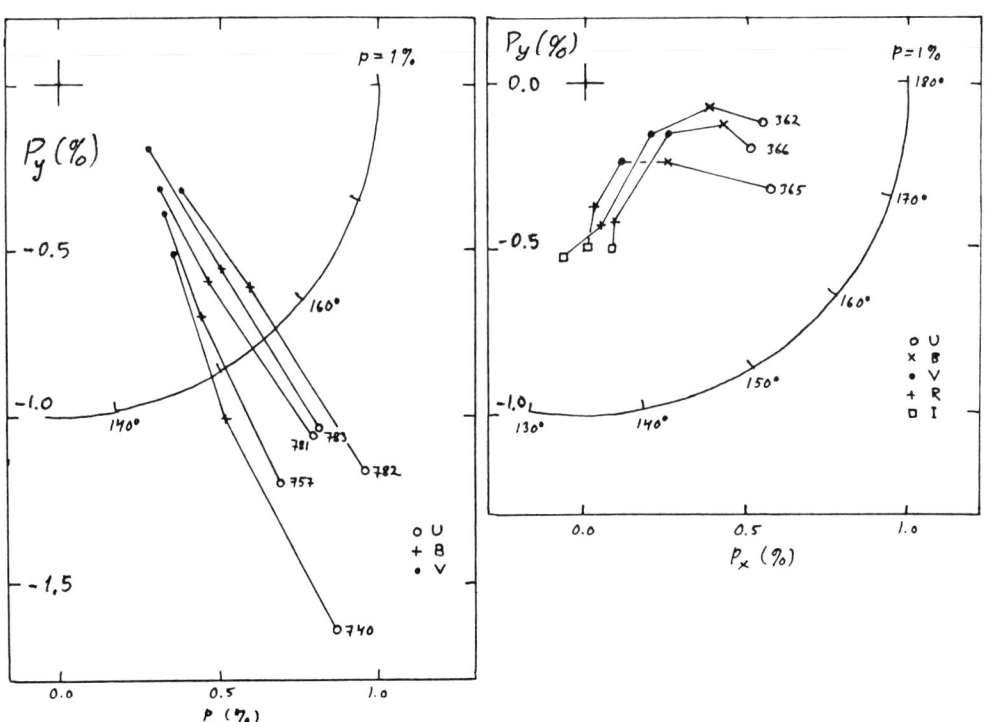

Figure 1. Polarization observations of CH Cyg in the autumn 1978 (left) and in May 1980 plotted in polar coordinates. The numbers in the figures give the last three digits of Julian Date.

Variable linear polarization has been detected in CH Cyg. Observations in 1977-79 (UBV) and 1979-80 (UBVRI) are presented. The amount of polarisation increases towards ultraviolet and the observed values range from 0.34 to 1.85 % in the ultraviolet, and 0.04 to 0.68 % in the yellow light. In the red and infrared observations the polarization was typically less than 0.1 %. During the interval September 1977 - February 1978 and in May 1980 a second component of polarization, with different direction and wavelength dependence, was present, resulting in a strong rotation of the position angle as a function of wavelength. The R and I observations of May 1980 ($P_R \cong 0.4$ %, $P_I \cong 0.5$%) showed that the second component was increasing towards the infrared. The peculiar wavelength dependence of polarization and position angle could be explained by variations in particle size and scattering geometry in a complicated dust envelope around the M giant. Another hypothesis is that the component of polarization increasing towards the ultraviolet is produced by electron scattering in an extended envelope of a hot companion and the second component by transient dust envelope of the M giant.

DISCUSSION ON CH CYGNI

Plavec: What is the shape on the UV continuum?

Hack: In 1978 and 1979 it was almost flat from 1240 to 1700 A. At 1700 A there is a sudden increase of flux, and then a slow increase from 1725 to 2200 A. In September 1980 the overall flux was about 8 times larger than that of 1978. In 1978 and 1979 the shape was about the same. We have no good data for the near ultraviolet continuum.

Friedjung: Can you give an upper limit to possible orbital radial velocity variations of the red component?

Hack: The radial velocity given by the photospheric line (neutral elements and TiO bands) fluctuates between -55 and -60 km s^{-1}. We do not have a continuous sequence of observations permitting us to decide between irregular or orbital radial velocity variations.

Houziaux: I would like to ask if other [OI] lines have been observed in addition to the 6300 and 6363 lines. It frequently happens that the red lines are observed while companion lines are absent. It should be noted that from the $2p^4$ ^1S level there is a coincidence with Lyα which could pump atoms to an upper level with possible autoionization/dielectronic recombination phenomena.

Hack: The 5577, 6158 and 6456 lines of [OI] have not been observed.

UV ECLIPSE OBSERVATIONS OF CI CYG

A.G. Michalitsianos[*], M. Kafatos[**], R.E. Stencel[***] and
A.A. Boyarchuk[+]
[*]NASA Goddard Space Flight Center, Greenbelt, Maryland USA
[**]George Mason University, Dept. of Physics, Fairfax, Va. USA
[***]JILA-University of Colorado, Boulder, Colorado USA
[+]Crimean Astrophysical Observatory-U.S.S.R.

INTRODUCTION

Low spectral resolution observations ($\sim$ 6 A) were obtained with the International Ultraviolet Explorer (IUE) during its eclipse phase. Additional data obtained by other IUE groups have been included in our eclipse observations, enabling us to examine the UV spectral properties of this system over nearly an entire orbit that spans early 1979 through mid-1981. Data obtained over this time interval suggest an overall decline in UV emission consistent with the decline of optical emission following the outburst of 1975, where CI Cyg attained an increase of $\sim$ 3 magnitudes in the visual. The short wavelength spectrum $\lambda\lambda 1200-2000$ A is characterized by numerous intense high excitation emission lines that become more prominent out of eclipse. The LWR wavelength range $\lambda\lambda 2000-3200$ A exhibits a few more additional lines of O III, Mg II and He II that are superimposed on continuum that rises gradually with increasing wavelength. Additionally, OH emission bands are identified at $\lambda\lambda 3064$, 2875 A (cf. Diecke and Crosswhite 1962). Collaborative ground-based observations of CI Cyg with W. Blair of McGraw Hill Observatory suggest the presence of the Balmer continuum jump at $\lambda 3646$ A, and enables us to ascribe the UV continuum observed with IUE to mainly Balmer free-bound recombination emission.

The variation of UV line intensity with optical phase divides our ensemble of emission lines into two broad groups that are differentiated by ionization potential. Higher excitation permitted lines of N V, for example tend to not weaken in deep eclipse as much as the more moderate excitation emission lines such as Mg II and Si III]. Unblended intercombination lines of N IV], N III], Si III], O III] and C III] indicate clear evidence for eclipse. Permitted lines of N V, C IV and He II appear to increase with time in a secular manner. Mg II appears to have been systematically declining over the observing period. However, very recent spectra obtained well outside of eclipse on 14 August 1981 indicates a substantial enhancement in intensity, for which the doublet has attained intensities comparable to pre-eclipse levels. Variations in C IV $\lambda\lambda 1548$, 1550 A emission during eclipse are confused owing to saturation of these lines. However, the emission lines investigated overall exhibit a relationship between secular change and excitation (see Figure 1).

Except for Balmer continuum and He II λ1640 A, the eclipse in the UV was deeper in the optical intercombination lines, but shallower than optical permitted lines. This is most apparent in the Mg II and N V lines which seemingly did not exhibit eclipse variations as compared to the Balmer continuum that vanished at mid-eclipse. Additionally, absence of variations in the Bowen flourescent excited lines of O III argues for an extended circumstellar nebula in the system that is not significantly affected by eclipse. The general absence of strong forbidden emission and strong emission from permitted lines argues for electron densities in the range 10^4 to 10^9 cm^{-3}. However, our recent IUE observations of 14 August 1981 that show enhanced Mg II emission also show high excitation ~ 140 eV emission from forbidden Mg V. The sudden appearance of high excitation forbidden lines well after the hot companion has emerged from eclipse is difficult to explain at present. He II λ1640 A presents different behavior than other permitted lines since it exhibits a rather deep eclipse, contrary to lines such as N V λλ1238, 1242 A (see Figure 2).

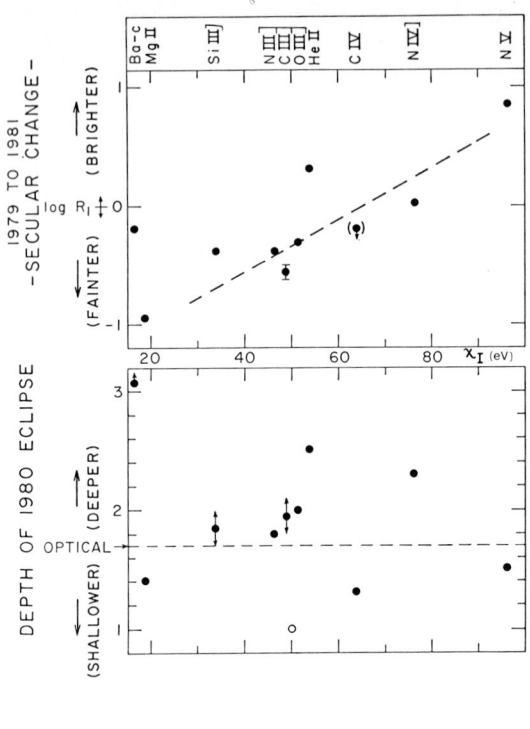

FIGURE 1

DISCUSSION

These observations are consistent with a binary star model that involves mass transfer from the extended cool envelope of the primary to the compact secondary. The formation of an accretion disc is a transitory phenomena in which viscosity eventually results in the dissipation of the disc over a timescale comparable to the binary orbital period. Emission from a low density circumstellar shell likely explains forbidden lines, for which the illumination of the M giant atmosphere by the intense radiation field of the secondary gives rise to moderate excitation emission, e.g. Mg II λλ2795, 2802 A, and a warm stellar wind from the primary. Secular variations can be used to rule out a simple expanding shell because the higher ionization lines are observed to increase in strength during eclipse, while intercombination lines decrease, contrary to expectation for a radially decreasing density nebula. Rather, we associate the resonance lines of C IV and N V with a large volume emitting region, possibly formed through shock collision from interacting stellar winds from the primary and secondary.

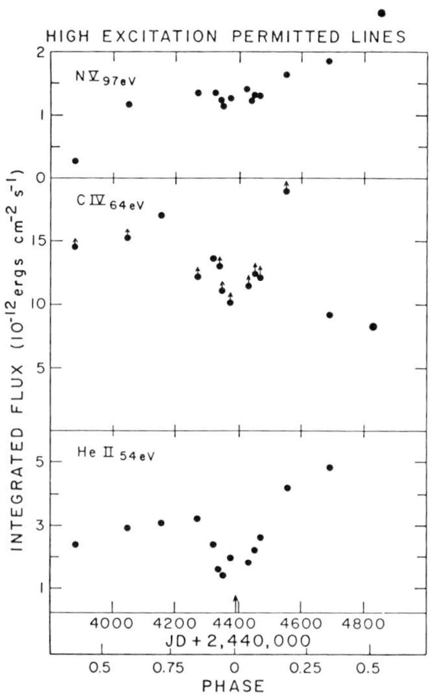

The intercombination lines in CI Cyg and Z And show systematic wavelength shifts of +30 ± 10 km s^{-1} relative to the permitted lines, suggesting possible P-Cygni type outflow from the M star (Freidjung et al. 1981; Altamore et al. 1981). A warm wind arising from the M star is reasonable if the giant does not even fill its Roche lobe. If in fact the M star does extend to its Roche limit the stellar wind would become enhanced. The intercombination lines appear to arise from an extended region of increasing ionization toward the hot subdwarf in a Stromgren-like manner. He II λ1640 A, however, strongly reflects eclipse in a manner similar to high excitation intercombination lines, thus departing in behavior from other permitted lines (Figure 2).

FIGURE 2

The Balmer continuum likely arises from the exterior of a thick accretion disc first demonstrated to exist by Webbink et al.(1981) following the 1975 outburst. We estimate flux of 8×10^{-11} erg cm^{-2} s^{-1} A^{-1} for the Balmer continuum (λλ2500 to 3646 A). Adopting a distance d ∼ 1500 pc and E(B-V) = 0.5 (Mikolajewska and Mikolajewski 1980), we estimate the Balmer continuum luminosity ∼ 67 L$_\odot$, which does not include effects of disc inclination.

Scaling laws that relate disc dimensions and luminosities to accretion rates have been developed (cf. Bath et al. 1974, Tylenda 1977 and Mayo et al. 1980), and are

$$L_{disc} \sim 10 \, \dot{M}_{-8} \, M \, R_9^{-1} \quad , \qquad (1)$$

and

$$T_{bl} \sim 3 \times 10^5 \, M^{\frac{1}{4}} \, \dot{M}_8^{\frac{1}{4}} \, R_9^{-0.75} \sim 5 \, T_{Disc} \quad , \quad (2)$$

where R_9 is the inner boundary with radius 10^9 cm, M is in units $M/M_\odot$ and M_{-8} in units of 10^{-8} $M_\odot$ yr^{-1}. $T_{bl} \sim 50,000$ K is obtained from estimates for the Balmer continuum. Equations (1) and (2) are combined to obtain R_9 which we find $\sim 10^8$ cm suggesting that 1 $M_\odot$ white dwarf is present. We stress that this is only a lower limit because we may not have included the disc luminosity, while a main sequence type star $R_9 = 100$ requires 10^{-5} $M_\odot$ yr^{-1}. The latter corresponds to super-Eddington luminosities if a white dwarf is present, thus mimicing the 1975 outburst. Further analysis is necessary to determine if in fact a white dwarf is present rather than an early main sequence type star, which has important ramifications for the evolution of the system and the mass transfer properties of the binary.

REFERENCES

Altamore, A., Baratta, G., Cassatella, A., Freidjung, M., Giangrande, A., Ricciardi, O and Viotti, R. 1981, Astrophys. J., 245, 630.
Diecke, G. and Crosswhite, H. 1962, J.Q.S.R.T., 2, 97.
Freidjung, M., Stencel, R.E. and Viotti, R. 1981, Astron. and Astrophys. in preparation.
Mayo, S., Wickramasinghe, D. and Whelan, J. 1980, M.N.R.A.S., 193, 793.
Mikolajewska, J. and Mikolajewski, M. 1980 Acta Astron., 30, 347.
Tylenda, R. 1977, Acta. Astron., 27, 235.

IUE OBSERVATIONS OF CI CYGNI DURING 1979-1981

G.B. Baratta[1], A. Altamore[1], A. Cassatella[2], M. Friedjung[3], D. Ponz[2], O. Ricciardi[4]
1. Osservatorio Astronomico e Università di Roma, Italy
2. Astronomy Division ESTEC, Villafranca, Spain
3. Institut d'Astrophysique, Paris
4. Istituto Astrofisica Spaziale (CNR), Frascati, Italy

High and low resolution IUE spectra of CI Cyg were obtained at VILSPA during 1979-81 allowing for an analysis of the spectral variations related to the decreasing activity of the star, and to the eclipse (June 1980).

In the high resolution spectra the emission lines have a width slightly larger than the instrumental one. This fact is particularly evident in the HeII 1640 A line and could be related to the peculiar behaviour of this line at low resolution as reported by Michalitsianos et al. (this volume). A systematic radial velocity difference between permitted and intercombination lines was found; this difference should be connected with the structure of the emitting region(s). "Secular" and eclipse variation was found in particular in the intercombination line intensities (Viotti et al. 1980). An electron density of $\sim 0.3-1.5 \times 10^{10}$ cm^{-3} was evaluated from the intensity ratios of the NIII] emission lines. No significant difference in these ratios and in the CIII]/NIII] ratio as well was found during 1979, 1980 (eclipse) and 1981 suggesting no large N_e variation with both the activity phase and eclipse of the star. This result should imply a low density gradient in the partially eclipsed NIII] and CIII] regions. A more detailed analysis of the high resolution data is in course to better clearfy these points, and their implications on the possible models for CI Cyg.

An interstellar extinction of E(B-V) = 0.40 was derived from the low resolution IUE spectra. The dereddened spectra during June 1979 - August 1980 are shown in figure 1. Differently from other symbiotic stars like Z And, V1016 Cyg, AG Dra and RR Tel, the continuum of CI Cyg does not show in the far ultraviolet the steep increase due to the presence of a hot black-body component. On the contrary the continuum appears flat and almost constant until January 1980. In March 1980, during ingress of eclipse the continuum found by us was eclipsed at 2500-2800 A, but almost unaffected in the far UV. At mid eclipse the whole continuum was lower by

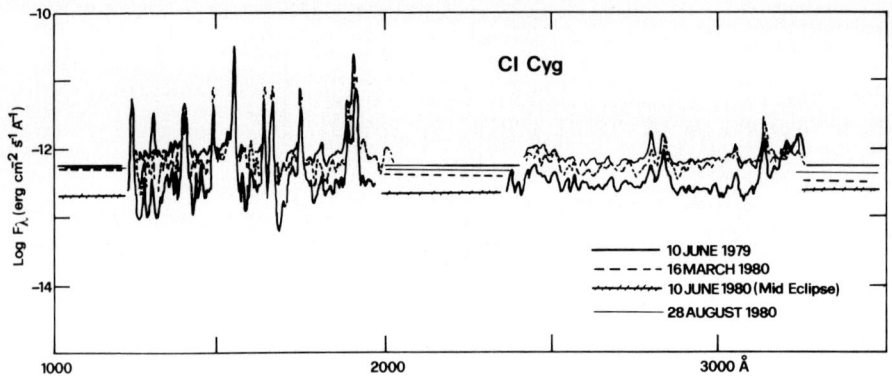

Figure 1. The low resolution spectra of CI Cyg between 1240 and 3200 A, during 1979-80. The mean slope of the continuum in different epochs is indicated.

a factor of about 2.5 with respect to the level before and after eclipse, when it nearly recovered the previous level.

REFERENCE

Viotti, R., Altamore, A., Baratta, G.B., Cassatella, A., Ponz, D., Friedjung, M., Muratorio, G.: 1980, IAU Circular No. 3518.

THE ECLIPSE OF CI CYGNI IN 1980 ON THE OBJECTIVE PRISM SPECTRA

J. Mikołajewska and M. Mikołajewski
Institute of Astronomy Nicolaus Copernicus
University Toruń, Poland

The observations of CI Cyg were carried out with the Toruń 60/90 cm Schmidt telescope with the objective prism giving a dispersion of 250 A/mm at H-gamma. Intensity traces of 20 selected spectra, made on the Kodak IIa-O plates, were made. The results were corrected for all instrumental effects and the interstellar extinction. The relative emission lines intensities have been obtained from the tracings after normalization to H-beta = 100.

The spectra were typical for the quiet symbiotic star. So, all changes in the spectrum were caused by geometrical effects (eclipse).

The preliminary analysis of obtained spectra leads to following results:

1) The high-temperature emission lines HeII 4686 and NIII 4641 disappeared in the eclipse. The changes of the brightness in the HeII 4686 line presents fig. 1 and fig. 2 shows the changes of the intensity of this line in relation to H-beta. We have estimated the rough phases of first and second contact as 0.93 and 0.97 respectively what gives the minimal radius of HeII line formation region about 0.4 r_c (where r_c is the radius of the cool component).

2) The Balmer emission lines were about 3 times weaker in relation to the same lines outside of the eclipse (fig. 1). Basing on that we have estimated the minimal radius of the HII region as 1.22 r_c.

3) The nebular lines of [OIII] and [NeIII] in principle have not changed their brightnesses (fig. 1).

4) On the ground of 10 spectra made outside of the eclipse we have derived [OIII] 5007+4959/4363 = 1.41±0.08. Using the formula given by Boyarchuk et al. (1963) and assuming as usual T_e=17000 K we have obtained N_e about $2 \times 10^7 cm^{-3}$.

5) The same 10 spectra were used to the temperature determination of the exciting star. From relative intensities of H-beta, HeI 4471 and HeII 4686 lines (Iijima, 1981a) we have obtained $T_{hot} = 15.2 \times 10^4$ K.

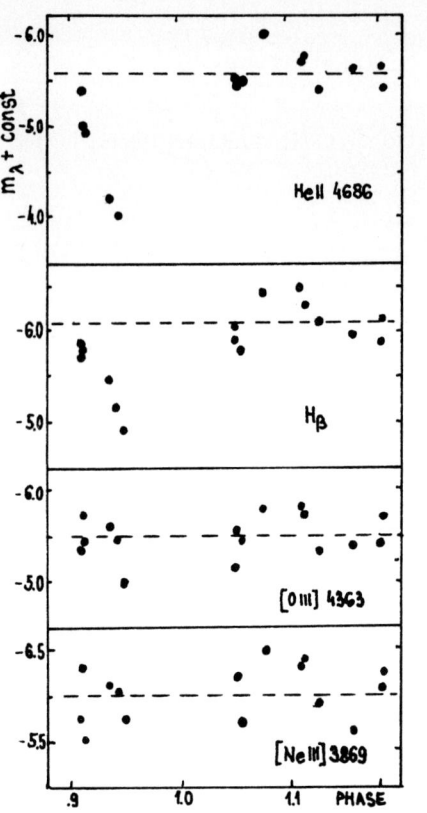

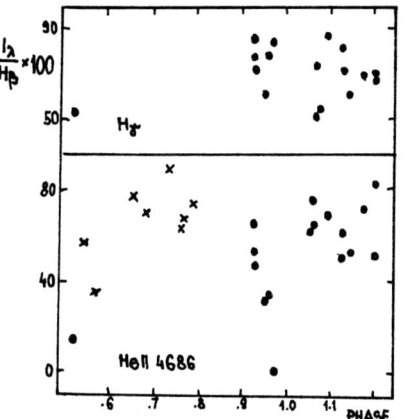

Fig. 1 (left). Changes of brightness in selected emission lines. The dashed line marks the average brightness outside of the eclipse.

Fig. 2 (bottom). Changes of HeII 4686 relative intensity. Crosses mark Iijima (1981b) observations. The relative intensities of $H\gamma$ are given for comparison.

References:

Boyarchuk, A. A., Gershberg, R.E., Pronik, V.I., 1963, Isvestia Crimea, 29, 291

Iijima, T., 1981a, in Photometric and Spectroscopic Binary Systems D. Reidel Publ. Co., 517-534

Iijima, T., 1981b, Astron. Astrophys., 94, 290

VARIATION SPECTRALE DE CI CYG EN 1975

Ch. Fehrenbach
Observatoire de Haute Provence   04870 St Michel l'Observatoire

En 1971 et 1972, cette étoile présentait un spectre d'émission comprenant des raies de l'hydrogène et les métaux ionisés CaII, FeII, etc... se superposant sur un fond continu sans raie d'absorption, [FeII] est présent par deux raies faibles, les raies de l'hydrogène, de FeII et de CrII présentaient un profil P Cyg.

Le 4 Juillet 1975, Madame Y. Andrillat a signalé la disparition de la raie de HeI 10830. Nous avons obtenu immédiatement un spectre à 20 A mm^{-1} au télescope de 152 cm. Ce spectre est un spectre d'absorption où seules les raies de l'hydrogène présentent des émissions à structures complexes.

Sur ce spectre, la vitesse radiale des raies d'absorption présente une forte variation avec l'ionisation et l'excitation. La vitesse est négative et l'atmosphère est en implosion. Le type spectral peut être estimé à A8 Ibp mais il présente de nombreuses particularités.

Le spectre obtenu le 8 Août 1975 montre une nette évolution avec une diminution de la température d'excitation et un renforcement de certaines raies. Cinq jours plus tard, le 13 Août 1975, le spectre a subi une modification considérable. Il peut être maintenant classé F5 Ib mais il présente aussi de très nombreuses particularités. Le phénomène essentiel est l'apparition des raies des éléments BaII et surtout des raies des terres rares GdII, PrII, YbII, NdII, SmII, CeII, EuII.

La présence de ces éléments dont l'identification a été examinée avec soin semble montrer l'existence d'une réaction du type "process s" et apparente cette étoile à FG Sagittae.

Ceci n'est qu'un résumé de l'article publié sous le titre "A Spectroscopic Study of CI Cygni : The S-process Episode" par J. Audouze, P. Bouchet, Ch. Fehrenbach and A. Wosczyck, dans Astron. Astrophys. 93, 1-7 (1981).

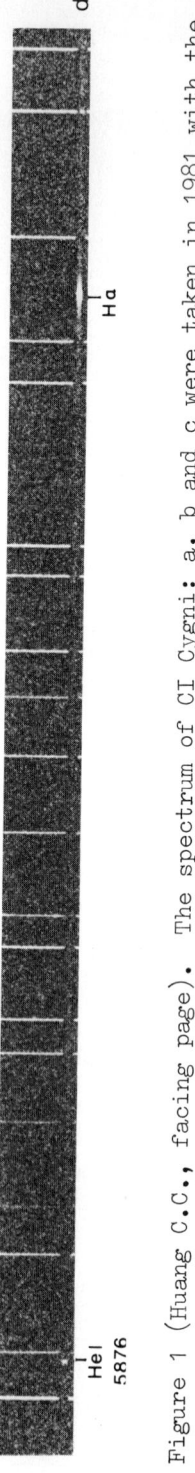

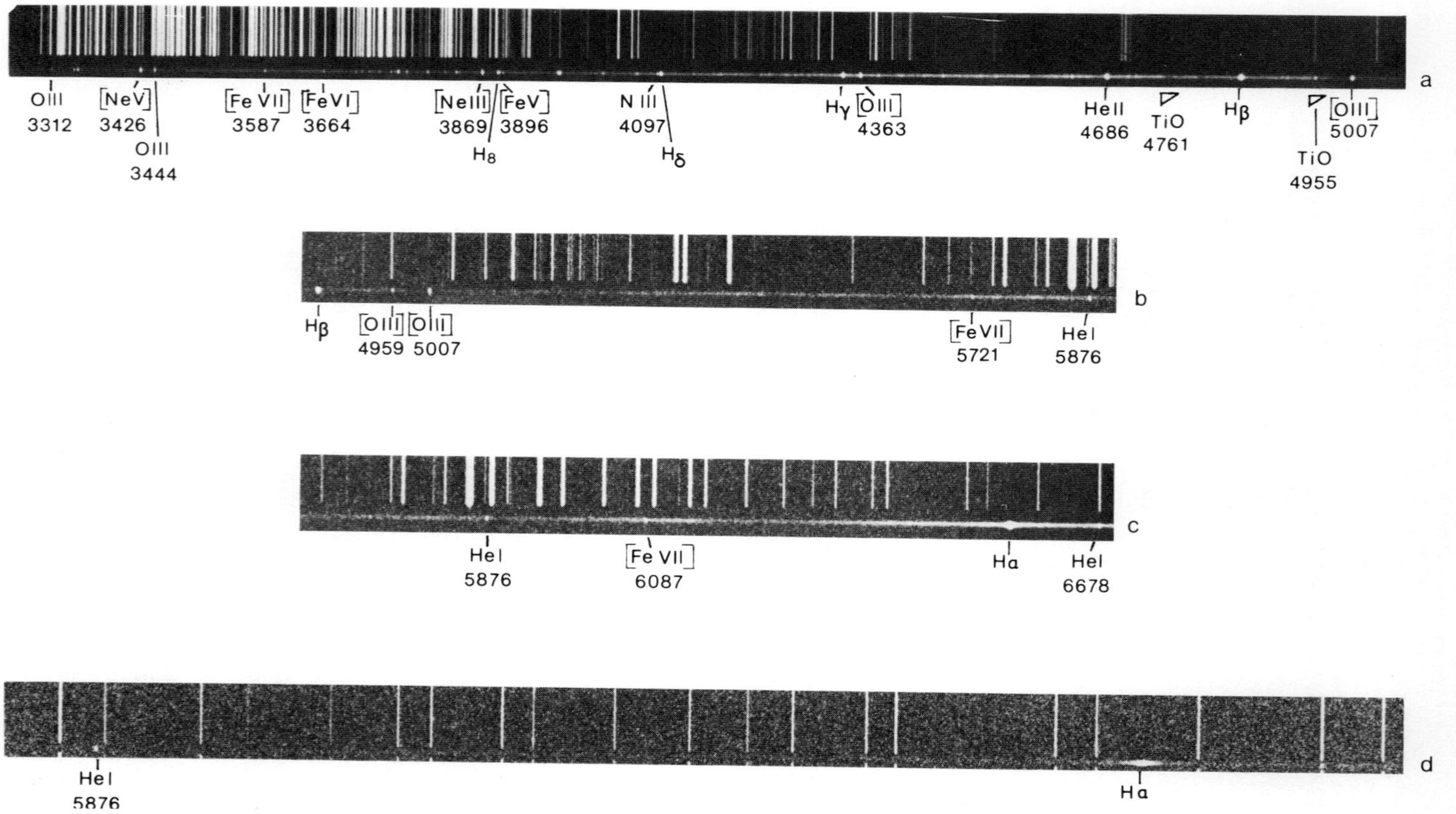

Figure 1 (Huang C.C., facing page). The spectrum of CI Cygni: a, b and c were taken in 1981 with the Marly spectrograph (original dispersion 80 Å mm^{-1}). d was taken on 18 August 1980 with the same spectrograph (original dispersion 40 Å mm^{-1}). The fading of Hα in 1981 and the change in structure are conspicuous. There were no [FeVII] emission lines in the 1980 spectrum.

SPECTRAL VARIATIONS OF CI CYG BETWEEN 1980 AND 1981

HUANG CHANG-CHUN
Purple Mountain Observatory, Academia Sinica

In our paper (Ch. Fehrenbach and C.C. Huang, 1981) we have described the details of our spectroscopic observations of CI Cyg in 1980. In that year, the star showed a fair presence of bright forbidden lines of [FeV] and very strong emission lines of high and low excitation, including forbidden [O III], [Ne III], [Fe II] and [S II] and permitted O III, N III as well as He I and Fe II ... etc ... The He II 4686 A was very strong. The Balmer lines were the strongest emission features in the spectrum and H_α and H_β were double. There were two groups of velocities in 1980, one was positive for forbidden lines and the other was negative for the permitted lines. The displacements of Balmer lines showed a regression.

Continuing our spectroscopic observations of CI Cyg, during summer 1981, a number of spectra of the star were obtained at the Haute Provence Observatory, using the Marly spectrograph with a dispersion of 80 A mm^{-1}.

In 1981, CI Cyg has clearly changed excitation and the spectral appearence is very different from that of 1980. The forbidden [Fe VII] which sometimes had been present in CI Cyg but was missing in 1980 have reappeared while the [Fe V] lines observed in 1980 have clearly weakened. In Fig. 1 a, b and c show the strength of [Fe VII] lines in ultraviolet region as well as in red region which our 1980 observations also covered. The [O III] lines are still intense but much weaker than that in 1980. There are great changes in the Balmer lines. The components shortward of the double lines of $H \alpha$ and $H \beta$ which definitely had been seen during 1980 have disappeared. Fig. 1, c and d show the fading of $H \alpha$ in 1981 and its structure changes between 1980 and 1981. In 1981, the regression of Balmer lines found in 1980 has disappeared. As to ionized helium, He 4686 A is still strong, but the Pickering lines present in 1980 are missing. We detected also a trace at 3203 A of He II (3-5) in 1981. It would be interesting to mention that the relative intensity of He II 4686 A in CI Cyg is as strong as that in AG Dra in our observations, but the He II 3203 is much stronger in AG Dra (C.C. Huang, 1981). In ultraviolet, the strongest lines are due to high I.P. lines of [Ne V] and fluorescence O III.

The continuum of the spectrum of CI Cyg much enhanced in 1981. The TiO bands are clearly visible in the photographic region. Fig.1, a shows the spectral appearence of the star in 1981.

There was an eclipse of CI Cyg in the summer of 1980, our 1980 observations from July 12 th to October, covered just the end of the eclipse. Although there were great spectral variations in the star between 1980 and 1981, the changes in the spectrum were not radical in our observations.

I am deeply indebted to professor Ch. Fehrenbach who has given me very great help every-way. I wish to express my thanks to M.L. Rolland who has done a great part of the measurements.

References

Fehrenbach, Ch., and Huang, C.C., :1981, submitted for publication in Astron.Astrophys. Suppl., in press.
Huang, C.C., : 1982, IAU Colloquium No. 70, The Nature of Symbiotic Stars, this volume, p. 151 and 185.

OPTICAL OBSERVATIONS OF CI CYGNI

N.A. Oliversen, C.M. Anderson and K.H. Nordsieck
Washburn Observatory, University of Wisconsin

ABSTRACT

We report optical emission line variations in CI Cygni during the 1980 eclipse.

I. OBSERVATIONS

We have obtained high (H-alpha) and low resolution spectra of CI Cygni during and after the 1980 photometric minima. The high resolution program at Washburn Observatory is described in Anderson, Oliversen and Nordsieck (1980), while the low resolution program is desribed in Oliversen (1981).

The balmer decrement was used to correct the emission line intensities for reddening due to interstellar extinction. The case B recombination balmer line ratio of I(4340)/I(4861) is .47, while the observed line ratio was .40±.03. Comparison of the observed to the case B line ratio implies a total visual extinction of $A_v = 1.1 \pm .5$.

		I(5007+4959)/I(λ)				
JD	PHASE	λ 3869	λ 4363	λ 4686	λ 4861	λ 5876
2444375	.97	2.1	2.8	4.7	1.2	3.7
2444399	.01	3.2	2.6	5.5	1.3	3.5
2444463	.08	4.6	3.1	1.4	.75	2.4
2444485	.10	3.8	2.9	1.3	.67	1.5
2444721	.38	2.7	2.6	.83	.55	1.6
2444823	.50	1.9	2.6	.80	.40	1.5

Table 1. CI Cyg Observed Emission Line Intensity Ratio's

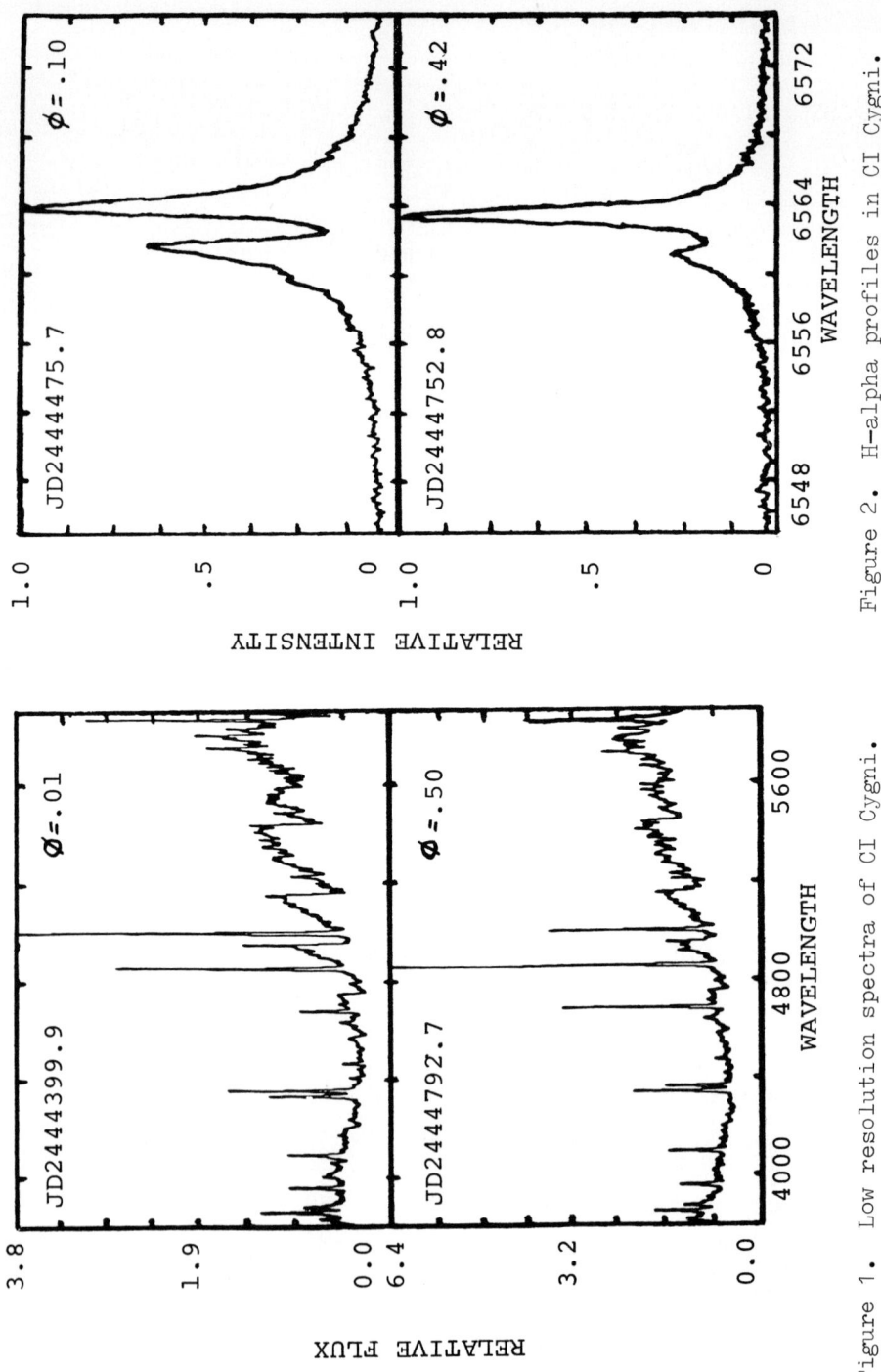

Figure 2. H-alpha profiles in CI Cygni.

Figure 1. Low resolution spectra of CI Cygni.

The observed line ratios of I(5007+4959)/I(λ) are given in table 1. The [OIII] λ4959 line occurs on top of a band head of TiO. The two features could not be resolved on our spectra. The line ratios quoted in table 1 assume a value of 2.93/1, which is the ratio of the spontaneous emission probabilities of 5007 and 4959A. Sample low resolution spectra (3800- 5900A) both in and out of eclipse are given in figure 1.

II. EMISSION LINE RATIO'S

The observed [OIII] line ratio varys irregularily where, I(5007+4959)/I(4363)=2.6±.3. Under conditions of high density ($N_e > 10^8$ cm^{-3}) the [OIII] line ratio becomes insensitive to density changes. This places a lower limit on the temperature of about 10000K based upon the [OIII] emissivities as given by Nussbaumer and Storey (1981), and Keyes (1981).

The emission line ratio's of I(5007+4959)/I(λ) all correlated with the optical eclipse. The [OIII] emission comes from a more extended region than the emission lines of [NeIII] λ3869A, He I λ5876, He II λ4686 or H-beta, confirming Stencel's (1981) result. The spectra record relative flux, therefore we can not distinguish between no change in the [OIII] emission versus a slower decline compared to the other emission lines. The [NeIII] λ3869 emission reached its minimum slightly after the optical eclipse at phase .08.

III H-ALPHA EMISSION LINE PROFILES

Sample H-alpha profiles both in and out of eclipse are shown in figure 2. CI Cyg exhibits the usual asymetrical profile along with a blue shifted reversal. During the eclipse the blue-shifted absorption feature became deeper. The deepest reversal occured slightly after the optical eclipse at phase .08. The absorption feature's heliocentric velocity also appears to correlate with phase. At phase 0 the velocity of the central reversal was about -15 km/s and at phase .5 the velocity is about -31 km/s. At phases 0 and .5 the radial velocity due to binary motion of the hot source should be zero. Perhaps these velocity shifts arise in absorbing matter streaming from the red giant to the hot object.

REFERENCES

Anderson, C.M., Oliversen, N.A., and Nordsieck, K.H. 1980, Ap.J., 242, p. 188.
Keyes, C. 1981, private communication.
Nussbaumer, H., and Storey, P.J. 1981, Astron. Astrophys., 99, pp. 177-181.

Oliversen, N.A. 1981, Ph.D. Thesis, Univ. of Wisconsin.
Stencel, R.E., Kafatos, M., Michalitsianos, A.G., and
  Boyarchuk, A.A. 1981, ApJ. In press.

DISCUSSION ON CI CYGNI

Boyarchuk: I would like to call your attention to the fact that CI Cyg is always brighter at mid of the eclipse than just after second contact or before the third one.

Slovak: A comment on the eclipse in CI Cyg as compared to the eclipse in AR Pav. CI Cyg appears to reach the same minimum light at mid-eclipse, whereas AR Pav varies by approximately 0.5 mag from "bright" eclipses to quiescent eclipses. Thus the accretion disk in CI Cyg appears very small and with no visible hot spot. The accretion disk in AR Pav, however, varies significantly from eclipse to eclipse.

Mikolajewska: HeII $\lambda 4686$ was not visible in 1971 and 72, but was very wide and strong in 1976. These variations are not related to the eclipses.

Rudak: I would like to ask Dr. Michalitsianos to briefly summarize the main parameters of CI Cyg, and to mention those observational facts for which an accretion disk geometry of matter around the hot component is necessary.

Michalistianos: For a 3 to 7 $M_\odot$ system the semi-major axis for the orbit is $a \approx 3.8$ to 5.0×10^{13} cm. The duration of the eclipse is approximately 0.2 of the orbital period (~ 880 days). Thus the radius of the primary is $\sim 0.38a$. The Balmer continuum absolute flux that we assume arises from a disk is $f_\lambda(2500-3646\text{Å}) \approx 8 \times 10^{-11}$ erg cm^{-2}s^{-1}Å^{-1} after correcting for $E(B-V)=0.5$. This gives $L(\text{disk}) \approx 67\ L_\odot$. About 1/3 of the observed luminosity arises from the disk. The primary may just fill its Roche lobe.

Kafatos: We have found from the duration of the eclipse that if the mass ratio between the two stars is 1:1, then the red giant is just filling its Roche lobe, but if the mass ratio is larger than that, it does not. Very probably, during quiescence the red giant does not fill its Roche lobe.

INTRODUCTORY REPORT ON V1016 CYGNI

A. Cassatella
Astronomy Division ESTEC, Villafranca Satellite Tracking
Station of ESA, Madrid, Spain

The formation of a new planetary nebula should be a rare event, and the case of V1016 Cyg may represent such an example. This is the reason of the considerable interest of the astronomers in this star since the discovery of its outburst in 1965. The light history of this star has been described in detail by several authors (e.g. Ciatti et al. 1971, FitzGerald 1971, etc.), and can be divided into three distinct phases:

1. <u>The preoutburst phase</u> (before 1964): the B-magnitude of V1016 Cyg has been probably always fainter than 15^m. In particular, during 1948-1963 it was close to 15.0-15.6 mag (FitzGerald et al. 1966).
On a 1947 objective prism plate it was classified as a late M type star (FitzGerald et al. 1966), but this is the only reported observation of a late type spectrum in this star before outburst. A strong H_α emission was first detected by Merrill and Burwell (1950) who denoted the star MH_α 326-116.

2. <u>The phase of increasing luminosity</u> (about 1963 to 1968): In 1963-64 the star underwent a large increase in its luminosity. This outburst was discovered in 1965 (FitzGerald et al. 1966) when the pre-outburst M-type spectrum was already disappeared, and an emission line spectrum and a hot continuum had appeared (FitzGerald et al., O'Dell 1967). According to Mammano and Ciatti (1975) and to Andrillat (this conference figure 2), TiO and VO bands are still detectable in the near infrared. Between 1965 and 1968 the mean ionization energy of the emission lines increased continuously, as shown by the successive appearance of lines of higher and higher ionization energies (see e.g. FitzGerald and Pilawaki 1974). The spectral and light evolution of V1016 Cyg during this phase clearly resembles that of a very slow nova and of other symbiotic stars like RR Tel (Thackeray 1977). But the lack of the spectral coverage for the earliest phases of the outburst prevents us to go further into this comparison.

3. **A quasi stationary phase** was reached after 1968 with no large luminosity changes, and small spectral variations mostly characterized by the further increase of the ionization range and the appearence of higher ionization stages (e.g. Ciatti et al. 1971, 1978). At high resolution the emission lines show a considerable structure: the strongets lines were characterized by two components separated by about 70 km s^{-1} and by a broad weaker blue shifted component (FitzGerald 1973).

V1016 Cyg is particularly interesting outside the optical spectrum. In the infrared, a large excess due to dust emission was found by Knake (1972) and Swings and Allen (1972). Baratta et al. (1974) from a study of the energy distribution and light curve of the star concluded that the dust envelope was present before the outburst. The IR spectrum probably originates from a nebula with at least two dust components with temperatures of $\sim$ 250 and 1000°K (Seaquist and Gregory 1973, Kwok 1976). Harvey (1974) discovered that the IR flux of V1016 Cyg is variable on time scales of the order of 450 days, quite appropriate for a Mira variable. No more recent systematic photometric observation in the IR has been so far published to confirm, or disprove, the periodicity of these variations.

Strong radio emission from the star was discovered by Purton et al. (1973) and confirmed by successive observations (see Kwok, this conference). Historically, this was the first astrophysical object where the slope of the radio spectrum ($S_\nu \propto \nu^{0.6}$) suggested emission from an optically thick nebula with a density gradient of the type $N_e \propto r^{-2}$ as in the case of continuous mass outflow from the central star (Seaquist and Gregory 1973; see also Panagia and Felli 1975). According to Ahern et al. (1977) this outflow started $\sim$ 600 years before the optical "outburst". When the mass loss ceased, the radiation from a remnant hot core of the star started to ionize the nebula producing the observed emission line spectrum and the visual brightening.

The ultraviolet spectrum of V1016 Cyg was extensively studied by Nussbaumer and Schild (1981) on the basis of low and high resolution IUE spectra. The low resolution, de-reddened UV spectrum of the star is shown in figure 1. Nussbaumer and Schild, to interpret the UV spectrum, developed a model of single star, with $T_* \cong 160000°K$ and $R_* \cong 0.06 R_\odot$, surrounded by a high density planetary shell with a mass of $\sim 2.8 \ 10^{-4} M_\odot$, an electron density of $3 \ 10^6 cm^{-3}$ and $T_{el} = 27000$ to 8000°K from the inner to the outer boundary. The interstellar extinction was estimated from different arguments to be $E(B-V) \cong 0.28$.

X-ray emission from V1016 Cyg has been detected by Allen (1981 and this conference) with the Einstein satellite and has been interpreted as

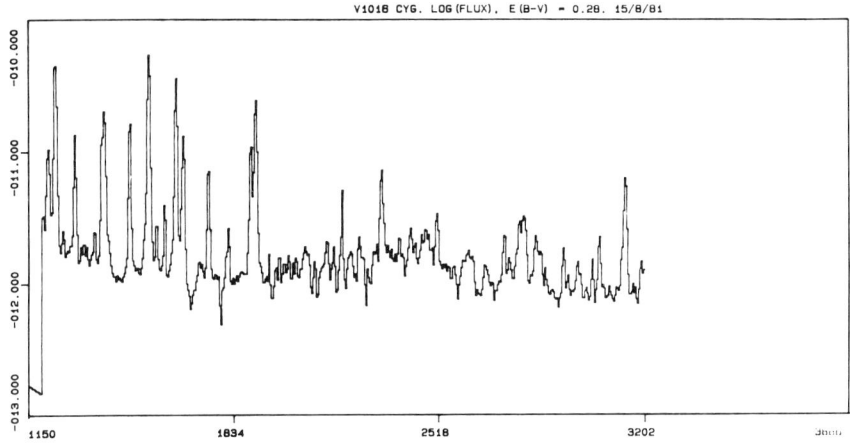

Figure 1. The low resolution IUE spectrum (1200-3200 Å) of V1016 Cyg.

an evidence for thermonuclear events on a white dwarf accreting from an M companion. Certainly, further X-ray observations spanned on a long time scale with the EXOSAT satellite will give a better insight on the origin of the X-ray emission and its relation with the other phenomena, like the mass outflow (and radio emission), the IR variations, etc.

Conclusions. Single star models of V1016 Cyg have been proposed by several investigators (Baratta et al. 1974, FitzGerald and Houk 1970, Kwok 1977, Ahern et al. 1977 etc.). Mammano and Ciatti (1975) suggested a binary model consisting of an M giant and a hot star ionizing the surrounding gaseous shell. The arguments in favour of this model are the presence of TiO and VO absorption bands (that in the single-star hypothesis are supposed to be formed in high density inhomogeneities of the expanding shell) and the IR variability, which suggests the continued presence of a late type (Mira variable) star.

We think that the problem of the nature of this star is still open. Certainly V1016 Cyg is a very interesting astrophysical object in all spectral ranges, and it would certainly require further studies in the future expecially with space experiments.

REFERENCES

Ahern, F.J., FitzGerald, M.P., Marsh, K.A., Purton, C.R. 1977, Astr. Ap. 58, 35
Allen, D.A. 1981, Mon. Not. R. astr. Soc. 197, 739
Baratta, G.B., Cassatella, A., Viotti, R. 1974, Astrophys.J. 187, 651

Ciatti, F., Mammano, A., Rosino, L. 1971, IAU Coll. No.15, Bamberg, p.64
Ciatti, F., Mammano, A., Vittone, A. 1978, Astr. Astrophys. 68, 251
FitzGerald, M.P. 1971, IAU Colloquium No. 15, p.73
FitzGerald, M.P., Houk, N. 1970, Astrophys.J. 159, 963
FitzGerald, M.P., Houk, N., McCuskey, S.W., Hoffleit, D. 1966, Astrophys. J. 144, 1135
FitzGerald, M.P., Pilawaki, A. 1974, Astrophys.J. Suppl. 28, 147
FitzGerald, M.P. 1973, Nature Phys. Sci. 245, 58
Harvey, P.M 1974, Astrophys.J. 188, 95
Knake, R.F. 1972, Astrophys. Letters 11, 201
Kwok, S. 1977, Astrophys.J. 214, 437
Mammano, A., Ciatti, F. 1975, Astr. Astrophys. 39, 405
Merrill, P.W., Burwell, C.G. 1950, Astrophys.J. 112, 72
Nussbaumer, H., Schild, H. 1981, Astron. Astrophys. 101, 118
O'Dell, C.R. 1967, Astrophys.J. 149, 373
Panagia, N., Felli, M. 1975, Astr. Astrophys. 39, 1
Purton, C.R., Feldman, P.A., Marsh, K.A. 1973, Nature Phys. Sci. 245, 3
Seaquist, E.R., Gregory, P.C. 1973, Nature Phys. Sci. 245, 85
Swings, J.P., Allen, D.A. 1972, Pub. astr. Soc. Pacific 84, 523
Thackeray, A.D. 1977, Mem. R. astr. Soc. 83, 1

RECENT STUDIES OF THE SPECTRUM OF V1016 CYGNI

G. Muratorio
Observatoire de Marseille, France

M. Friedjung
Institut d'Astrophysique, Paris, France

Two coudé spectra of V1016 Cyg taken on June 24 and 27, 1979 were reduced, using a computer programme developed in Marseille. Radial velocities and full widths at half maximum were measured for the emission lines, and are summarized in the following table were VR is the mean radial velocity in km s^{-1}, DV the velocity corresponding to the mean FWHM and Xi the effective ionization potential for the ion.

Ion	Xi	Plate 4230 (June 24) VR	DV	Plate 4239 (June 27, 1979) VR	DV
ScII	6.54	−71.5	64	−70	74.5
TiII	6.82	−80.1±1.9	41.4±13.2	−78.8±3.7	34.4±14.1
FeII	7.87	−80.5±8.3	44.2±11.0	−82.6±5.8	32.5±14.0
H I	13.60	−82.6±3.8	77.7±12.4	−82.5±5.3	72.1±23.2
SiII	16.34	−74	63	−99	35
HeI	24.54	−78.4±4.5	60.4±17.4	−78.8±6.1	45.4±15.5
NIII	47.45	−78.7±3.2	56.3±18.4	−78.7±5.1	41.7±12.1
CIII	47.84	−82	58	−77	54
HeII	54.42	−74.0	71.5	−74.5	77.0
OIII	54.93	−76.5	63	−89	61.5

forbidden lines

Ion	Xi	VR	DV	VR	DV
[NiII]	7.64	−37	25	−29	32
[FeII]	7.87	−77.6±4.1	45.8±7.5	−80.9±4.3	37.9±9.8
[FeIII]	16.16	−77.0±21.5	116±36	−66.7±8.5	133±32
[A IV]	40.74	−69	112	−110	136
[NeIII]	40.96	−67	134	−73	125
[FeV]	54.8	−72.6±5.4	140.8±17.7	−78.2±10.0	125.0±13.2
[NeIV]	63.45	−74	151	−81	152
[FeVII]	100	−85.5±6.1	132.5±9.7	−85.2±16.3	118.0±24.3

It will be seen that the mean width of <u>all</u> permitted lines of different ions, and of forbidden lines for effective ionization potentials needed to form ionization stage dominant in the line formation region below 10eV, are not more than 80 km s^{-1}. For forbidden lines of ions with Xi above 10eV, DV is however greater than 110 km s^{-1}. This may be understandable in an interacting wind model.

In addition observations of [FeVI] emission line equivalent widths with the Multiphot detector system of the Haute Provence Observatory in August 1980, can be used to estimate the electron density of its region of formation. Three different line ratios indicate electron densities of at least 10^7 or perhaps $10^8 cm^{-3}$ for assumed electron temperatures of $2 \; 10^{4} °K$ or less. The physical data used for this ion are from Nussbaumer and Storey (1978, Astr. Astrophys. 70, 37).

DISCUSSION ON V1016 CYGNI

<u>Swings</u>: In a paper by Y. Andrillat, F. Ciatti and J.P. Swings submitted to Astrophys. and Space Sci. we describe the recent spectral evolution of three peculiar emission line objects with IR excess: V1016 Cyg, HM Sge and MWC 349, on the basis of data obtained in the blue visible, red and near infrared (Reticon data for instance) regions. The main variations are the following (Andrillat and Swings 1980, and OHP and Asiago data): (a) V1016 Cyg: a conspicuous intensity increase of OIλ 8446 between 1979 and 1980; simultaneous presence in 1980 and 1981 of strong permitted emissions due to a hot star and of forbidden lines of FeII, V, VII.
(b) HM Sge: appearance of HeIIλ 10123 in 1980.
(c) MWC 349 A: important increase of OIλ 8446 and slight strengthening of HeIλ10830 between 1979 and 1980.

<u>Nussbaumer</u>: I have questions to the observers in the optical, infrared and radio domains: 1) What can be said about the TiO bands in V1016 Cyg; 2) Harvey's IR observations do not clearly establish a period. Have these observations been continued? 3) Have the radio data been extended beyond 100 GHz to establish clearly at which frequency the flux density becomes flat.

<u>McCarthy</u>: As an old M star observer I wish to point out that in the data displayed by Swings the strengths of TiO and VO bands were most indicative of the presence of a late type M star (> M6). I would also comment that since most late type M stars are variable in spectral type as well as in apparent magnitude, we should be ready to think that the M giant may be doing "its own act" in addition to interacting with the hot object.

<u>Ciatti</u>: In reply to the question by Nussbaumer on late-type features in V1016 Cyg: late-type features due to TiO and VO (like in M6-7

stars) are reported e.g. by Mammano and Ciatti (1975, Astr. Astrophys. 39, 405). They are in any case barely recorded in the last years on plates obtained with the same instrument (Ciatti et al. 1978, Astr. Astrophys. 68, 251). These variations are also discussed by Andrillat, Ciatti and Swings (submitted). It would be interesting to compare them with the Mira-type photometric variations reported by Harvey.

Slovak: No systematic program of broad-band infrared photometry has been pursued for V1016 Cyg since Harvey's work. Such observations for V1016 Cyg and HM Sge are a definite desiderata and observers are encouraged to add these objects to their infrared programs.

Whitelock: The IR period and amplitude determined by Harvey is comparable to those of the symbiotic stars containing Mira variables which have been extensively stidied from SAAO. The variation in the molecular bands might be due to filling in by the dust emission or swamping by increased emission from the hot binary component. Variations in molecular bands are of course to be expected from a Mira variable.

Houziaux: We are carrying on an observing program of the near ultra violet in symbiotic stars.

Cassatella: Observations near the Balmer Jump are certainly very important for the detection of the Balmer continuum emission which is useful for a correct diagnostic of the electron temperature which is often difficult to determine.

Kwok: The turnover of the radio spectrum of V1016 Cyg is now at 20 GHz. A question to IUE observers: from the emission lines one can evaluate the emission measures. What are the characteristic sizes of the ne-nebulae calculated from such emission measures?

Hack: Do you know which are the time scales for the transition stage from red giant to proto-planetary nebula?

Cassatella: 10^5 yr or less.

THE PECULIAR SYMBIOTIC OBJECT V1329 CYGNI: SINGLE-STAR VERSUS BINARY MODELS

Jiri Grygar
Institute of Physics, Czechoslovak Academy of Sciences, Rez
Drahomir Chochol
Astronomical Institute, Slovak Academy of Sciences, Tatranska Lomnica, Czechoslovakia

1. Recapitulation of photometric and spectroscopic observations.

The variable emission-line object V1329 Cyg (= HBV 475) was discovered by Kohoutek (1969). Crampton and Grygar (1969) identified more than 100 emission lines in the blue portion of the spectrum, while Andrillat (1969) found evidence for the late-type (M) spectrum in the near infrared. This justified the classification of the object among the symbiotic stars. The classification was subsequently confirmed by all authors who studied the spectroscopic evolution of the object.

Its optical variability was first noted by Kohoutek and Bossen (1970). Stienon et al. (1974) analysed the archive photographic data in the years 1891-1966 and established several modes of variability. From the absence of the object on some patrol plates they inferred that the object is probably an eclipsing binary with a period of 959 days. The duration of primary eclipse is about 60 days. In the years 1960-64 the object increased its photographic brightness from $15^m.1$ to $14^m.2$. Then started a sharp increase of the rate 0.02 mag/day up to $11^m.9$. Maximum brightness of $11^m.5$ was reached in the middle of the year 1966 (Arkhipova and Mandel 1973 and 1975). Since then the object slowly faded, but the original photometric period of about 960 days is preserved.

Radio emission of the object was discovered by Altenhoff and Wendker (1973) and Terzian and Dickey (1973).

2. Postdiscovery studies

A detailed description of the evolution of the postdiscovery spectrum was subsequently published by Crampton et al. (1970), Mammano and Righini (1973), Baratta et al. (1974), Stienon et al. (1974), Andrillat and Houziaux (1976) and Tamura (1977). The last author noted the appearance of the emission lines of [FeVII] and the strengthening of the HeII λ 4686 line in the years 1974-76. This tendency of the appearance of the emission lines of higher and higher excitation was found by the others, too (for a summary see Grygar et al. 1979).

The complex structure of some emission line profiles strongly suggests that an ejection process started during the main outburst in the year 1966. However, the [OIII] lines were either weak or absent until July 13, 1969 and Tamura (1981a) did not find P Cygni profile in the HeI λ6678 line in the years 1976 and 1977. Grygar et al. (1979) noted that the line profile of [OIII] resembles profiles typical of the nebular stage of novae. Thus, separate blobs of matter were apparently ejected along the equatorial as well as polar directions with respect to the parent star. The ejection were in the range from 50 to 350 km s^{-1} (Crampton et al. 1970; Tamura 1981a).

3. Binary hypothesis

Following the suggestion by Stienon et al. (1974) we investigated the possibility that the object is an eclipsing binary. Using the pre-outburst photometry we have searched for possible periodicities in the data by various mathematical techniques (Grygar et al. 1977). From the analysis we may infer that in the time interval from 20 to 1000 days there is only one plausible period of about 950 days. The phase of the pre-outburst light curve is perfectly preserved in the post-outburst photoelectriv measurements (Grygar et al. 1979).

The radial velocity data for emission lines of HI, HeI, HeII, FeI, FeII, OI, [OIII] and [NeIII] in the years 1970-76 yielded the elements of spectroscopic binary circular orbit with a period of 950.1 days and semiamplitude of 61 km s^{-1}. Again, the derived phase of the primary minimum is in excellent agreement with the photometric data (Hric et al. 1978). The spectroscopic orbit was confirmed by Iijima et al. (1981) who from more numerous data covering the time interval of 3814 days found an eccentric orbit (e=0.17) with the same period and semiamplitude of 63 km s^{-1}.

Nice agreement of the pre-outburst and post-outburst photometry as well as its match with spectroscopic orbit lead us to the formulation of the binary hypothesis on the nature of the object V1329 Cyg, in spite of some problems arising:

(i) The mass function $f(M) = 23\ M_\odot$ is pretty large. This infers the minimum mass of the late-type component $M_2 = 25\ M_\odot$; the value that is too high for the estimated bolometric magnitude $M_{bol} = -5$ implying $M_2 = 10\ M_\odot$ (Iijima et al. 1981).

(ii) The late-type component does not fill its Roche lobe by a wide margin. Thus all mass transfer may occur through the stellar wind only. Tamura (1981b) doubts whether the amount of matter transferred by the wind is sufficient enough to preserve the activity of the star.

(iii) The nature of the hot component is unclear. It should be a compact object (no secondary minima were detected) and its Roche lobe is small by 1-5 (!) orders of magnitude to accomodate the matter seen in emission (Tamura 1981b; Iijima and Mammano 1981). Thus, it is objected that the radial velocities as derived from the peaks of emission lines do not represent an orbital motion.

Still, we feel that all these objections could be overcome. Our binary model seems to be theoretically supported by evolutionary considerations published by Tutukov and Yungelson (1976), Bath (1977) and Paczynski and Rudak (1980).

4. Alternatives

Iijima and Mammano (1981) discussed the possibility that V1329 Cyg is a nova-like object with recurrent outbursts at intervals of about 950 days, were the compact component is a white dwarf with rapid mass accretion ($2\ 10^{-5} M_\odot yr^{-1}$), following the theoretical calculations of the hydrogen flashes by Nomoto et al.(1979).

Tamura (1981a, b) advocates a single-star model where a central star has a variable effective temperature in the range $1.2-1.8\ 10^5$°K and the expanding zones of [FeVII] and HeII extend to $7\ 10^{14-16}$cm from the star. He is supporting the interacting stellar winds model by Kwok et al.(1978).

5. Discussion

In spite of all observational as well as theoretical efforts no clear picture of the object V1329 Cyg seems to emerge. The only sure fact is that there was an outburst in the year 1966 followed by an ejection of matter. The interpretation of the rest of the data depends critically on the assumption about the binary or single-star nature of the object.

The relation to other symbiotic objects may give us a better clue. Iijima et al. (1981) noted the similarity of its optical spectrum to the spectrum of the optical counterpart of the X-ray binary GX 1+4. Morphologically, the stars V1016 Cyg, RT Ser, RR Tel, HM Sge and AS 295B are also similar.

We may be fortunate in getting more decisive results in the near future. Following the suggestion by Iijima and Mammano (1981), another period of activity of V1329 Cyg is expected since the end of the year 1981 and the next outburst should occur in November 1982. Thus, we advise the observers present at the Colloquium to run to their powerful instruments immediately after returning home in order to get the latest data in the end of the quiescent stage of this marvellous object.

REFERENCES

Altenhoff, W.J., Wendker, H.J.: 1973, Nature 241, 37
Andrillat, Y.: 1969, IAU Circular No. 2182
Andrillat, Y., Houziaux, L.: 1976, Astr. Astrophys. 52, 119
Arkhipova, V.P., Mandel, O.E.: 1973, Inf. Bull. Var. Stars No.762
Arkhipova, V.P., Mandel, O.E.: 1975, IAU Symposium No.67, p.391
Baratta, G.B., Cassatella, A., Viotti, R.: 1974, Astrophys.J. 187, 651
Bath, G.T.:1977, Mon. Not. R. astr. Soc. 178, 203
Crampton, D., Grygar, J.: 1969, IAU Circular Nos. 2174 and 2176
Crampton, D., Grygar, J., Kohoutek, L., Viotti, R.: 1970, Astrophys. Letters 6, 5
Grygar, J., Hric, L., Chochol, D.: 1977, IAU Coll. No. 42, p.383
Grygar, J., Hric, L., Chochol, D., Mammano, A.: 1979, Bull. astr. Inst. Czechosl. 30, 308
Hric, L., Chochol, D., Grygar, J.: 1978, Inf. Bull. Var. Stars No. 1525
Iijima, T., Mammano, A.: 1981, Astrophys. Space Sci. 79, 55
Iijima, T., Mammano, A., Margoni, R.: 1981, Astrophys. Space Sci. 75, 237
Kohoutek, L.: 1969, Inf. Bull. Var. Stars No. 384
Kohoutek, L., Bossen, H.: 1970, Astrophys. Letters 6, 157
Kwok, S., Purton, C.R., FitzGerald, P.M.:1978, Astrophys.J. 219, L125
Mammano, A., Righini, G.M.: 1973, Contr. Oss. Asiago No. 285
Nomoto, K., Nariai, K., Sugimoto, D.: 1979, Pub. astr. Soc. Japan 31, 287
Paczynski, B., Rudak, B.: 1980, Astr. Astrophys. 82, 349
Stienon, F.M., Chartrand, M.R., Shao, C.Y.: 1974, Astron. J. 79, 47
Tamura, S.: 1977, Astrophys. Letters 19, 57
Tamura, S.: 1981a, Astrophys. Letters in press
Tamura, S.: 1981b, Pub. astr. Soc. japan in press
Terzian, Y., Dickey, J.: 1973, Astron. J. 78, 875
Tutukov, A.V., Yungelson, L.R.: 1976, Astrofizika 12, 521

PHYSICAL PROPERTIES OF V1329 CYGNI

T. Iijima
Asiago Astrophysical Observatory, Italy
(On leave from Nagoya University, Japan)

In this paper, I shall present another model for V1329 Cyg. Grygar et al. (1979) proposed a long period eclipsing binary model consisting of a massive M giant and a hot component. Some problems, however, have remained on the binary model. Namely, the mass function is too large, the amplitude of the light variations changes from period to period, the light curve differs from that of Algol type systems and is similar to long period binaries, and so on (Iijima et al. 1981). It has been believed that some peculiarities in the system could explain those problems. Recently however, some phenomena which are inconsistent with the binary model were observed (Iijima and Mammano 1981).

Figure 1 shows the intensity ratios of high and low excited lines and the temperatures of the exciting star in 1979 and 1980. The temperatures were derived from the relative intensities of H_β, HeI 4471 and HeII 4686 by a modified Ambartsumyan's method (Iijima 1981). The intensities of high excited lines and the temperature of the exciting star increased at first. They reached the maxima at phase 0.3, then decreased suddenly. Their minima were found on phase 0.5 - 0.6; namely on the phase of the maximum optical brightness. Following the minima those values are recovering gradually. These phenomena are well known in outbursts of novae (McLaughlin 1960) and nova-like variables (Swings 1970). On the other hand it is difficult to explain these new phenomena with the binary model. This result suggests that the light curve in this period arised from a mild outburst. Meanwhile, the light variation during this period was almost the same as those observed in the last 10 years (Iijima and Mammano 1981). Therefore, it may be reasonable to consider that also all the other light variations in the last 10 years arised from recurrent outbursts.

As seen in figure 1, the temperature of the exciting star changed from 180000°K to 130000°K in this period. If this temperature decrease occurs holding its total bolometric luminosity, which is possible in nova

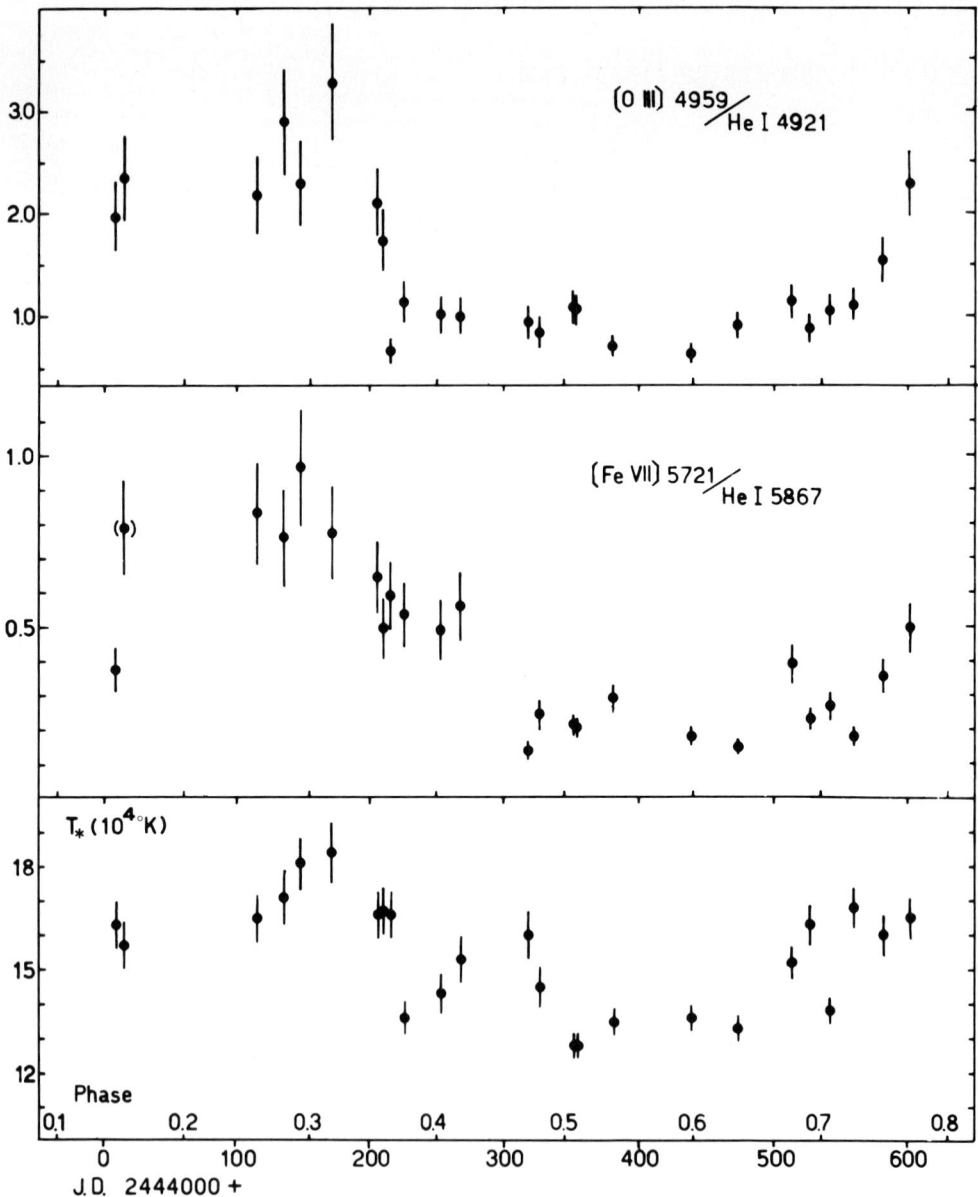

Figure 1. The intensity ratio of the emission lines and the temperature of the exciting star in V1329 Cygni during 1979-80. Phases are reckoned from a photometric minimum (from Iijima and Mammano 1981).

outbursts, an increase of optical brightness of $\Delta m_V \cong 1^m$ is generated (Bath 1977). This value well agrees with the observed light variations (Grygar et al. 1979, Iijima et al. 1981). Recently, Tamura (1981) suggested that the change of the radial velocity of emission lines may not be due to the orbital motion of the binary system, because the dimensions of the emission line regions are too much larger than that of the binary system.

White dwarfs with rapid mass accretion may be one of the possible models for V1329 Cyg. Nomoto et al. (1979) studied rapid mass accretion ($2 \times 10^{-5} M_\odot yr^{-1}$) onto a white dwarf. They found that mild hydrogen shell flash arised with little accreted mass ($\sim 7.8 \cdot 10^{-5} M_\odot$). The effective temperature of the white dwarf ($15 \times 10^{4} °K$, Nomoto et al. 1979) well agrees with that of the exciting star of V1329 Cyg. Though only one outburst is registered in their model calculations, the mass accretion rate can supply hydrogen for one flash in about 1000 days. It will be possible to reproduce recurrent outbursts with a little modification of the initial parameters. If this is the case, it is reasonable to consider that V1329 Cyg is a binary. However, its orbital period may differ from the period of 950 days, because the M type component must fill its Roche lobe.

REFERENCES

Bath, G.T.: 1977, Mon. Not. R. astr. Soc. 178, 203
Grygar, J., Hric, L., Chochol, D., Mammano, A.: 1979, Bull. astr. Inst. Czech. 30, 308
Iijima, T.: 1981, in "Photometric and Spectroscopic Binary Systems, E.B. Carling and Z. Kopal (eds.), D. Reidel Publ. Co., Dordrecht, p. 517
Iijima, T., Mammano, A.: 1981, Astrophys. Space Sci. 79, 55
Iijima, T., Mammano, A., Margoni, R.: 1981, Astrophys. Space Sci. 75, 237
McLaughlin, D.B.: 1960, in Stellar Atmospheres, J.L. Greenstein (ed.), The University of Chicago Press, Chicago, p.585
Nomoto, K., Nariai, K., Sugimoto, D.: 1979, Publ. astr. Soc. Japan 31, 287
Swings, P.: 1970, in Spectroscopic Astrophysics, G.H. Herbig (ed.), University of California Press, p. 189
Tamura, S.: 1981, Publ. astr. Soc. Japan, in press

VARIABILITY OF HBV 475 IN THE NEAR INFRARED

Y. Andrillat
Observatoire de Haute Provence
04870 St Michel l'Observatoire (France)

In the spectral range $\lambda\lambda 5800-8750$, HBV 475 shows important spectral variations between 1969 and 1974. Sometimes the "hot component" spectrum dominates (many emission lines), sometimes the "cool component" is preponderent (many molecular absorption TiO bands) (Andrillat 1973 - Andrillat, Houziaux 1976).
On August 4 1974, June 6 1975 and August 9 1981, we have extended the observations up to 1.1μ (fig.1) (Haute Provence Observatory 193 Telescope - Spectro ROUCAS + Image tube - dispersion 230 A.mm^{-1}).
Some emissions are present :
He I $\lambda 10830$ - He II $\lambda 10123$ - P7λ 10049 - O Iλ 8446.
The Ca II triplet $\lambda\lambda$ 8498-8543-8662 is not visible on August 9 1981 because it is blended with the strong TiO molecular band λ 8432.
For the same reason, O I λ8446 appears very weak.
At this date, other TiO bands are visible at λ 8868.
These features are characteristic of M4-M5 spectral type (Andrillat 1981). The near infrared observations confirm the symbiotic nature of HBV 475 and allow to specify the spectral type of the cool component.
It is well known that the spectral type of the hot component is WN5.
There are a few symbiotic stars which have WR stars as their hot components:
AG Peg : WN6 + M3 III
HBV 475: WN5 + M4
RX Pup : WN7- WN8 + M5 - HM Sge : WN5 + M4
We have observed HD 229227 WN9-WN10 in the LMC during november 1980 (ESO 3,60 m telescope - Boller Chivens Spectrograph + RETICON - dispersion 228 A.mm^{-1}).
In the infrared region, the spectrum shows the characteristics of a symbiotic star (Ca II triplet lines in absorption and many absorption molecular bands of TiO $\lambda\lambda 7054-7126-7589-8432$ which permit to assign a spectral type M4 III.)
Thus, we have found another symbiotic star having a WR star as its hot component (Andrillat, Vreux, Dennefeld 1981).
It is interesting to note that all those WR stars which are members of symbiotic stars belong to the nitrogen sequence.

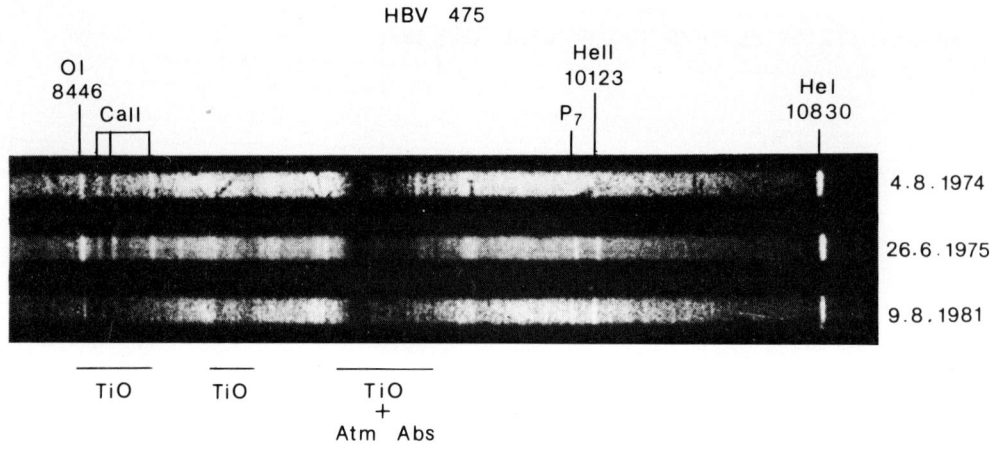

Figure 1

REFERENCES

Andrillat, Y.: 1973, Mém.Soc.Roy.Sci.Liège, T.5, 371.
Andrillat, Y., Houziaux, L.: 1976, Astron.Astrophys.52, 119.
Andrillat, Y., Vreux, J.M., Dennefeld, M.: 1982, IAU Symp. 99
    (forthcoming).

SPECTRAL EVOLUTION OF HBV 475 (= V1329 CYGNI) IN THE
ULTRAVIOLET

Ch. Kindl and H. Nussbaumer
Institute of Astronomy, ETH-Zentrum,
CH-8092 Zürich, Switzerland

Figure 1 shows the low resolution IUE spectra of HBV 475 during 1979-81. Whereas the May 1981 spectrum looks very similar to that of June 1979, there is a clear difference in the August 1980 spectrum when the star was brighter. The strength of the NV multiplet at 1240 A surpasses those of the preceeding and succeeding years by at least a factor 4. A first glance at the continuum and the strongest lines suggests a strong activity increase between June 1979 and August 1980, followed by a decrease to the 1979 level by May 1981. Closer inspection shows that neither CIII] λ1908 nor NIII]λ1750 followed that activity pattern, but rather decreased steadily in their fluxes. Thus in June 1979 CIII]λ1908 showed the same flux as HeII λ1640, whereas in May 1980 CIII] was weaker by a factor two. The OI λ1304 and MgII λ2800 multiplets show the same behaviour as HeII, CIV and NV.

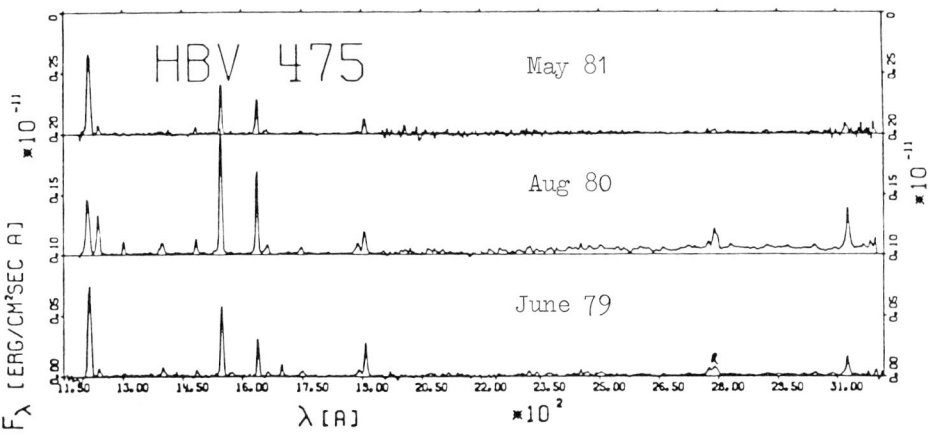

Figure 1. The low resolution ultraviolet spectrum of HBV 475.

DISCUSSION ON V1329 CYGNI

Plavec: I think that the binary model of V1329 Cyg is fundamentally correct. The only problem is the large disparity of masses, 25 $M_\odot$ for the red star, 1 $M_\odot$ for the hot star. Normally, components of binary systems tend to have much more equal masses. Naturally, here we have an old and evolved system, in which the hot star probably initially was the more massive one, and subsequently lost most of its mass. But we have an additional problem, namely the 25 $M_\odot$ for an M giant. All this is based on an unusually large mass function, which in turn is based on the radial velocity curve derived from the emission lines. Probably the lines do not reproduce the velocity curve of the hot star. Both the large range and the eccentricity may be spurious, as they often are in interacting binaries.

Grygar: In our last solution (Grygar et al. 1979) we assumed a circular orbit and obtained $K_1 = (62 \pm 8)$ km s^{-1}. Iijima et al. (1981) found the eccentricity $e = (0.17 \pm 0.09)$, but their radial velocity semi amplitude remained the same, i.e. $K_1 = (63 \pm 6)$ km s^{-1}.

Iijima: In my recurrent outburst model, the change of the radial velocity of the emission lines are explained with a hypothesis of moving ionization front. The 950 days is the period of the outbursts. The temperature of the exciting star changed owing to the outbursts. Therefore, the ionization front moves with the period of 950 days.

Slovak: Which are the magnitudes of V1329 Cyg at the time of the IUE observations?

Viotti: According the the AAVSO Circulars, V =12.95 on August 1980 near maximum, as compared to V = 13.4 (June 1979) and V = 13.2 (May 1981) was it was close to light minima.

INTRODUCTORY REMARKS ON THE SYMBIOTIC STAR AG DRACONIS

Nancy A. Oliversen and Christopher M. Anderson
Washburn Observatory, University of Wisconsin-Madison

AG Draconis contains a relatively early cool component. Quoted spectral types range from dG7 (Wilson 1943) to K3III (Boyarchuck 1969). Most recently Smith and Bopp (1981) adopt G7IIIe. The optical spectrum shows in emission the permitted lines of hydrogen and helium, both neutral and singly ionized. At various times the spectrum of the late type component has been heavily veiled by a blue continuum and numerous lines of various stages of ionization of iron have been seen, including the 6830 A feature which Allen (1980) attributes to Fe VI. On the other hand, on a KPNO echelle spectrum taken by one of the current authors (CMA) in 1977 August, the late giant spectrum was very sharp and no iron or forbidden lines were present. The lack of the forbidden lines identifies AG Dra as a high density case and thus it conforms to the spectral type-density correlation found by Allen (1980).

The Hα profile of AG Dra has a shape which is typical of many of the symbiotic stars. The strong, sharp central peak is bordered by wings of unequal strength. The blue wing of the emission is decidedly stronger than the red wing, a situation commonly referred to as a "blue assymetry". Occasionally, as shown in figure 1, the blue wing develops a distinct shoulder and perhaps even a reversal or separate blue shifted component. Similar phenomena have been reported by Smith and Bopp (1981). They show that the blue component is most pronounced at photometric minimum and suggest that it comes from a wind off the hot component which at that time is less dominated by the redward component which originates around the cool component. Our spectra in which the continuum is well detected and which are linear and unsaturated over their entire intensity range, show that the red peak remains nearly constant relative to the continuum while a distinct absorption cuts into the blue wing.

Anderson, Cassinelli and Sanders (1981, and, with Oliversen, elsewhere in this conference) have reported the detection of AG Dra by the HEAO 2 Imaging Proportional Counter. AG Dra is a remarkably strong source of soft X-rays.

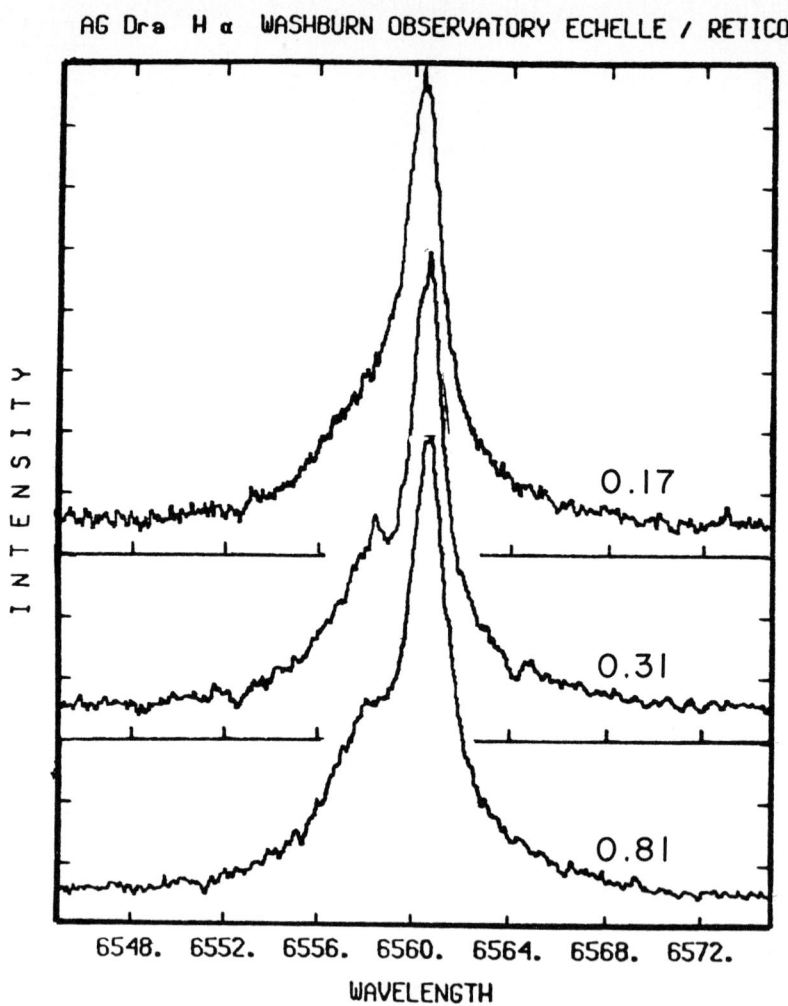

Figure 1. The H-alpha profile of AG Dra at three different phases. Photometric maximum defines phase zero. A shoulder or reversal on the blue wing appears from phase .3 to .8, i.e. during times of depressed U-brightness.

Meinunger (1979) has shown that AG Dra exhibits variations in the U-band of about one magnitude with a period of 554 days. Variations in the B- and V- bands are also present but their amplitudes are much smaller and there are no deep minima obviously correlated with the deep and unmistakeable U-minima. On the other hand, the most recent maximum, which was an unusually high one, shows up as an obvious outburst in the visual data of the AAVSO. Meinunger's data have been considerably augmented by Kaler (private communication). In figure 2 we show the combination of these two data sets, phased according to the Meinunger ephemeris. Although there is considerable scatter, it is clear from the figure that there is an underlying consistency to the light curve which would not be suspected in Meinunger's plot. The light curve is characterized by somewhat rounded maxima, sharp minima and a rise to maximum which is slightly faster than the decline to minimum.

If one ignores the slight asymmetry in the light curve, the shape of the curve is similar to that of a close, eclipsing binary. However the long period and the relatively small size of the cool component makes such a geometry highly unlikely. If an eclipse is to produce a continuous variation, the size the eclipsed object must be of the same order of size as the orbit. Clearly such a component can not be stellar or it would dominate the spectrum of the entire system. An extended disk might be such a component.

An alternate model for AG Draconis is a star with active regions of enhanced surface brightness. Rotation of the star produces the modulation in the U-magintude. In order to produce continuous variations the active region must cover a substantial fraction of the stellar surface. The erratic nature of the light curve would occur if the average lifetime of an active region or part thereof is less than or comparable to the rotational period of the star. Models with differing spot characteristics were calculated by following a generalized version of the method of Bopp and Evans (1973). A region symmetric in latitude about the equator and covering a large extent in longitude (about 220 deg) reproduces the rounded shape of the U light curve. A U amplitude of about .9 magnitude was produced with a region having a brightness enhancement of 1.4 times over the non-enhanced region and covering 43% of the stellar surface. It was found that a symmetric region is unable to reproduce the slight asymmetry in the U light curve. An asymmetry can be introduced by allowing for longitudinal variations in region size and/or for surface brightness gradients. Figure 3 illustrates symmetric and non-symmetric active region models. No attempt was made to find a 'best fit' model since the solution is by no means unique. We conclude that if the U light curve is due to a rotating mottled star, then the active region must cover a large part of the star and that it is probably non-uniform in size and/or brightness.

Models which utilize the Planck function for surface brightness

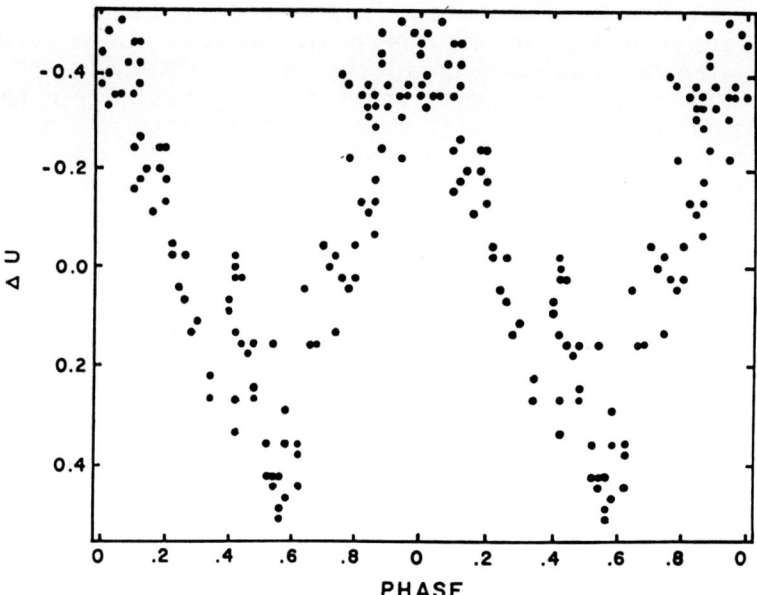

Figure 2. Composite light curve of AG Dra

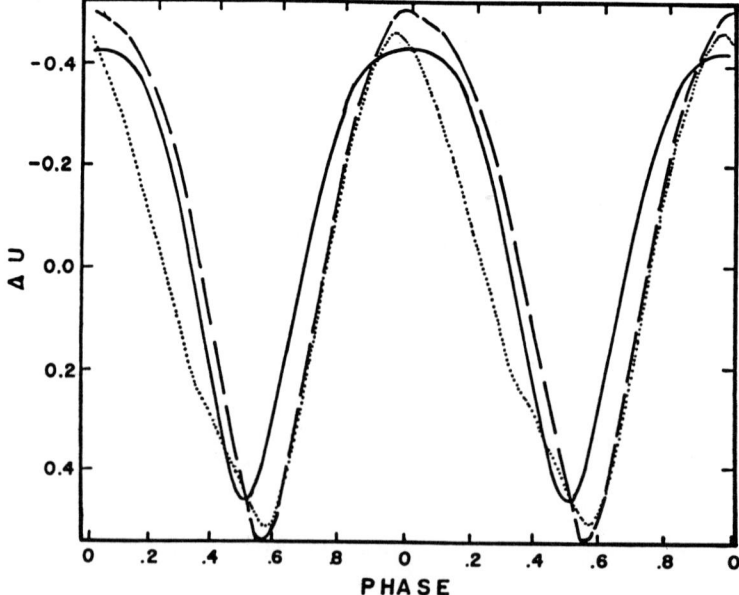

Figure 3. Active region models: T_B=brightness temperature, Δl=longitude extent, $\Delta \beta$=latitude extent, Normal surface T_B=3275K
Solid line: single region T_B=3525, Δl=220, $\Delta \beta = \pm 40$.
Dashed line: 2 regions T_B=3525, Δl=100, $\Delta \beta = \pm 40$ leading, $\Delta \beta = \pm 70$ trailing. Dotted line: 3 regions, $\overline{\Delta l}$=120, $\Delta \beta = \pm 50$, T_B=$\overline{3400}$ leading, T_B=3600 trailing.

and which yield the appropriate U-magnitude amplitude uniformly predict V-magnitude amplitudes near .6 magnitudes whereas the observed variations are of the order of 0.3. A simple thermal emission model, i.e. hot "spots", can not be the physical mechanism producing the surface brightness enhancement.

Enhanced Balmer line and continuum enhancement is often associated with stellar activity. The Hβ equivalent width is observed to be correlated with the light curve reaching a maximum at about the same time as does the U-band brghtness. Thus we suggest that the photometric behaviour of AG Dra may be the result of chromospheric activity modulated by rotation.

REFERENCES

Allen, D. A. 1980 M.N.R.A.S., 190, 75 .
Anderson, C. M., Cassinelli, J. P., and Sanders, W. T. 1981,
    Ap.J. 247, L1.
Bopp, B. W. and Evans, D. S. 1973 M.N.R.A.S., 164, 343.
Boyarchuk, A. A. 1969 I.A.U. Coll. "Non-Periodic Phenomena in
    Variable Stars", p. 395.
Meinunger, L. 1979 I.A.U. I.B.V.S. #1611 .
Smith, S. E. and Bopp, B. W. 1981 M.N.R.A.S., 195, 733 .
Wilson, R. E. 1943 Publ.A.S.P., 55, 282.

THE ULTRAVIOLET SPECTRUM OF AG DRACONIS

A. Altamore[1], G.B. Baratta[1], A. Cassatella[2], A. Giangrande[3], D. Ponz[2], O. Ricciardi[3], R. Viotti[3]
1. Osservatorio Astronomico e Università di Roma, Italy
2. Astronomy Division ESTEC, Villafranca, Spain
3. Istituto Astrofisica Spaziale (CNR), Frascati, Italy

The high resolution ultraviolet spectrum of AG Dra was observed with IUE in April and August 1981 at phases 0.50 and 0.69 according to the Meinunger (1979, Inf. Bull. Var. Stars No.1611) U-light curve. The UV spectrum of the star appears rather different from that of the other classical symbiotic stars. The low resolution IUE spectrum of AG Dra shown in figure 1. The continuum is rather strong with respect to the emission lines and detectable at high resolution. Many intense interstellar lines are present, in spite of the low reddening of the star (E(B-V)= 0.06, according to the depth of the 2200A interstellar band).

The emission line spectrum has a high ionization temperature, showing permitted and intercombination lines of CIV, NIV-V, OIV-V, SV, SiIV, while CIII] and NIII] are very weak. OI and MgII are also present. The radial velocity difference, emission minus absorption lines is about -110 km s^{-1}, in agreement with the optical observations.
An electron density of $\sim 3 \cdot 10^9$ cm^{-3} is derived from the OIV] lines. HeII λ 1640 is very prominent in emission with broad wings. If this line is produced by photoionization by ultraviolet radiation, a 100000°K blackbody temperature for the hot source is required.

We have compared the 1981 observations with the pre-outburst 1979 (phase 0.46) and 1980 (phase 0.90) IUE spectra.
The continuum and the emission lines before outburst appeared much weaker than in 1981, and the 1979-80 variations are in agreement with the Meinunger light curve, suggesting that the whole UV spectrum of AG Dra before outburst is variable in phase with the U-curve.

In figure 2 we show the intensity variations of the emission lines during 1979-81, as compared with the visual light curve as derived from the AAVSO Circulars. This light variation was followed by most of the emission lines except SiIII] that was almost steady. For the high ionization lines, in spite of the large variations in intensity, there is an indication of no change in the relative intensities of the NV, CIV, SiIV

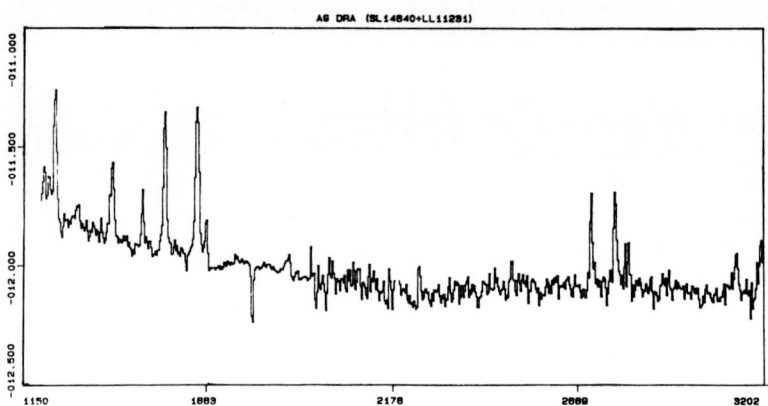

Figure 1 (Altamore et. al., facing page). The low resolution ultraviolet spectrum of AG Draconis after outburst, in August 1981.
The most prominent emission lines are NV λ 1240, CIV λ 1550, HeII λ 1640, HeII λ 2732 and MgII λ 2800. A weak continuum depression near 2200 A is present. The line near 1900 A is SiIII], while CIII] λ 1909 is very weak.

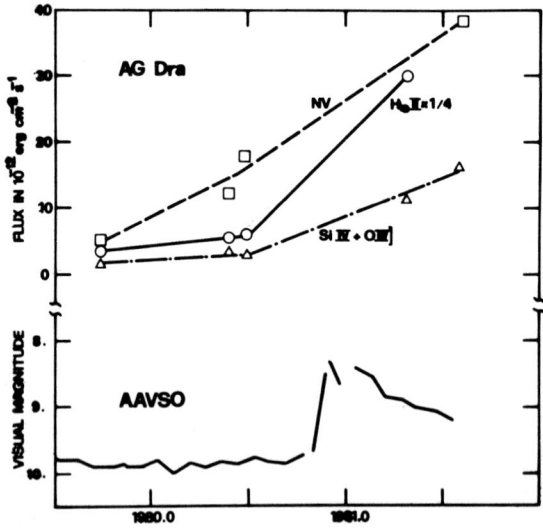

Figure 2. Spectral variations of AG Draconis during 1979–81 from high and low resolution IUE spectra. The visual light curve (from the AAVSO Circulars) is shown. The 1979 and 1980 data are referred to a U-light minimum and maximum respectively.

and OIV] lines which may indicate that the mean ionization level has not varied after the outburst.

SPECTROSCOPIC OBSERVATIONS OF AG DRA

HUANG CHANG-CHUN
Purple Mountain Observatory, Academia Sinica

AG Dra is an interesting symbiotic star, on account of its very high negative velocity and its earlier spectral type among the symbiotic stars. This star has been classified as dG7 (Wilson,1943), K1 II (Roman,1953) and K3 III (Boyarchuk,1966). It has a variable radial velocity.

During summer 1981, spectroscopic observations of AG Dra were performed at the Haute-Provence Observatory using the Marly spectrograph with a dispersion of 80 A mm^{-1} at the 120 cm telescope and using the Coudé spectrograph of the 193 cm telescope with a dispersion of 40 A mm^{-1}. Professor Ch. Fehrenbach very kindly given me a plate of the star which he had taken in July,1966, using the coudé spectrograph of the 193 cm telescope with a dispersion of 40 A mm^{-1}.

The actual outlook of the spectrum of AG Dra is very different from what it was in 1966 in the sense that only a few intense absorption lines remain, the heavy emission continuum masking the absorption spectrum, while on the 1966 plate, about 140 absorption lines have been measured. They were due to FeI, TiII, TiI, CaI, CrI, SrII, CeII and BaII,etc. Perhaps SmII was present. The lines of TiII, SrII and BaII were relatively strong, showing very high luminosity star features. There was also a sharp intense absorption at λ 3933 of the interstellar K line with a radial velocity of about -38 Kms^{-1} which still exists with the same radial velocity. A number of coudé plates of giant and supergiant standard stars of early K and late G types were obtained in 1981, the TiII, SrII and BaII lines clearly show luminosity effects in these stars. The 1966 spectrum of AG Dra matches well enough that of the K0Ib Star, 12 Peg. (Fig.1).

Numerous emission lines have been measured, most of them, present in 1981, could also be detected in 1966. They are due to H, HeI and HeII. The Balmer lines are recorded in emission as far as H24. The HeII 4686 A is always strong. A few Pickering lines have been detected. We have found also a very strong line at 3203 A of HeII (3-5) on the Marly spectrograph plates. There is evidence that the HeI and fluorescence lines of O III are more intense in 1981. On the 1981 plates, we have found also emission lines of OII multiplet (1). We have not seen any one of the three nebular lines of [OIII].

The emission continuum which heavily veils the absorption spectrum in 1981 in the photographic region was much weaker in 1966. The Balmer emission continuum was as strong in 1966 as in 1981. On the Marly spectrograph plates, the stellar spectrum in ultraviolet can be traced beyond λ 3200 A (Fig.1). However, it seems there was not much changes in the ultraviolet region between 1966 and 1981 as shown by the comparison of two coudé plates taken in 1966 and 1981 in Fig.1.

The velocities obtained by us are summarized in Table 1, n is the number of the lines. The absorption mean velocities were derived essentially from lines of FeI.

Table 1. Radial Velocities for AG Dra

	1966			1981			1966		1981	
em.	V	n	em.	V	n	absorp.V	n	absorp.V	n
H	-158,2±2.6	21	H	-146,0±2.8	20	-149,8±1.5	41	-157,9±7.2	5
HeI	-163,0±3.4	7	HeI	-154,4±3.0	15				
HeII	-149,9±2.7	3	HeII	-164,8±4.9	5				
OIII	-139,7±1.1	2	OIII	-153,0±1.7	5				
			OII	-165,1±5.1	3				

From the data of Wilson (1945) and Roman (1953) and ours, there is no reason to have doubts about the variability of the radial velocities of the star. The emission variations seem to be periodic with a period of about 35 years. This is shown in Fig.2. The absorption line velocity which has been derived from the very few available absorption lines when the spectrum of the star was too much veiled by the emission continuum is not sure.

In conclusion, AG Dra could be classified as supergiant K0Ib, according to the relative intensities of TiII, SrII, BaII and CaI on the 1966 plate. The velocity variations of emission lines suggest a period of about 35 years. But more spectroscopic and photometric observations are necessary to confirm the period. It would be interesting to learn more about the blue emission continuum which greatly enhanced in 1981, while there was no conspicuous changes in ultraviolet continuum between 1966 and 1981. As to the strong emission line HeII 3203 A, it may be recalled that Y. Andrillat (1979) drew attention to the emission line HeII 10124 A (4-5) in some Of stars. Like HeII 10124 A in Of stars, the formation

SPECTROSCOPIC OBSERVATIONS OF AG DRA

Figure 1 : AG Dra and comparison stars

The spectrograms of AG Dra, taken on July 23rd 1966 and on August 16th 1981, as well as the spectrograms of comparison stars 12 Peg K0 Ib and ηCyg K0 III, were obtained with Coudé Spectrograph of 193cm telescope. The others were taken with Marly spectrograph.

of HeII 3203 A in AG Dra also could involve chromospheric phenomena.

I am deeply indebted to professor Ch. Fehrenbach who has given me very valuable help every-way in my work and for his loan of the plate taken in 1966. I would like very much to express my thanks to Dr.Y. Andrillat for her discussions and helpful suggestions in this work. I wish to thank M.L. Rolland who has done a great part of the measurements.

References

Andrillat,Y.,:1979, I.A.U. Colloquium n° 47,137. Vatican Observatory.
Boyarchuk, A.A.,:1966, Astrofizika, Vol.2,n° 1,50.
Roman, N.G.,:1953, AP.J., 117, 467.
Wilson,R.E., Pub. A.S.P.,55, 282, 1943 ; 57,309, 1945.

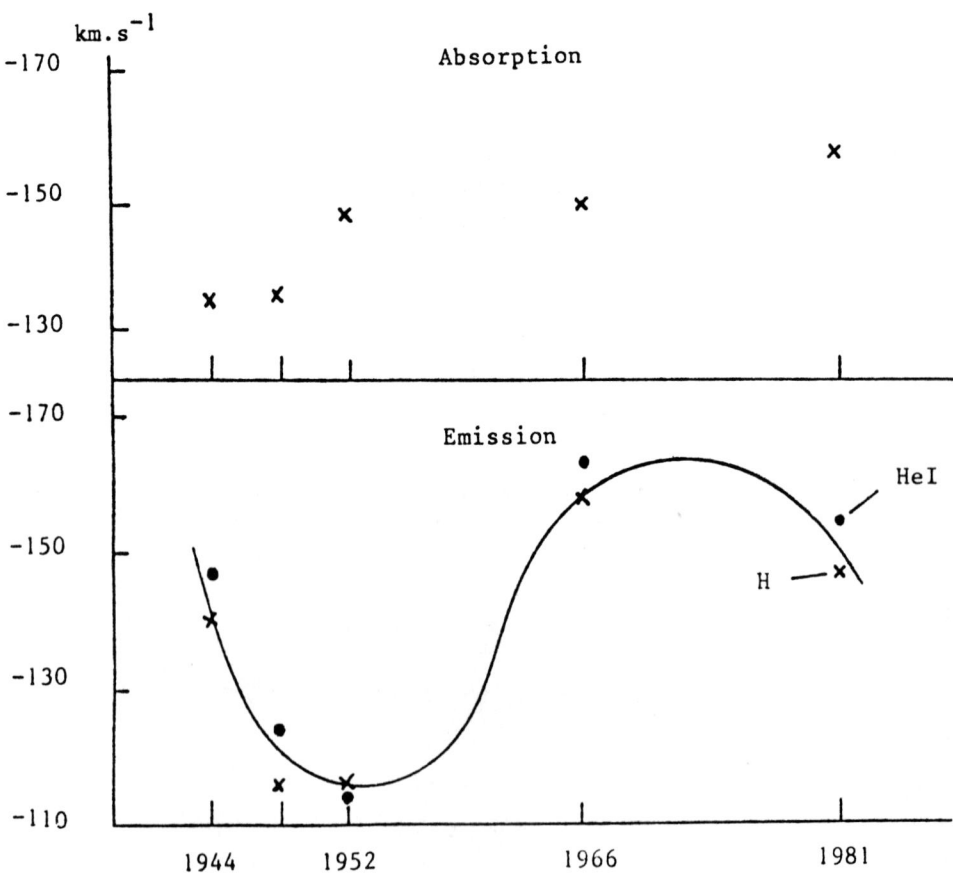

Figure 2.

DISCUSSION ON AG DRACONIS

<u>Plavec</u>: (to Oliversen) I notice that the light curve you are using is in the U filter. I believe that one of the biggest blunder in astronomy was the introduction of the U filter defined so that its sensitivity extends to both sides of the Balmer jump or, rather, the use of this filter for binary star work. In particular, if continuous bf + ff hydrogen radiation is present, variable Balmer emission will dominate the U magnitude. From the scans Keyes and I have, we concluded that this H emission is strong. Therefore the U-light curve will reflect predominantly the behaviour of circumstellar gas. Thus, you have justly concluded the the light curve cannot be interpreted to mean that you are dealing with a contact binary.

<u>Fehrenbach</u>: AG Dra has a strong Balmer discontinuity, whose variations should explain that of U. The radial velocities of the emission and absorption lines are of the order of -150 km s^{-1}, and unchanging. The period of 20 years suggested by Huang is the result of an attempt.

<u>Slovak</u>: AG Peg shows a 820^d sinusoidal variation in its light, ascribed to the heating of the secondary by the hot primary. The variations in AG Dra may arise from a similar situation, if indeed it is a binary system.

<u>Oliversen</u>: Our model does not rule out a binary model. Modulation of U could be produced by rotation of the cool star <u>or</u> by the binary period of the hot object. Let me just state that the U-light curve is not probably produced by an eclipse solely of the hot object, since the separation is too great.

<u>Keyes</u>: Our simultaneous IUE and ground-based scans covering 1200-7000 A in May and October 1980 (both pre-outburst; the latter within two weeks of outburst) show no continuum variation; however H, HeI and HeII have increased by ~20 % in October, possibly an outburst precursor. It should be noted that for the densities quoted by Altamore et al. the lack of [OIII] 4363 + 5007 is most likely only a contrast effect because of the optical continuum which is considerably brighter at 4300 A relative to λ 1660 A (OIII]) than in most symbiotic stars.

<u>Fehrenbach</u>: There are lines with P Cygni profile in the ultraviolet?

<u>Viotti</u>: Only NV λ1238 presents at high resolution a P Cygni profile with an absorption component shortward shifted by about -100 km s^{-1}. CIV appears symmetric only in emission. Note that the presence strong interstellar lines in MgII with a radial velocity difference of $+110$ km s^{-1} with respect to the emissions could be erroneously interpreted as inverse P Cygni profile.

UV TIME-DEPENDENT EMISSION IN SY MUSCAE

A.G. Michalitsianos[*], M. Kafatos[**]
[*] NASA Goddard Space Flight Center, Greenbelt, Md. 20771 USA
[**] George Mason University, Department of Physics
    Fairfax, Virginia 22030   USA

INTRODUCTION

Ultraviolet spectra acquired with the International Ultraviolet Explorer (IUE) of SY Mus = HD 10036 on 20 September 1980 and 11 June 1981 indicate a substantial enhancement of UV emission over a nine month period. The general UV flux level appears to have increased by approximately one order of magnitude between the first and second observing epochs. The strong ultraviolet continuum evident throughout the entire IUE spectral range $\lambda\lambda 1200$-3200 A on 11 June 1981 is closely approximated by a star with T_{eff} = 40,000 K, where previously on 20 September 1980 the continuum distribution presented a more complex structure that is possibly explained by a combination of thermal emission from an early type main sequence star, and nebular recombination emission (Michalitsianos et al. 1981).

The redistribution of continuum energy flux with wavelength is accompanied by an increase in intensity in permitted high excitation emission lines. High excitation lines of N V $\lambda\lambda 1238.8$, 1242.8 A, O V $\lambda 1371.5$ A, Si IV $\lambda 1393.8$ A, O IV] $\lambda\lambda 1397.2$, 1399.8, 1401.2 A, C IV $\lambda\lambda 1548.2$, 1550.2 A, He II $\lambda 1640.4$ A, N III] $\lambda\lambda 1746.8$, 1748.6, 1749.7, 1752.2, 1754.0 A, Si III] $\lambda 1892.0$ and C III] $\lambda\lambda 1906.7$ and 1908.7 A are among a number of lines that increased in absolute intensity by a factor 1.5 to 5. In contrast, forbidden Mg V $\lambda\lambda 2783.2$, 2928.7 A, [Ne V] $\lambda 1574.9$ A decreased in absolute intensity by approximately 15%. A decrease of [Mg V] line emission during an enhanced UV emission phase suggests that SY Mus increased overall in electron density. This is substantiated by high dispersion spectra of the C III] doublet (available for 11 June 1981 only) where $I(\lambda 1906.7 \text{ A})/I(\lambda 1908.7 \text{ A}) \ll 1$ ratio suggests $n_e \sim 10^{10}$ cm^{-3}. Alternatively, if created in a region very far removed from the UV continuum forming region, forbidden Mg V might appear to diminish in intensity if the continuum increased substantially. As such, contrast between emission line and continuum might explain the apparent decrease in forbidden line intensities in both [Mg V] as well as [Ne V].

Mg II resonance doublet emission increased in absolute intensity by a factor ~ 60 between observing epochs. It is of interest to note at this point that the line profiles in low dispersion on 20 September 1980, and low as well as high dispersion on 11 June 1981 do not exhibit anomalous profile structure. There is no evidence for P-Cygni type profiles in any of the high or low ionization lines, and evidence for significant turbulent broadening in any of the emission lines investigated in detail is lacking. As such, the overall increase in UV emission is best described as an enhancement of quiescent emission rather than an outburst or eruption. Based on this, we suspect that SY Mus is an eclipsing symbiotic system in which the hot secondary has recently emerged from behind the cool primary. On 11 June 1981 the UV spectral properties of SY Mus resembles that of RW Hya (Kafatos et al. 1980) and Z And (Altamore at al. 1981).

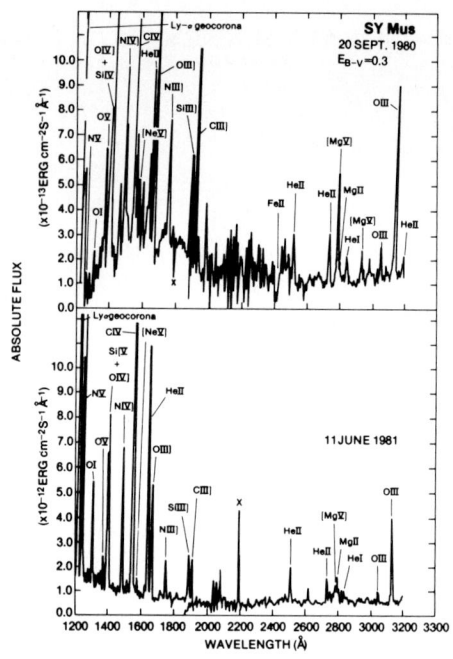

FIGURE 1

DISCUSSION

Feast et al. 1977 identify the cool primary in SY Mus as an M3 III star with absolute magnitude M_V = -0.4. They find an interstellar absorption corresponding to $E(B-V)$ = 0.23, consistent with our estimate for $E(B-V)$ = 0.3 derived from the $\lambda \lesssim 2200$ A continuum dip, and which has been applied to data shown in Figure 1. Given these uncertainties for the primary the distance modulus yields $d \sim 1.1_2$ kpc. If we consider the absolute UV continuum flux at $F(\lambda 1400A) = 10^{-12}$ erg cm^{-2} s^{-1} A^{-1}, we can extrapolate the continuum into the visible region by applying a Planckian blackbody curve for a T_{eff} = 40,000 K star. For d = 1.1 kpc we find the following parameters for the hot companion: L = 90 $L_\odot$, R = 0.2 $R_\odot$ and m_V = 13.6. This places the secondary in a region of the H-R Diagram occupied by central stars of planetary nebulae. The corresponding relative magnitudes between secondary and primary is Δm_V = 2.9. Similarly, as an upper limit if we consider a T_{eff} = 10^5 K secondary, we obtain L = 530 $L_\odot$, R = 0.08 $R_\odot$ and m_V = 14.4 (or Δm_V = 3.7).

The enhancement of moderate excitation emission lines such as as Mg II $\lambda\lambda$ 2795, 2802 A doublet suggests possible photon interaction between the secondary and primary. Here a hot spot formed from the intense radiation field of the companion that illuminates a small portion of the M giant atmosphere likely results in emission from medium excitation emission lines that generally characterize chromospheres. This is consistent with our model for an eclipsing system because the portion of the M giant atmosphere directly subjected to the radiation field of the companion becomes more visible as the secondary emerges from eclipse. Therefore, monitoring this object is extremely important to determine if in fact this is correct.

Alternatively, the hot companion may have ejected a thick ionizing shell that was created through mass transfer from the cool primary onto the compact secondary. Bath (1978) has suggested that such an object might outwardly give the appearance of being an early main sequence type star in terms of luminosity and surface effective temperature. Instabilities that could develop in the shell envelope due to continuous accretion might result in the dislodging of the envelope, thus exposing the underlying thermal emission from the hot subdwarf. However, high dispersion line profiles do not suggest motion of this character. Rather the line profiles appear narrow and are centered near the laboratory rest wavelength. Continued UV and optical observations of SY Mus are essential in order to determine the nature of this interesting object.

REFERENCES

Altamore, A., Baratta, G.B., Cassatella, A., Freidjung, M., Giangrande, A., Ricciardi, O. and Viotti, R. 1981, Astrophys. J., 245, p. 630.
Bath, G.T. 1978, M.N.R.A.S., 182, p. 35.
Feast, M.W., Robertson, B.S.C. and Catchpole, R.M. 1977, M.N.R.A.S., 179, p. 499.
Kafatos, M., Michalitsianos, A.G. and Hobbs, R.W. 1980, Astrophys. J., 240, p. 114.
Michalitsianos, A.G., Kafatos, M., Feibelman, W.A. and Hobbs R.W. 1981, Astrophys. J., in press.

DISCUSSION ON SY MUSCAE

<u>Kafatos</u>: This is a request to the southern hemisphere observers to follow this particular object in the visible to see if indeed it is an eclipsing system.

<u>Whitelock</u>: We are presently following many of these objects, and if we know dramatic changes we can follow up with observations.

AR PAVONIS: THE ROSETTA STONE OF THE SYMBIOTICS

Mark H. Slovak

University of Texas at Austin, USA

1. Early History and Observations

With Mayall's discovery of the eclipsing nature of AR Pavonis = CD-66°3307, a nearby unique opportunity to directly probe the structure of the symbiotic binary became available. Few eclipsing systems have been detected among the symbiotics and AR Pav is fortunately bright enough to be observed spectroscopically at Coudé dispersions.

Due to the dominance of the giant component over the optical region, it has been difficult to reliably derive the properties of the hot companion from optical studies alone. The light curves published by Mayall (1937) and more recently by Andrews (1974), in addition to the detailed spectroscopic studies undertaken by Thackeray (1959) and Thackeray and Hutchings (1974), were used to determine the binary properties. At a distance of approximately 3.8 kpc, the system appears to consist of an M3III secondary and a subluminous primary ($T_e \cong 30,000$ K) embedded in an accretion "shell". The binary model proposed for AR Pav by Thackeray and Hutchings is reviewed in light of recent ultraviolet, infrared, and radio observations.

2. The Light Curve

Using extensive Harvard plate collection, Mayall (1937) published a photographic light curve for AR Pav, spanning the period from 1889 to 1936. A skewed primary eclipse ($\delta m_{pg} = 2.5$ mag) is clearly present, lasting approximately 100 days from initial ingress to final egress. No convincing evidence for a secondary eclipse is discernible. The eclipse depth is variable, showing "bright" eclipses where the level at minimum light is nearly 0.5 mag above that seen at "quiescent" eclipses. The system brightness outside of eclipse is also variable, with a quasi-periodicity of approximately seven years. There is no apparent correlation between "bright" eclipse minima and an elevated light level between eclipses.

Andrews (1974) presented UBV observations over the period 1967-1973, refining Mayall's value for the eclipse period (605 days) to P_{ecp}= 604.6 days. Eclipses were observed in 1967, 1969, 1970, and 1972; the mean "quiescent" eclipse profile derived from the observations displays a marked asymmetry at ingress, exhibiting a "stillstand" near phase = 0.9. The eclipse minima are not flat, indicating the non-stellar appearance of the primary object. The U-B index changes abruptly by -0.3 mag following the stillstand into eclipse, and returns to the pre-eclipse value of -0.60 mag upon egress.

During a "bright eclipse", the stillstand apparently disappears and the time of mid-eclipse advances to pháse 0.98. While the eclipse duration remains essentially the same, the central depth of a bright eclipse is about 0.5 mad deeper than for a quiescent eclipse.

3. Spectroscopic Properties

In 1900-1901 the earliest spectra of AR Pav showed P Cygni line profiles, with HeII 4686 A and NIII 4640 A emission lines appearing in addition to the Balmer series (Mayall 1937). Sahade (1948/) obtained additional spectra and identified emission from MgII, FeII, HeI, CIII, [OIII], and [NeIII], superimposed on an early-type absorption spectrum.

The most detailed spectroscopic studies remain those of Thackeray (1959) and Thackeray and Hutchings (1974). The rich emission line spectrum and its behaviour throughout the eclipse cycle was the subject of the first paper. The Balmer lines of Hγ and Hδ were resolved into two components, whose separation decreased by 34 km/s during eclipse from 114 km/s seen outside of eclipse. Both the helium and hydrogen lines weakened in eclipse relative to the [OIII] 4363 A nebular line; HeII 4686 A virtually disappeared at mid-eclipse. TiO absorption was seen only during the eclipse. The stratified nature of the emission producing region is indicated by this behaviour, where the high excitation lines appear to be concentrated towards the central region.

In the later paper, Thackeray and Hutchings discuss the radial velocity and emission line intensity variations across the eclipse cycle, using an extensive set of spectra spanning a twenty year period. From the ratio of [OIII] (4959+5007)/4363 the mean electron density in the nebula was estimated to be log $N_e \sim 7$, substantially higher than found in the planetary nebulae.

Velocities were measured for the permitted and forbidden emission lines, as well as for the absorption lines of TiII, ScII, and SrII. Orbital solutions were independently calculated for each set of velocities. The permitted line (PE) solution, identified with the primary, gave a semi-amplitude of K = 13.0 $\pm$ 1.6 and an eccentricity of e = 0.11

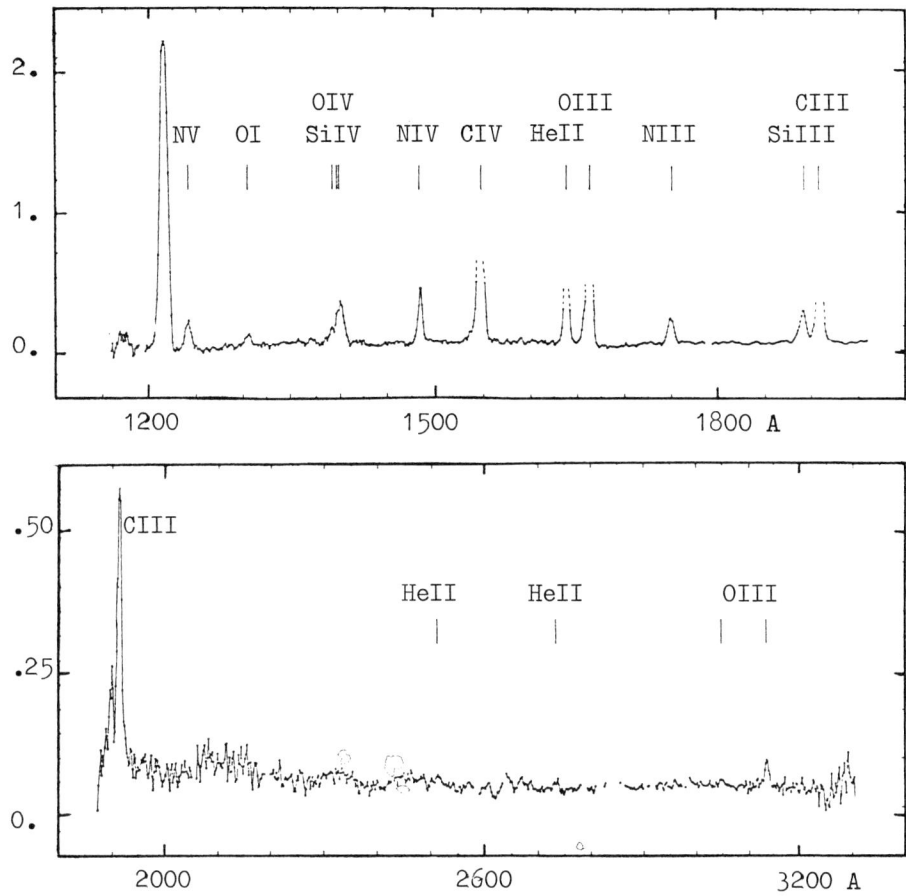

Figure 1. Low resolution IUE spectra of AR Pavonis. Ordinates are fluxes in 10^{-11} erg cm^{-2}s^{-1}Å^{-1} dereddened for $E(B-V) = 0.30$.

± 0.11. The absorption line (A) solution, associated with the late type secondary, yielded values of $K = 26.4 \pm 11.2$ and $e = 0.45 \pm 0.39$. The forbidden line (FE) solution indicates that these velocities only weakly reflect the orbital motion ($K = 6.4 \pm 2.1$ and $e = 0.53 \pm 0.20$), having a phase shift of 0.1 from the A and PE solutions. The large scatter of the individual velocity measurements (± 5.7 km/s) suggests that the streaming velocities in the line forming regions are significant compared to the orbital motion.

From the PE and A solutions, Thackeray and Hutchings (1974) estimate a mass ratio $q = M_1/M_2 \approx 2$, yielding probable masses of $M_1 = 2.5\ M_\odot$ and $M_2 = 1.2\ M_\odot$ for the primary and secondary respectively, assuming the eclipse is total.

4. Recent Observations

Over the last decade, observations of AR Pav at infrared, radio and

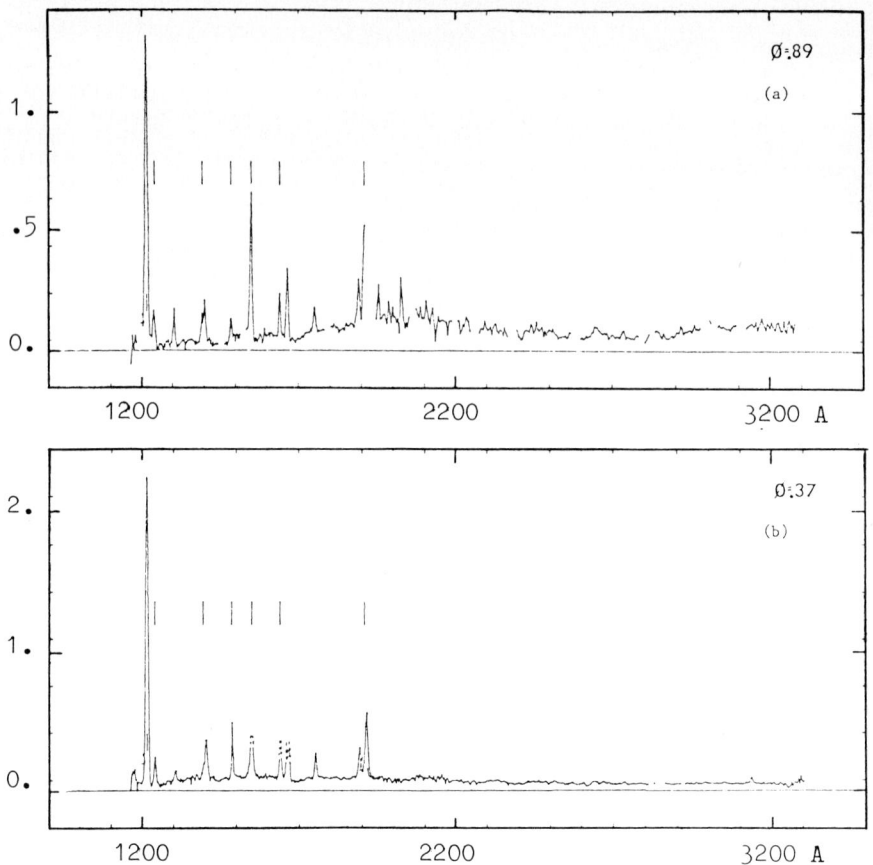

Figure 2. Combined SWP and LWR spectra of AR Pavonis at (a) eclipse ingress, and (b) outside of eclipse. Ordinates as in figure 1.

ultraviolet wavelengths have been obtained. In the infrared Glass and Webster (1973) and recently Allen (1978) have published broad-band Johnson JHKL photometry. The infrared properties do not differ markedly from normal non-variable giant stars, and Allen (1978) subsequently classified AR Pav as an S, or stellar type symbiotic. AR Pav was included, but not detected, in the 2.1 cm continuum survey of Wright and Allen (1978). The lack of detectable radio emission is consistent with its S type classification.

Low resolution ultraviolet IUE spectra have been obtained for AR Pav at eclipse ingress ($\phi = 0.89$) and outside of eclipse ($\phi = 0.37$) (Slovak 1981). The SWP and LWR spectra at phase 0.37 are seen in figure 1. Prominent lines of NIV], CIV, HeII, OIII], NIII] and CIII] dominate the SWP bandpass. The LWR spectrum shows a strong Balmer continuum with few emission features. The interstellar 2200 A feature was used to estimate a value of the reddening E(B-V) = 0.30 substantially higher than the value of E(B-V) < 0.1 determined by Andrews (1974) from UBV observations of surrounding stars in the field.

A comparison of the IUE observations at eclipse ingress and outside of eclipse is seen in figure 2 (a) and (b) respectively. The ratios of CIV/NIII] and HeII/NIII] weaken as the eclipse progresses while the ratio NIV]/NIII] remains relatively constant. This behaviour is consistent with the optical counterparts of these ratios. The peak continuum flux also moves from $\lambda \sim 2000$ A near eclipse to $\lambda \sim 1500$ A outside eclipse, reflecting the contribution of the hot primary.

5. Review of the Binary Model

The recent observations serve to confirm the validity of the binary model proposed for AR Pav by Thackeray and Hutchings (1974). The secondary appears as an M3 giant showing normal infrared properties. The nature of the hot primary, however, remains uncertain as it appears obscured by an accretion "shell" both during and outside of eclipse.

These comments are reinforced by the absolute energy distribution shown in figure 3. Absolute mean fluxes are compared to an intrinsic composite stellar distribution derived from a B0V star (T_e = 28,000 K) combined with an M3III star. A reasonable fit is achieved beyond one micron, but shortward of 3300 A the distribution is dominated by Balmer continuum radiation. Thus the accretion shell dominates the UV region, and significantly alters the eclipse appearance when bright.

The author gratefully acknowledges travel grants from the University of Texas at Austin, the American Astronomical Society, and the IAU organizing committee. This research was supported in part by the NASA Grant NSG 5379.

REFERENCES

Allen, D.A.: 1978, in IAU Coll. No.45, "Changing Trends in Variable Star Research", Hamilton, New Zeeland, p.125.
Andrews, P.J.: 1974, Mon.Not.R.astr. Soc. 167, 635.
Glass, I.S., and Webster, B.L.: 1973, Mon.Not.R.astr.Soc. 165, 77.
Mayall, M.W.: 1937, Harvard Annals 105, 491.
Sahade, J.: 1949, Astrophys.J. 109, 541.
Slovak, M.H.: 1981, Unpublished PhD Thesis, University of Texas.
Thackeray, A.D.: 1959, Mon.Not.R.astr.Soc. 119, 629.
Thackeray, A.D., and Hutchings, J.B.: 1974, Mon. Not.R.astr.Soc. 167, 319.
Wright, A.E., and Allen, D.A.: 1978, Mon.Not.R.astr.Soc. 184, 893.

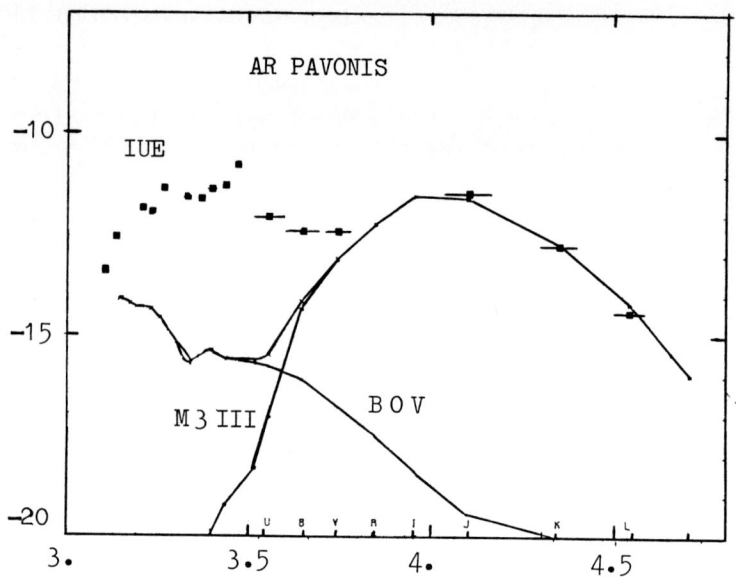

Figure 3. Absolute energy distribution for AR Pav. Abscissae: log wave length in angstrom. Ordinates: log flux in erg $cm^{-2} s^{-1} A^{-1}$. Filled squares (■): IUE, optical and IR photometry of AR Pav. The absolute energy distributions of M3III, B0V stars, and the combined distribution are also indicated.

DISCUSSION ON AR PAVONIS

Plavec: I would like to call your attention to the bulge seen in the continuum of AR Pav at approximately 1900A. This bulge is seen in all Serpentids (that are strongly interacting binaries with strong emissions of NV, CIV, SiIV, etc.), notably in β Lyrae. Since β Lyrae and the other Serpentids show strong emissions of FeIII, and since FeIII has usually many lines just in the region about λλ1850-2300 A, it was suggested that the bulge in the continuum is caused by a suprposition of numerous faint FeIII emission lines. However, there are no FeIII emissions visible in AR Pav. This support my opinion, based on my work on β Lyrae, that the bulge is not due to FeIII emission, but is of a different nature, completely (in AR Pav) or at least predominantly (as in β Lyrae), namely, that it is due to a peak in the continuous bf + ff radiation of a hydrogen cloud in the system.

INTRODUCTORY REPORT ON AG PEGASI*

The light history of AG Peg (HD 207757) is characterized by a ~ 3 mag brightening which took place during 1850-70 followed by a slow decline ever since. In recent times the V magnitude has shown variations between 7.7 and 8.8, and a semi-regular behaviour with maxima 600 to 700 days apart (Mattei 1981). The optical spectrum and its time variation was studied among others by Boyarchuk (1967) and Hutchings et al. (1975) who suggested the presence of a binary system consisting in an M type giant and a WN6 companion surrounded by a nebular region.

The IR spectrum is typical of a cool star (Allen 1979) and seems not to be variable. AG Peg is also a radio source; Gregory et al. (1977) and Ghigo and Cohen (1981) found a radio spectral index consistent with that expected for a continuous outflow model.

A strong UV flux was first observed with OAO-2 by Gallagher et al. (1979) and confirmed by the IUE observations. Keyes and Plavec (1980) observed AG Peg simultaneously in the optical and UV wavelength regions, and found a strong UV continuum with numerous strong permitted and intercombination lines. The high resolution IUE spectra revealed line profiles similar to those seen in WN stars, with broad components in the CIV, NV, NIV lines and in NIV] λ 1486, while the other intercombination lines display only a sharp emission component. According to Keyes and Plavec the hot continuum is most likely a combination of a true stellar continuum plus a nebular emission. Model fitting and Zanstra arguments indicate $T_e \sim 30000K$ for the hot star.

The cool spectrum dominates longwards of $\sim 4300A$. Keyes and Plavec found an effective temperature of 3570K, corresponding to an M 1.7 spectral type. If the luminosity class is III a distance of 500pc and a radius of 56 $R_\odot$ is derived for the cool star.

REFERENCES

Allen, A.D.: 1979, Changing trends in Variable Star Research, IAU Coll. No. 46, F.M. Bateson, J. Smak and I.H. Urch (eds.), University of Waikoto, Hamilton, p. 125

---

* Since the original report of C.D. Keyes has not arrived in time to be included in the Proceedings, a short abstract was prepared by the Editors partly based on notes taken during Keyes' talk.

Boyarchuk, A.A.: 1967, Soviet Astronomy AJ, 11, 8 (Astr. Z. 44, 12)
Gallagher, J.S., Holm, A.V., Anderson, C.M., Webbink, R.F.: 1979, Astrophysical J. 229, 994
Ghigo, F.D., Cohen, N.L.: 1981, Astrophys. J. 245, 988
Hutchings, J.B., Cowley, A.P., Redman, R.O.: 1975, Astrophys.J. 201, 404
Keyes, C.D., Plavec, M.J.: 1980, The Universe at Ultraviolet Wavelengths, R.D. Chapman (ed.), NASA CP-2171, p. 443
Mattei, J.A.: 1981, North American Workshop on Symbiotic Stars, R.E. Stencel (ed.), JILA and University of Colorado, p. 31

DISCUSSION ON AG PEGASI

Friedjung: Would you expect to see X-ray emission? Would it not be expected to be absorbed by the strong wind?

Keyes: It is true that the densities we calculate are rather high (10^9–10^{10} cm^{-3}), although these are for the intercombination line region and not the wind. May be one might expect some X-ray absorption at these densities, but we might expect to see the ionization effects (OVI and other high ionization species) of this absorption in the wind and we do not.

Kafatos: I would like to point out that the presence of absorption in the NIV] profile is very interesting because it would imply regions close to the source of the continuum, but where densities are not excessively high, e.g. $N_e \sim 10^{10}$ cm^{-3} for a scale size of $L \sim 2 \cdot 10^{12}$ cm.

Viotti: As far as the CIV doublet is concerned showing in AG Peg the sharp emission component at 1550 A much stronger than the 1548 A one, if you have an absorption region moving at ~ 500 km s^{-1} the strongest emission line at 1548 A is strongly depressed due to blending with the absorption component of the other line at 1550 A. The same effect has been for example seen in the MgII doublet of η Car.

Keyes: Yes, I agree that such a mechanism could give the observed diminution of the 1548 A emission. However, the 1548 A component of your absorption region should be observable in an image exposed long enough to reach the continuum. We have only one sufficiently deeply-exposed image and we shall certainly look for this blue shifted 1548 component.

THE PECULIAR STAR RX PUPPIS

M. Kafatos
Department of Physics, George Mason Univ., Fairfax, Va., U.S.A.
A. G. Michalitsianos
Laboratory for Astronomy and Solar Physics, NASA, Goddard Space Flight Center, Greenbelt, Md., U.S.A.

ABSTRACT

We have obtained the first high dispersion observations of RX Puppis in the wavelength region 1200 - 3200 A with the "International Ultraviolet Explorer" (IUE). The anomalies we observed in lines such as He II, C III], C IV, N III], N IV], O III], and Si III], that show split line profiles, Doppler displaced component(s) suggest dynamic activity in circumstellar material that probably has the form of rings and/or gas streamers between the cool giant and the hot companion. The Mg II lines show P-Cygni structure arising in the Mira primary. The continuum cannot be due to a star earlier than A0 II and it may arise in an accretion disk around the hot secondary. Moreover, the line emission requires photoionization either from a hot subdwarf or the inner accretion disk.

1. INTRODUCTION

RX Pup is classified as a symbiotic variable in the "General Catalogue of Variable Stars". As noted by Swings (1970), no evidence of a late-type star was found in its spectrum. Observations by Sanduleak and Stephenson (1973) revealed low-excitation emission lines. An M star could not be confirmed in slit spectra taken by Swings and Klutz (1976) and Klutz, Simonetto and Swings (1978). Swings and Klutz (1976) suggested that RX Pup could be compared to η Carinae or slow-novae like objects. Feast, Robertson and Catchpole (1977) detected variability in the 1-4μ region and argued from this that a Mira is present in RX Pup. Klutz et al. (1978), however, discounted the presence of an M-type star and argued instead that the optical continuum is that of a late B or early A-type giant. The existence of a Mira has been confirmed by the presence of the steam absorption band at 1.9μ (Barton, Phillips and Allen 1979). They concluded that RX Pup is a symbiotic star with a Mira primary. Multi-component P-Cygni profiles of the Balmer lines-and strongest Fe II lines-were detected in 1976 (Klutz et al. 1978). Sharp, blue-shifted absorptions ranging up to 1100 km s^{-1} were seen in 1975, 1976 and 1977.

2. OBSERVATIONS AND DATA ANALYSIS

We obtained IUE low and high resolution spectra of RX Pup in the SWP region of the spectrum (1200 - 2000 A) and low resolution LWR spectrum (2000 - 3200 A) on Sept. 20, 1980 and June 11, 1981. A high resolution spectrum in the LWR region of the spectrum was also obtained on June 11, 1981. The spectrum of RX Pup is dominated by strong UV lines superimposed on a weak continuum in the SWP part of the spectrum and a fairly rapidly rising continuum in the LWR part of the spectrum. The strongest SWP lines are Si IV and O IV] at ∼ 1400 A, N IV] at 1486 A, C IV at ∼ 1550 A, He II at 1640 A, O III] at ∼ 1666 A, N III] at ∼ 1750 A, Si III] and C III] at 1892 and 1909 A respectively. Weaker lines of O I and Si II at ∼ 1305 and ∼ 1816 A, respectively are also seen. The LWR lines are primarily He II at 2511, 2733 and 3203 A, the Bowen fluorescence lines of O III at 3024, 3047 and 3133 A as well as He I at 2829 and 3188 A. The Sept. 20, 1980 observations are discussed in greater detail elsewhere (Kafatos, Michalitsianos and Feibelman 1981). We found that the continuum remained essentially the same over the two dates. The lines, though, varied somehow the greatest changes being in the Mg II (∼ 3 times increase) and the O III 3133 A line (∼ 2 times increase in the latest observations) the overall fluxes being in the range of ∼ 10^{-11} ergs cm^{-2} s^{-1} for the strongest features. The

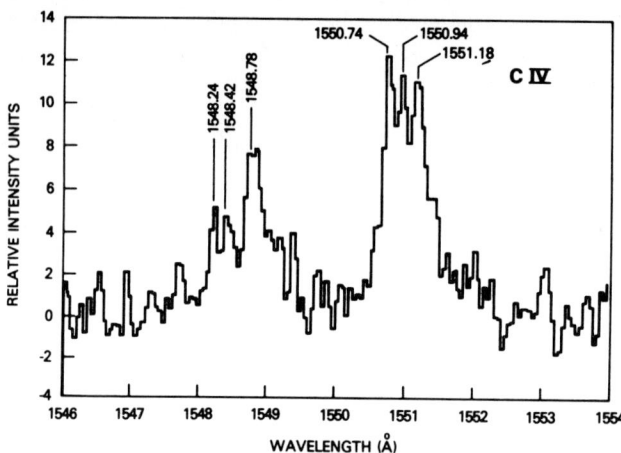

Figure 1. C IV profiles of Sept. 20, 80

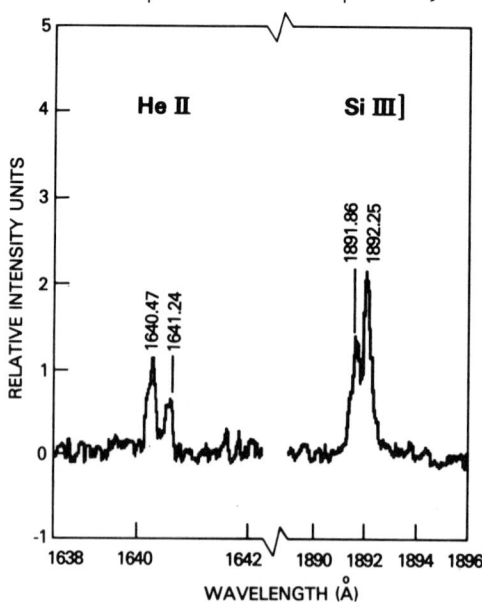

Figure 2. He II and Si III profiles of Sept. 20, 80

high resolution line profiles show complex structure and demonstrate the importance of such observations. Even though the low resolution line emission is similar to other symbiotics such as RW Hydrae (Kafatos, Michalitsianos and Hobbs 1980) the high resolution line profiles are unique. The C IV doublet observed on Sept. 20, 1980 shows components at the rest wavelengths of the two lines as well as two components redshifted from the rest wavelengths by $\sim$ 40 and 90 km s^{-1} (Figure 1). The June 11, 1981 profiles show the same overall flux but significant variations in the detailed profiles, the 1548 A component being much stronger at the latest observation. The He II line shows two components, the strongest centered near the rest wavelength, the weaker redshifted by $\sim$ 170 km s^{-1} (Figure 2). The O III] and Si III] (Figure 2) lines show similar structure, two components located more or less symmetrically on either side of the rest wavelength. The C III] shows a single component while the N III] and N IV] profiles are asymmetric, the red wing dropping abruptly. Larger velocity separations are associated with higher ionization stages. These results were substantially the same in the latest observations. The Mg II profiles obtained on June 11, 1981 are the only ones showing P-Cygni structure in the UV. The separation of the emission from the absorption component is $\sim$ 80 km s^{-1}, this velocity being of the magnitude expected in the Mira primary. Finally, the N V doublet was unseen in Sept. 1980 and became observable in June 1981, indicating an increase in the level of ionization.

Klutz et al. (1978) obtained a value of $\sim$ 1 kpc for this object and E_{B-V} in the range 0.7 - 1.0. Feast et al. obtained $E_{B-V} \sim 0.3$. We estimated values in the range 0.4 - 1.1 using theoretical line ratios of He II (Seaton 1978) and the O III fluorescence intensities (Osterbrock 1974). We find that the UV continuum mimics that of an early F to an early A-type depending on the value of E_{B-V}, but not earlier than A0, contrary to the contention of Klutz et al. (1978) about the existence of a B-type star. For E_{B-V} = 0.7, and parameters appropriate to the UV line emission region (T_e = 15,000 K, $n_e \sim 10^9 - 10^{10}$ cm^{-3}, $L \gtrsim 10^{13}$ cm) free-free continuum and bound-free continuum make up a substantial contribution to the total observed UV continuum.

We have carried out detailed theoretical analysis of the observed semiforbidden line ratios of C III], N III], N IV], and O IV] (cf. Nussbaumer and Schild 1979). We also used the results of Kafatos and Lynch (1980) for the forbidden lines, which-if present at RX Pup-are very weak. The results are presented in great detail in Kafatos et al. (1981) and are summarized here. We find that electron densities in the line emitting region are $n_e \sim 10^9$ - a few$\times 10^{10}$ cm^{-3} and the size of the line emitting region $L \gtrsim 10^{13}$ cm. The electron temperature is harder to estimate but we find upper limits of 30,000 K at least where the semiforbidden lines and He II $\lambda\lambda 1640$ arise. Very likely $T_e \sim 15,000$ K. We find that carbon is depleted by a factor of $\sim$ 7 similarly to R Aqr (Michalitsianos, Kafatos and Hobbs 1980) and RW Hya (Kafatos et al. 1980). The line analysis implies photoionization in RX Pup, similarly to what Nussbaumer and Schild (1981) find for V1016 Cygni.

3. DISCUSSION AND CONCLUSIONS

The Bowen resonance-fluorescence lines of O III can be used to estimate the number of photons below 228 A (He II Ly continuum photons). We find this number to be 1/10 - 1/30 of the hydrogen Ly continuum photons. The source of the photoionization may be a hot subdwarf or an accretion disk/boundary layer, or a combination of the two.

We find that when we compare the stellar flux F_ν-estimated from our UV line analysis-with theoretical stellar models (e.g. Hummer and Mihalas 1970) that the effective temperature of the secondary star is in the approximate range 75,000 K $\lesssim T_{eff} \lesssim$ 100,000 K and 4.5 $\lesssim \log g \lesssim$ 6.0. Such a star would be located in the central stars of planetary nebulae region (Kafatos et al. 1980).

The photoionizing photons may alternatively arise from an accretion disk and its inner boundary layer region. The velocity structure of the line profiles is suggestive of rings of material around and/or streamers onto a secondary. An accretion disk model may not be applicable at all times in RX Pup but it presents an interesting theoretical limit. We have derived a number of useful relations that can be used to obtain the relevant disk accretion parameters from a number of quantities which are in principle observable (Kafatos et al. 1981). We find that the boundary layer temperature (cf. Bath et al. 1974) is in the approximate range 90,000 K - 110,000 K. Accretion rates onto the secondary are in the range $\sim 10^{-6}$ - a few$\times 10^{-5}$ $M_\odot$/yr and the secondary is very likely a main sequence star. To this date the best candidate of a main sequence secondary accreting material from the M-type primary is CI Cygni (Bath and Pringle 1980; Stencel et al. 1981). Future observations of RX Pup will help in determining whether an accretion disk is required.

REFERENCES

Barton,J.R.,Phillips,B.A.,and Allen,D.A.:1979,M.N.R.A.S. 187,p.813.
Bath,G.T.,et al.:1974,M.N.R.A.S. 169,p.447.
Bath,G.T.,and Pringle,J.E.:1980,M.N.R.A.S. 194,p.967.
Feast,M.W.,Robertson,B.S.C.,and Catchpole,R.M.:1977,M.N.R.A.S. 179,p.499.
Hummer,J.B.,and Mihalas,D.:1970,M.N.R.A.S. 147,p.339.
Kafatos,M.,and Lynch,J.P.:1980,Ap.J.Supp. 42,p.611.
Kafatos,M.,Michalitsianos,A.G.,and Hobbs,R.W.:1980,Ap.J. 240,p.114.
Kafatos,M.,Michalitsianos,A.G.,and Feibelman,W.A.:1981,Ap.J. submitted.
Klutz,M.,Simonetto,O.,and Swings,J.P.:1978,Astron.Astroph. 66,p.283.
Michalitsianos,A.G.,Kafatos,M.,and Hobbs,R.W.:1980,Ap.J. 237,p.506.
Nussbaumer,H.,and Schild,H.:1981,Astron.Astroph. in press.
Nussbaumer,H.,and Schild,H.:1979,Astron.Astroph. 75,p.L17.
Osterbrock,D.E.:1974,Astrophysics of Gaseous Nebulae (San Francis.Freeman).
Sanduleak,N.,and Stephenson,C.B.:1973,Ap.J. 185,p.899.
Seaton,M.J.:1978,M.N.R.A.S. 185,p.5P.
Stencel,R.E.,et al.:1981,Ap.J.Lett. submitted.
Swings,P.:1970,Spectroscopic Astrophysics (Los Angeles,Univ. Calif.press).
Swings,J.P.,and Klutz,M.:1976,Astron.Astroph. 46,p.303.

LONG TERM TRENDS IN THE 3.5μ LIGHT CURVE OF RX PUPPIS

Patricia A. Whitelock and R.M. Catchpole
South African Astronomical Observatory

SAAO photometry of RX Pup at 3.5μ between 1972 and 1981 shows a very well defined period of 580 days with an amplitude of 0.35 mag, which is attributed to a Mira variable component. The infrared variability of RX Pup was demonstrated by Feast, Robertson and Catchpole (1977), who also pointed out that the infrared colours were indicative of the presence of a Mira variable with a dust excess. Further evidence for the Mira comes from the observations of H_2O in the infrared spectrum by Barton, Phillips and Allen (1979).

In addition to the 580^d period, secular variations are also evident in the 3.5μ data. Between 1972 and 1980 RX Pup brightened steadily at a rate of 0.09 mag per year. The more recent results do not follow this trend and in fact RX Pup got fainter by $0.^m15$ between 1980 and 1981. The origins of this variation are not obvious but could be associated with changes in the conditions of the dust cloud, the nature of which may become clear when these observations are compared with those at other wavelengths. It is obviously of considerable importance to continue observations of this and other symbiotic objects over long time-periods and at all wavelengths in order to clearfy the nature of the secular changes. The observations discussed here are to be published in full elsewhere. We are indebted to SAAO colleagues for use of data in advance of publication.

REFERENCES

Barton, J.R., Phillips, B.A., Allen, D.A.: 1979, Mon. Not. R. astr. Soc. <u>187</u>, 813

Feast, M.W., Robertson, B.R.C., Catchpole, R.: 1977, Mon. Not. R. astr. Soc. <u>179</u>, 499

DISCUSSION ON RX PUPPIS

<u>Swings</u>: I wish to discuss the spectrum of RX Pup in the 1970's (B[e] - type), and its evolution from March to December 1979. RX Pup seems indeed to be returning rapidly to the conditions exhibited four decades ago; the strong and broad emission lines of HeII, NIII, OII, [OIII] that appeared in that short period indicate an appreciable increase of the degree of excitation. The Balmer continuum is now strongly present in emission. For details see Klutz and Swings (1981, Astr. Astrophys. <u>96</u>, 406).

<u>McCarthy</u>: How do the trends of the H, J, and K magnitudes for RX Pup compare with that shown with time by the L magnitudes?

<u>Whitelock</u>: The trends are similar but there is no sign of the long term variation.

<u>Viotti</u>: I think that the model on RX Pup based on the UV observations is quite reddening dependent. Did you determine $E(B-V)$ from the strength of the 2200 A interstellar band?

<u>Kafatos</u>: The continuum changes a bit when you go from $E(B-V) = 0.3$ to $E(B-V) = 0.7$; for values outside this range you get unreasonable looking UV continua.

HM SAGITTAE - A MOST REMARKABLE STAR

Sun Kwok
Herzberg Institute of Astrophysics
National Research Council of Canada
Ottawa, Canada K1A 0R6

HM Sagittae is one of the most unusual objects in the Galaxy for it displays activity in every spectral band from x-ray to radio. Its present variable-star designation was given after the discovery of its optical brightening from 16^m to 12^m between April and September 1975 (Dokuchaeva and Balazs 1976). It was soon found to have a rich emission-line spectrum similar to that of a planetary nebula (Stover and Sivertsen 1977). Post-brightening monitoring of the object by Ciatti, Mammano and Vittone (1977, 1978) found the B and V magnitudes to be variable with amplitudes of at least one magnitude. Evidence for increasing excitation was found by Ciatti, Mammano and Vittone (1979) with HeII 4686 emerging in October 1978. Wolf-Rayet features of velocities up to 2000 km s^{-1} have also been seen (Belyakina, Gershberg and Shakhovskaya 1978, 1979; Brown et al. 1978; Ciatti et al. 1978; Wallerstein 1978; Allen 1980; Andrillat and Swings 1982). Analysis of the forbidden line ratios gives an estimated nebula density of 10^6-10^7 cm^{-3} and a gas temperature >10^4K (Ciatti et al. 1977; Arkhipova and Dokuchaeva 1978; Davidson, Humphreys and Merrill 1979; Arkhipova, Dokuchaeva and Esipov 1979).

Photospheric absorption bands of CO and H_2O have been detected by Puetter et al. (1979), suggesting the presence of an M giant. Thronson and Harvey (1981) confirmed these absorption features in 1979 but found the CO feature missing entirely in 1980. Thronson and Harvey also detected 8 members of the Brackett series in the 1.5-2.3 μm region. Using standard extinction curves, they derive a visual extinction (A_v) of 12^m. Most of the extinction is likely to be circumstellar, however. Davidson et al. (1978) found a strong infrared continuum with a colour temperature of ~1000 K, together with a 9.7 μm silicate feature. This suggests an oxygen-rich system and the existence of a stellar wind from the cool component. Since most of the energy is emitted in the infrared, the total luminosity of the object can be estimated to be ~4×10^3 (D/1 kpc)$^2 L_\odot$. Variations of the infrared continuum were noted by Slovak (1978) and further photometric observations by Taranova and Yudin (1980) show the variations to be consistent with a Mira variable with a period > 1 year.

HM Sge also has a rich emission line spectrum in the ultraviolet (Feibelman *et al.* 1980). Widths of the CIII] lines indicate expansion velocities >120 km s^{-1} (Flower, Nussbaumer and Schild 1979). Narrow spikes observed on the broad CIII] line suggest emission from two separate components (Feibelman *et al.* 1980). Analysis of the optical line profiles by Wallerstein (1979) shows line widths of large disparity, ranging from 35 km s^{-1} for [NII] to 1700 km s^{-1} for Hα. This was interpreted by Kwok and Purton (1979) using an interacting-winds model where the wide permitted lines originate from a wind from the hot component, forbidden lines from the cool-star wind and forbidden lines of high critical densities from a dense shell formed by the interaction of the two winds.

Soft x-ray emission from HM Sge was detected by Allen (1981). He suggests that the x-ray emission has a thermal surface comparable to the size of a white dwarf and is probably excited by accretion or a thermonuclear runaway during the outburst.

Radio emission was first detected by Feldman (1977) in May, 1977. Since then the radio emission has undergone a continued brightening at frequencies between 1 and 15 GHz (see Kwok 1982). The radio spectrum is still optically thick in 1981 but it is expected to become optically thin (first at high frequencies) in the near future. High resolution maps made at the VLA show the object to be elongated along the NE direction (Kwok, Bignell and Purton 1982).

What is the cause of the optical brightening? With increasing evidence for the actual presence of an M giant, it seems unlikely that it is the result of the unveiling of the hot core of an M giant as suggested by Kwok and Purton (1979). It is possible that the brightening is the result of ignition of nuclear burning after a period of sustained accretion due to an increase in mass loss rate from the M giant. Willson (1981) suggests that such a sudden change in mass loss rate can be caused by the onset of Mira pulsation while the star is ascending the asymptotic giant branch. If wind accretion is the dominant mode of accretion then one must also be concerned about whether the accretion process would be disrupted by the possible initiation of another stellar wind from the compact component. If the accretion process is to remain steady, then the wind from the hot component cannot be spherically symmetric. A possible model is that accretion occurs on the plane of the accretion disk while the hot-star wind is concentrated only in the polar directions.

In summary, HM Sge is probably best described by a binary system consisting of a late M giant and a hot compact object which is similar to central stars of planetary nebulae. The presence of a wind from the M giant implies that Roche-lobe overflow is not a necessary condition for mass transfer. The complex structure of the circumstellar nebula is possibly the result of wind interactions. The ongoing spectral evolution of HM Sge after its recent outburst makes it an ideal candidate to test models of the symbiotic phenomenon.

REFERENCES

Allen, D.A. 1980, *Mon. Not. Roy Astron. Soc.*, 190, 75.
—— 1981, paper presented at the North American Workshop on Symbiotic Stars, Boulder.
Andrillat, Y. and Swing, J.P. 1982, this volume.
Arkhipova, V.P. and Dokuchaeva, P.D. 1978, *Sov. Astron. Letters*, 4(1), 48.
Arkhipova, V.P., Dukuchaeva, P.D. and Esipov, V.F. 1979, *Sov. Astron.*, 23(2), 174.
Belyakina, T.S., Gershberg, R.E. and Shakhovskaya, N.I. 1978, *Sov. Astron. Letters*, 4(5), 219.
—— 1979, *Sov. Astron. Letters*, 5(6), 349.
Brown, L.W., Feibelman, W.A., Hobbs, R.W., McCracken, C.W. *Astrophys. Letters*, 19, 75.
Ciatti, F., Mammano, A. and Vittone, A. 1977, *Astron. Astrophys.*, 61, 459.
—— 1978, *Astron. Astrophys.*, 68, 751.
—— 1979, *Astron. Astrophys.*, 79, 247.
Davidson, K., Humphreys, R.M. and Merrill, K.M. 1978, *Astrophys. J.*, 220, 239.
Dokuchaeva, P.D. and Balazs, B. 1976, *Astron. Tsink.*, No. 929, 1.
Feibelman, W.A., Boggess, A., Hobbs, R.W. and McCracken, C.W. 1980, *Astrophys. J.*, 241, 725.
Feldman, P.A. 1977, *J. Roy. Astron. Soc. Canada*, 71, 386.
Flower, D.R., Nussbaumer, H. and Schild, H. 1979, *Astron. Astrophys.*, 72, L1.
Kwok, S. 1982, this volume.
Kwok, S. and Purton, C.R. 1979, *Astrophys. J.*, 229, 187.
Kwok, S., Bignell, C. and Purton, C.R. 1982, in preparation.
Puetter, R.C., Russell, R.W., Soifer, B.T., Willner, S.P. 1978, *Astrophys. J. (Letters)*, 223, L93.
Slovak, M.H. 1978, *Astron. Astrophys.*, 70, L75.
Stover, R.J. and Sivertsen, S. 1977, *Astrophys. J. (Letters)*, 214, L33.
Taranova, O.G. and Yudin, B.F. 1980, *Sov. Astorn. Letters*, 6(4), 273.
Thronson, H. and Harvey, P.M. 1981, *Astrophys. J.*, in press.
Wallerstein, G. 1978, *Publ. Astron. Soc. Pacific*, 90, 36.
Willson, L.A. 1981, paper presented at the North American Workshop on Symbiotic Stars, Boulder.

SPECTRAL EVOLUTION OF HM SAGITTAE IN THE ULTRAVIOLET

Ch. Kindl and H. Nussbaumer
Institute of Astronomy, ETH-Zentrum,
CH-8092 Zürich, Switzerland

We have observed the ultraviolet spectrum of HM Sge with the International Ultraviolet Explorer during 1978-1980. The low resolution spectra are shown in Figure 1.

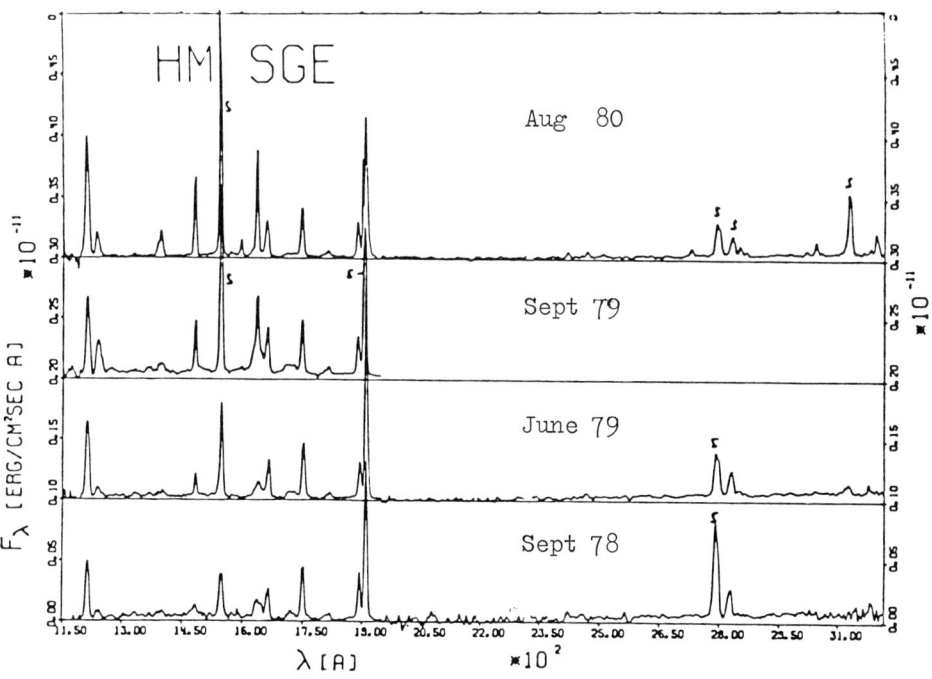

Figure 1. The low resolution ultraviolet spectrum of HM Sge during 1978-1980.

There are strong changes over the two years. The fluxes of OIII] $\lambda 1664$, NIII] $\lambda 1750$, SiIII] $\lambda 1892$, CIII] $\lambda 1908$, and the CII/OIII blend at $\lambda 2323$ have remained approximately constant, but the blend of OIV] and SiIV at $\lambda 1400$, NIV] $\lambda 1485$, CIV $\lambda 1549$, and HeII $\lambda 1640$ increased their fluxes by approximately a factor 3, so did the Bowen line at $\lambda 3133$. NV $\lambda 1240$ decreased after an initial increase. The change was particularly rapid between June and September 1979.

The interpretation of these spectra has only started, it will rest heavily on high resolution spectra which we took as well. We intend to continue these observations; we hope that well calibrated spectra in other wavelength domains will be taken as well.

DISCUSSION ON HM SAGITTAE

<u>Ciatti</u>: Since the many similarities between HM Sge and V1016 Cyg have been often remarked upon, it is also to be recalled that:
V1016 Cyg showed a very smooth photometric and spectroscopic evolution; HM Sge has presented several phases of higher and lower excitation (in the general trend leading to higher ionization stages; now HeII and some times [FeVII]), and light variations around "maximum" light.

Recent variations in the optical spectrum of HM Sge are reported by Ciatti et al. (1979, Astr. Astrophys. <u>79</u>, 247) and Andrillat, Ciatti and Swings, Astr. Astrophys. submitted. Furthermore, [FeVII] is not recorded during July-August 1981 on the Asiago plates.

<u>Keyes</u>: Dr. Kwok mentioned that Wolf-Rayet characteristics existed in HM Sge. Are these well-defined, do they indicate a WN or a WC type? (This has a potentially important bearing on the evolutionry state of the system from the standpoint of what types of nuclear processing may have occurred).

<u>Kwok</u>: HM Sge has been classified as WN6 by Ciatti et al. (1978, Astr. Astrophys. <u>68</u>, 751), whereas Brown et al. (1978, Ap. Letters <u>19</u>, 75) call it WC.

RR TELESCOPII

Patricia A. Whitelock, R.M. Catchpole and M. W. Feast
South African Astronomical Observatory

Introduction

In 1948 the South African amateur astronomers de Kock and Kirchhoff were the first to notice that RR Tel had brightened to 7th magnitude. Subsequent examination of the Harvard Patrol plates showed that it had in fact reached a maximum in 1944 after rapidly brightening 7 magnitudes (Mayall, 1949). Prior to 1944 the object showed variations with a period of 387 days and an amplitude of up to 2 magnitudes in the blue. After the outburst the spectrum evolved in a manner characteristic of very slow novae: an F5 supergiant absorption spectrum gave way to strong permitted and forbidden emission lines. The ionization levels characterising the emission increased with time. The spectral development has been extensively studied and has been well summarised for the period up to 1973 in the Thackeray's (1977) monograph on the subject. More recent spectra (e.g. Penston et al, 1981) show that although the trend towards species of increasingly higher ionization has probably stopped, the visual and near ultraviolet light is still completely dominated by strong emission lines with a weak blue continuum.

Most if not all slow novae and symbiotic stars are now thought to be interacting binary systems, in which a hot subdwarf (probably a white dwarf) accretes matter from a cool giant. In some cases, including RR Tel, the cool giant is a Mira variable. Such systems have been modelled by various authors (e.g. Bath 1977).

Infrared Photometry

The SAAO infrared (JHKL) photometry of RR Tel from 1975 to 1980 clearly shows variations with a period of about 387 days. This period, which has an amplitude of 0.6 mag at 2.2μ, confirms the suggestion of Feast and Glass (1974) that the RR Tel system contains a Mira. The suggestion is further supported by the observation of TiO in the photographic spectrum (Webster 1974) and H_2O in the infrared (Allen et al. 1978).

Mayall's pre-outburst photographic photometry is quite consistent with the idea that a normal Mira was present in the system before the outburst. Where the amplitude of the variations was low we can simply assume that the photographic region was dominated by emission from the hot binary component. It thus seems probable that the Mira has been present in the RR Tel system throughout its history and that it was not fundamentally affected by the nova outburst.
As well as being a slow nova RR Tel appears to be a typical member of the subgroup of symbiotics which contain a Mira variable and an infrared dust excess. Other symbiotics of this type for which periods have been derived at SAAO are RX Pup (580 days), He 2-38 (431 days) and He 2-106 (176 days). It is notable that these objects have a large range in periods, indicating that Miras are not of a very restricted population type.

It has been suggested (e.g. Feast et al. 1977) that the infrared excess in symbiotic systems such as RR Tel implies the existence in them of larger quantities of dust than are associated with solitary Miras. An alternative explanation, which is consistent with the photometry, is that the hot binary component provides an extra source of excitation, giving rise to a larger infrared flux for the same quantity of dust. Photometry at longer wavelengths (5-20μ) is required to clearfy this problem. The observations discussed here are to be published in full elsewhere. We are indebted to SAAO colleagues for use of data in advance of publication.

REFERENCES

Allen, D.A., Beattie, D.H., Lee, T.J., Stewart, J.M., Williams, P.M.: 1978, Mon. Not. R. astr. Soc. 182, 57p
Bath, G.T.: 1977, Mon. Not. R. astr. Soc. 178, 203
Feast, M.W., Glass, I.S.: 1974, Mon. Not. R. astr. Soc. 167, 81
Feast, M.W., Robertson, B.R.C., Catchpole, R.M.: 1977, Mon. Not. R. astr. Soc. 179, 499
Mayall, M.W.: 1949, Harvard Bull. 919, 15
Penston, M.V., Benvenuti, P. Cassatella, A., Heck, A., Selvelli, P., Beekmans, F., Macchetto, F., Ponz, D., Jordan, C., Cramer, N., Rufener, F., Manfroid, J.: 1981, Mon. Not. R. astr. Soc. in press
Thackeray, A.D.: 1977, Mem. R. astr. Soc. 83, 1
Webster, B.L.: 1974, IAU Symposium No. 59, 123

ULTRAVIOLET OBSERVATIONS OF RR TELESCOPII

D. Ponz, A. Cassatella
Astronomy Division ESTEC, Villafranca Satellite Tracking
Station of ESA, Madrid, Spain

R. Viotti
Istituto Astrofisica Spaziale, CNR, Frascati, Italy

The ultraviolet spectrum of RR Tel was extensively studied with the IUE satellite since 1978 in both the high and low resolution modes. A comprehensive study of these observations was made by Penston et al.(1981) who measured more than 400 emission lines. As it is clearly shown in figure 1, the UV spectrum of this symbiotic star is very rich in emission lines. Like in the optical spectrum (Thackeray 1977), the UV spectrum presents emission lines of ions belonging to a wide range of ionization energies, from neutral up to four and five times ionized species (OV, MgV, CaVI, etc.). Permitted, intercombination and forbidden transitions were found which may allow a diagnosis of the physical conditions of the emitting regions. Electron temperatures of $1.2-1.9\ 10^4$ °K and densities of 10^6-10^8 cm^{-3} were derived by Penston et al. (1981), but higher densities could be inferred from the relative intensities of the NIII] lines (Alta-

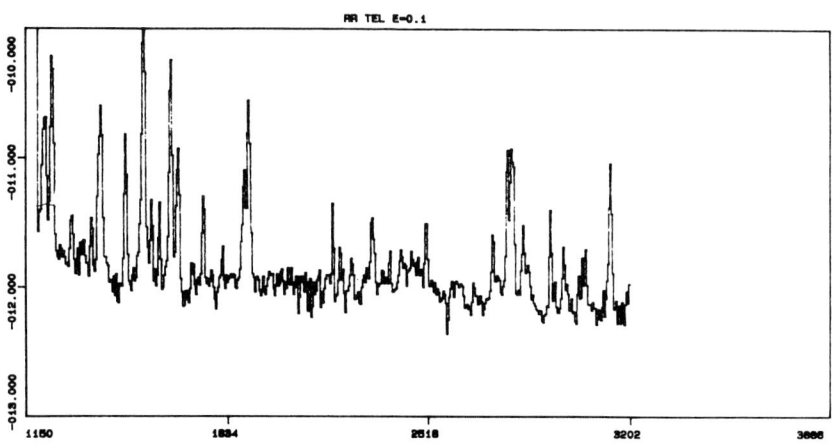

Figure 1. The low resolution ultraviolet spectrum of RR Tel.

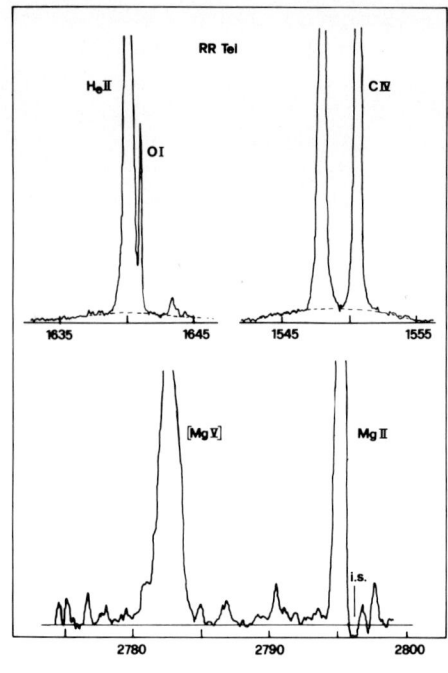

Figure 2.

more et al. 1981). Several FeII emission lines were identified by Penston et al. belonging to resonance and highly excited transitions as well. From the high resolution spectra they also found a systematic increase of the line widths with the ionization energy. A similar result was previously found by Friedjung (1966) and Thackeray (1977) for the optical lines. The optical spectra also show broad emission features resembling the WR line spectra.

Weak shallow emissions are also present in correspondence of the strongest emission lines in the most exposed IUE SW images, while they are less evident in the LW region probably because of the high density of weak emission lines (figure 2). These wings could be of instrumental origin but an inspection of the most exposed calibration spectra taken at VILSPA showed no trace of wings in the most intense lines. The width and relative intensity of the wings seem to be different from line to line, being larger for CIV and HeII and smaller for CIII] λ 1909 line.

Broad emission wings have been observed by Keyes and Plavec (1980) in AG Peg and by Altamore et al. (these proceedings) for the HeII λ 1640 emission line in AG Dra. They could be the result of Thomson scattering in an extended ionized envelope, or of emission from a very hot region or rotating disk. A detailed study of these features in the UV and optical spectra of symbiotic stars is required to improve the current models on these stars.

REFERENCES

Altamore, A., Baratta, G.B., Cassatella, A., Friedjung, M., Giangrande, A., Ricciardi, O., Viotti, R.: 1981, Astrophys.J. <u>245</u>, 630
Friedjung, M.: 1966, Mon. Not. R. astr. Soc. <u>133</u>, 401
Keyes, C.D., Plavec, M.J.: 1980, The Universe at UV Wavelengths, NASA, 443
Penston, M.V., Benvenuti, P., Cassatella, A., Heck, A., Selvelli, P.L., Beeckmans, F., Macchetto, F., Ponz, D., Jordan, C, Cramer, N., Rufener, F., Manfroid, J.: 1981, Mon. Not. R. astr. Soc. in press
Thackeray, A.D.: 1977, Mem. R. astr. Soc. <u>83</u>, 1

V 4049 SGR

M.F. McCarthy S.J.
Vatican Observatory, SCV
B.M. Lasker
CTIO, La Serena, Chile
T.D. Kinman
KPNO, Tucson, Arizona, USA

V 4049 Sgr was detected by the authors as a possible symbiotic star on Curtis Schmidt objective prism plates. This star is Nova Sgr 1978 and was first discovered by Stenholm and Lindstrom. When observed in 1979 with SIT specgrograph on CTIO 4 m telescope the outstanding features were the presence of very strong lines of [FeVII] chiefly at 6085 A where its flux relative to H_α is 1.6. Details have been published in Pub. astr. Soc. Pacific, 93, 470 (1981).

DISCUSSION ON RR TELESCOPII

Slovak: Have Warner or Walker pursued high speed photometry of RR Tel, and if so, what has been found?

Whitelock: I do not think that Walker detected flickering in RR Tel. Penston et al. discussed flickering detected by Stromgren photometry.

Cassatella: I think that these observations need confirmation before we can say that flickering has definitely detected.

Friedjung: I would like to make three comments:
1) The high velocity wings seen in the UV remind me of the Wolf Rayet profiles seen by Webster in the visual. I would suspect this to be due to a wind from the hot component.
2) I object slightly to one talking of a normal D type symbiotic star. The kinematics studied by Thackeray is complex. This is a unique animal!
3) L. A. Willson in a talk given in Boulder concluded that the Mira components of D type symbiotics were quite normal.

Michalitsianos: There appears to be a relationship between mass loss rate and pulsation period in single Miras from infrared work. Accordingly, if symbiotics contain Miras of longer periods, say $P > 350$ days, the interaction between components in the system may be enhanced due to higher mass loss rates.

Kafatos: It is puzzling that in RX Pup in particular, where the 1551/1548 ratio is reversed (about 2:1) with respect to the theoretical ratio, this would indicate the only evidence of broad P Cygni profiles in the IUE range (~ 600 km s^{-1}), i.e. the 1551 P Cygni absorption pulls down the 1548 emission.

THE KUWANO PECULIAR OBJECT (PU VULPECULAE)

T.S. Belyakina, R.E. Gershberg, Yu.S. Efimov, V.I. Krasnobabtsev
E.P. Pavlenko, P.P. Petrov, K.K. Chuvaev
Crimean Astrophysical Observatory
V.I. Shenavrin
Crimean Station of the Sternberg State Astronomical Institute

U, B, V, J, H, K photometry, polarimatry and spectroscopy of this peculiar object was made from August 1979 to December 1980. During 1979 it was at a phase of stable maximum brightness with variations of less than $0^m.2$ in magnitude and less than $0^m.1$ in colour. A change occurred at the beginning of 1980; the object was $0^m.5$ fainter in mid March and then faded rapidly. During the rapid fading the object became redder in U-B and B-V, and there was an intrinsic polarization of about 1 %. At minimum in September 1980, B-V become even bluer than at maximum, while the 1.2–2 μ flux was twice lower in October 1980 than in April 1979. At maximum the spectrum was of type F; the BaII lines indicating a luminosity higher than that of a normal supergiant. In autumn 1979 H_α had a narrow absorption core superposed on a broad shallow base, not present for H_β. During the rapid fading, the spectrum became of M type with clear emission in H_α. The spectral type was M5 or M6 in summer 1980, and M4 in autumn 1980. From our spectra we cannot exclude the dwarf and supergiant luminosity classes. The equivalent width of H_α decreased by a factor of two from summer to autumn 1980, while H_β emission became absorption. [NII] and [OIII] lines appeared, and NaI emission increased by a factor of 3. Radial velocities had not changed by more than 20–30 km s^{-1} from 1979.

We consider PU Vul to be binary with a late M giant and a component of probably low luminosity. The latter flared in 1979, and the former also changed a little. From the photometric, spectroscopic and polarimetric data we estimate $E(B-V) \lesssim 0.4$ and locate the star at 5–7 kpc from the sun and 0.7–1.0 kpc from the galactic plane. The photometric behaviour, duplicity, and nature of the components are similar to those of novae and symbiotic stars, but no evidence for an ejected envelope is seen.

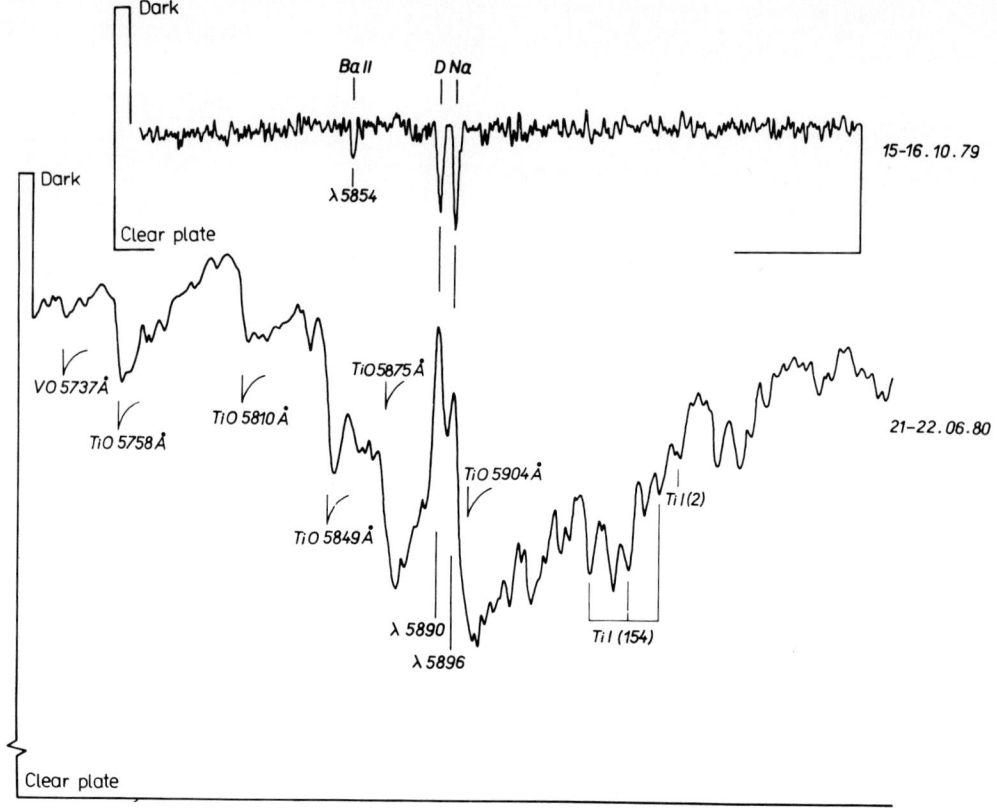

Figure 1. The spectrum of PU Vul at maximum and minimum luminosity. Note the strong emissions of NaI doublet during minimum.

DISCUSSION ON PU VULPECULAE

Friedjung: Infrared observations made at maximum by Bensammar et al. (1980 Astr. Ap. <u>83</u>, 261) showed the presence of a cool component even then. Actually, this is a very funny object. People who observed it keep thinking they have seen the wrong star, because its spectrum is so normal compared with those of novae and symbiotic stars.

Viotti: IUE observations of the LW spectrum of PU Vul made by Cassatella and Ponz at VILSPA on August 3 this year (when the star was again at maximum) show a normal A9 spectrum, while they failed to detect it in August 1980, during minimum phase.

SESSION III

INTERPRETATION

Chairman: H. Nussbaumer

Introductory reports on:

THE TERM SYMBIOTIC STAR (A.A. Boyarchuk)
BINARITY (M. Plavec)
MODELS (M. Friedjung)

Symbiotic Stars: Outburst of one Component.

DETERMINATION OF THE TERM SYMBIOTIC STAR

A. A. Boyarchuk
Crimean Astrophysical Observatory
U.S.S.R. Academy of Science

Symbiotic stars have been studied for many years. But the common determination of the term "symbiotic star" does not exist even now. Merrill (1958) introduced this term in order to emphasize unusual spectral features - absorption TiO bands and emission lines belonging to highly ionized ions.

If we look on the eruptive stars spectra we well see absorption and emission features in the spectra of many stars which we cannot certainly consider as symbiotic stars. For example the U Gem type stars have absorption and emission features in their spectra. A similar situation exists in the case of the old novae. The T Tau type stars have many emission lines, and their spectral type corresponds to G-K. Even the long period variables have some emission lines though their spectral type is M. On the other hand, there are also the BQ[ ] type stars which are characterized by the presence in their optical spectrum of forbidden emission lines and of a rather hot absorption spectrum.

I believe that most of the astronomers that are studying non-stable stars, do think that U Gem, T Tau, old novae and BQ[ ] stars do not belong to the category of the symbiotic stars. The main reason is the fact that symbiotic stars have emission lines with higher ionization degree.

It appears therefore suitable to propose the following criteria for the use of the term symbiotic star:

> The symbiotic stars must have a spectrum which simultaneously present the cool star features (TiO bands or G-band, etc.), and the emission lines of HeII and/or [OIII], and/or [NeIII], and lines which require even higher ionization level.

Of course there are other types of observations which could be considered, but they can be used to divide symbiotic stars into different subclasses. As a first approximation, it can be proposed the following classification of symbiotic stars according to different types of observations:

1. According to UBV photometry:

    Z-stars, with light curves as Z And
    A-stars, with light curves as AG Peg
    R-stars, with light curves as R Aqr

2. According to the infrared observations:

    S-star, whose infrared spectrum corresponds to that of a cool star
    D-star, whose infrared spectrum corresponds to that of dust emission
    N-star, with constant infrared radiation ($\Delta K \lesssim 0^m.1$)
    V-star, with variable infrared radiation ($\Delta K \gtrsim 0^m.1$)

3. According to the radio observations:

    E-star, having a detectable radio emission ($\sim > 10$ mJy)
    Q-star, with no radio emission ($\sim < 10$ mJy)

4. According to the absorption spectrum:

    M-star, with M-type absorption spectrum
    Y-star (yellow), with F, G, K-type absorption spectrum

5. According to the emission spectrum:

    a) the degree of excitation which is indicated by the average ionization potential of the emission lines (Allen 1979)
    b) relative intensity of the recombination and forbidden lines which is an indication of the electron density. Allen (1979) has proposed four classes: l - low, m - middle, h - high, and e - extremely high electron density.

The amount of the ultraviolet and X-ray observations is not enough at present to permit any classification.

I think that the above groups of symbiotic stars are not independent. For example D-type stars (having dust) show infrared variations (V-type). They are also radio sources (E-type) and usually have lower electron density (l-type). I think that the future investigation will permit to establish a small number of independent classes of symbiotic stars.

REFERENCES

Allen, D.A.: 1979, Proc. IAU Coll. No.46 "Changing Trends in Variable star Research", F.M. Bateson, J. Smak, I.H. Urch (eds), of Waikato, Hamilton, p.125.
Merrill, P.W.: 1958, "Etoiles à Raies d'Emission", 8th Coll. Liège, p.436.

DISCUSSION ON THE TERM SYMBIOTIC STAR

Whitelock: When we are trying to ascertain the presence of a cool star, then we should not restrict ourselves to the photographic region. The presence of H_2O or CO in the infrared spectrum is just as good an indication of the presence of a cool star as is TiO or the G Band.

Slovak: We need a luminosity class criterion in addition to a spectral class criterion, since most late type components in symbiotics are believed to be giants, as opposed to dwarfs or supergiants.

Boyarchuk: The determination of the luminosity class is not so simple a problem in the case of the symbiotic stars. The main reason is that those stars are faint and their spectrograms have low dispersion. Moreover, the presence of both line and continuous emission gives additional difficulties. As a result only a few symbiotic stars were classified according the luminosity.

Friedjung: The recurrent nova T Coronae Borealis would fit these criteria. The cool component is even a giant. The trouble is that we do not know enough the physics. Ground based spectra also cover too narrow a range.

Huang: Are there any dwarf symbiotic stars? Now we have got giant and supergiant symbiotic stars, being in my opinion AG Dra a supergiant star.

Viotti: I want to recall the fact that AG Dra recently underwent a nova-like outburst after a long period of relative quiescence. Its historical light curve resmbles that of Z And, although the stars are different in many aspects.

Hack: I think we should speak of the symbiotic phenomenon rather than of symbiotic stars (e.g. AG Dra at some epochs was more similar to Z And, and at others to VV Cep). We should base our classification also on the properties of the outbursts like the number of shells observed, the expansion velocity, the mass loss per outburst, the energy emitted during an outburst, the ratio between the energy emitted during an outburst and that emitted between two successive outbursts, the average time interval between outbursts, and the average duration of an outburst.

Slovak: We need to restrict the symbiotic definition, at least spectroscopically, to quiescent phases, since the spectral features change remarkably during and following outbursts.

Kafatos: Should we add the term <u>variability</u> to the definition? If so, it seems to me that symbiotics are characterized by timescales of years or so rather than the much shorter ones which characterize other types of stars, like dwarf novae, etc.

Slovak: The only thing we can say is that variability over timescales of minutes does not exist.

Kafatos: At least in the IUE range we can exclude timescales of hours. We have made a point to observe various symbiotics at the beginning and end of one or two shifts (8-16 hours) and we find no variations over these timescales.

Whitelock: When discussing infrared variability we shoyld refer to large amplitude variation. Our infrared observations indicated that those objects that are often called 'non variable' - the S type symbiotic stars are variable with a low amplitude, sometimes periodically.
In reply to the question: 'is this fundamental, or a matter of degree', I would say that it is fundamental in the sense that the large amplitude variables are Miras, while the others are not Miras.

McCarthy: Regarding the variability of Red Stars, I agree with Dr. Whitelock that it is important to distinguish in the infrared between large amplitude (Mira variability) stars and non large amplitude stars. I know no M type giant where stability greater than 0.1 mag has been established. I add that spectral types seem to show much more constancy.

Michalitsianos: Symbiotics in the ultraviolet present a spectrum very similar to planetary nebulae. The only thing that tells a UV observer that he is in fact looking at a symbiotic star is that someone has found evidence for a cool component in the optical or infrared.

Whitelock: In the infrared the reverse is often the true: they are not different from M giants.

Fehrenbach: The two features: TiO and G band, and the emission lines of NeIII etc. appear during time variations, and do not need to be simultaneous.

Kwok: How many of Allen's 112 objects satisfy our criteria? As I recall, the only criterion he used was an emission line object with some emission.

McCarthy: Here we are at an important stage in our study of these fascinating Symbiotic Stars. Let us try not to be overrestrictive now in assigning "defining" criteria. I agree with Boyarchuk's basic definition - which elaborates correctly Merrill's definition and does not impose excessive limitations on the kinds of objects we may study. In future years wide angle H_α emission surveys will give us more and more objects for study, including Be stars, SS433 objects, T Tauri stars, carbon + M emission stars, Planetary Nebulae, Emission Line Galaxies and Symbiotic Stars. These surveys will reach fainter and fainter limits of apparent red magnitudes. We should not now overclassify these symbiotic objects. I suggest we look back to our predecessors in spectral classification about 100 years ago. Let some of us work as Huggins did in England on a very restricted number of stars whose spectra he studied in great detail; let others of us work as Secchi did discovering numerous emission objects and separating them out on a broader scale and in less detail.
So as we come towards the conclusion of this most interesting and productive conference, we realize more than we did before that we are only at the beginning.

Nussbaumer: We have tried to define what makes a symbiotic star. From the desperate remarks (which alas will not enter the published version) that emerge it is obvious that these discussions will not result in anything like a final definition of symbiotic stars. Could we try to arrive somewhere by way of elimination. Is there some strong feeling among the partecipants that some objects which have been called symbiotic stars should refuse that name?

Michalitsianos: We want to have a formal definition of a symbiotic star because many observers place peculiar emission stars in such a category when they know of no other category. As such, symbiotics have tended to be a collection of odd objects that clearly do not belong to any other astronomical category, somewhat like a rubbish bin!

Houziaux: From the discussion, it appears that a star at a symbiotic phase is a composite object with cyclic variations whose line spectrum looks like that of a planetary nebula in the ultraviolet, and like a K or M giant star in the infrared.

Viotti: I think that the present discussion has a broader astrophysical interest than it would appear from the small number of stars called "symbiotics" here discussed.
Clearly, classification - in Astronomy, or in Biology or else - is a question of methodology: one collects a number of astronomical events and puts them in a few boxes - the O, B, A... stars, novae, T Tauri

stars, etc. — to make some kind of "order" in the multiform astronomical phenomenology. The common error is to assume that stars in the same box have the same physical properties and should belong to the same evolutionary stage. This is certainly not the case of the symbiotic stars which may well represent a collection of objects of different nature.

It is true, on the other hand, that the symbiotic phenomenon is not peculiar to only these stars. It is related to other physical processes, like mass loss/transfer/accretion, stellar coronae etc., that are presently of large astrophysical interest, and that probably are particularly effective in the stars discussed here.

Hence their detailed study may give the key to better understand these processes, rather than to define a new class of stars.

<u>Friedjung</u>: I would like to make a historical note. I proposed one should have a meeting on symbiotic stars 2 years ago in Montreal. However Roberto did not believe symbiotic stars existed and I was at first discouraged. He changed his mind some months later, and then we started to organize this meeting.

<u>Nussbaumer</u>: Perhaps we might after all still add the property which I suggested half jokingly:
It cannot be classified as something else.

THE SYMBIOTICS AS BINARY STARS

Mirek J. Plavec,
Department of Astronomy, University of California
Los Angeles, CA 90024, U.S.A.

ABSTRACT

Symbiotic stars have become an important testing ground of various theories of binary star evolution. Several physically different models can explain them, but in each case certain fairly restrictive conditions must be met, so if we manage to identify a definite object with a model, it will tell us a lot about the structure and evolutionary stage of the stars involved. I envisage at least three models that can give us a symbiotic object: I have called them, respectively, the PN symbiotic, the Algol symbiotic, and the novalike symbiotic. Their properties are briefly discussed. The most promising model is one of a binary system in the second stage of mass transfer, actually at the beginning of it: The cool component is a red giant ascending the asymptotic branch, expanding but not yet filling its critical lobe. The hot star is a subdwarf located in the same region of the Hertzsprung-Russell diagram as the central stars of planetary nebulae. It may be closely related to them, or it may be a helium star, actually a remnant of an Algol primary which underwent the first stage of mass transfer. In these cases, accretion on this star may not play a significant role (PN symbiotic). Perhaps more often, the subdwarf is a "rejuvenated" degenerate dwarf whose nuclear burning shells were ignited and are maintained by accretion of material coming from the red giant in the form of a stellar wind. Eruptions are often inevitable: this is the novalike symbiotic. A third alternative is a system in the first stage of mass transfer, where the photons needed for ionization of the nebula come from an accretion disk surrounding a main sequence star: an Algol symbiotic.

In spite of considerable observational effort, the symbiotics are known so poorly that it is hard to decide between the models, or even decide if all three can actually exist. The theorists seem to be ahead of the observers. Every effort should be made to obtain better information on the components stars of any of the symbiotic systems.

1. INTRODUCTION: LET'S ACCEPT THE HETEROGENEITY OF THE SYMBIOTICS

We hear frequent complaints that the symbiotic stars are a very inhomogeneous group of objects. The first prize in this respect goes to Roberto Viotti, who last year in Trieste (discussion following Plavec, 1981, p. 455) came forward with the statement that symbiotics do not exist, while this year he has organized an excellent three-day colloquium on them with a rather crowded program. His argument, of course, was that the symbiotics do not exist as a homogeneous class of objects. I do not understand why anything like this should be held against them; in fact, heterogeneity makes them much more important, since they can tell us a lot about the evolution of binary stars. I do not want to claim that single-star models are excluded; the heterogeneity may easily go that far. My task is to talk about binary models for the symbiotics, and I believe that any binary system that contains a red giant and at the same time displays evidence of the presence of a much hotter object in the system is worth looking at, whether it satisfies any formal criteria or not.

According to Merrill's original definition, symbiotic stars have combination spectra, in which the high-excitation emission lines regularly found in planetary nebulae are superposed on a low-temperature absorption spectrum. We usually are a little more specific and postulate the co-existence of a He II emission with the TiO absorption bands. By this specification we have selected the most extreme cases from a much more general phenomenon, namely the existence, in a binary system spectrum, of emission lines requiring a hotter radiation source than the one suggested by the underlying stellar continuum. An M star spectrum with superposed emission lines of He II and [O III] is a puzzle; but it is only an extreme case of the puzzle presented by the existence of Balmer emission lines in, for example, the eclipsing binary SX Cassiopeae, whose components have been classified as A6 III + G6 III. This potential relationship of the symbiotics to a much larger group of stars with emission lines should be kept in mind.

In order to construct a symbiotic object, we need a) a source of a late-type continuum, b) a source of circumstellar gas, c) a source of ionizing photons. The two former requirements are satisfied simultaneously if we postulate the presence of a late-type giant or supergiant. Observations of single stars of high luminosity have provided ample evidence of stellar winds from late-type giants or supergiants. The rate of mass outflow generally increases with increasing luminosity and decreasing effective temperature. This makes the M giants prime candidates for membership in the symbiotics, and this is what we actually observe. Only M supergiants produce an even higher mass outflow, and indeed in related stars like VV Cephei, red supergiants are present. There must be some reason why giants predominate in the symbiotics.

2. A SUMMARY OF OBSERVED PROPERTIES.

We agree that most, if not all, symbiotics are binary systems. But coming to the properties of the stellar components of these presumed binaries, how many hard facts about them do we know? Very few indeed, and that must be kept in mind whenever we are tempted to generalize.

For several systems, we know orbital periods. The eclipsing systems are most reliable: AR Pavonis has P = 605 days, CI Cygni has 855 days. Then there are well-observed spectroscopic orbits: AG Pegasi with P ≅ 820 days, T Coronae borealis with 227 days. For other systems the radial-velocity data are less reliable, but in general, periods between 1 year and perhaps 20 years or more are indicated.

Concerning masses, we are even worse off. For AG Peg, Hutchings, Cowley, and Redman (1975) <u>assumed</u> the cool star mass to be 3-4 $M_\odot$, and obtained about 1 $M_\odot$ for the hot star. For AR Pav, again from the radial velocity curve of only the cool component, Thackeray and Hutchings (1974) suggest $M_c \cong 2.5\ M_\odot$, $M_h \cong 1.2\ M_\odot$. In the recurrent nova T CrB, both radial velocity curves were determined by Kraft (1958), and the rediscussion by Paczynski (1965) indicates $M_c \gtrsim 2.2\ M_\odot$, $M_h \gtrsim 1.6\ M_\odot$.

Actually, no object has been analyzed succesfully enough to make it possible for us to present it here as a model case. Since we need to have a model before our eyes, I will describe AG Pegasi. It is questionable if it is a representative symbiotic object, or if a representative symbiotic object exists at all. At least, AG Pegasi may have a representative orbital period, 2.25 years. From a number of discussions of AG Peg, I will attempt a synthesis. If we assume $M_h = 1\ M_\odot$, $M_c = 3\ M_\odot$, then their separation is probably $A \cong 600\ R_\odot = 2.7$ AU. The radius of the critical Roche lobe around the red giant is then about $A \cong 280\ R_\odot$. The temperature of the cool star appears to correspond to about a spectral type M2, and this in turn permits us to express its radius and luminosity as functions of the distance D (in kpc): $R_c = 108$ D, $L_c = 1.7 \times 10^3\ D^2$, $MBOL_c = -3.3 - 5 \log D$. The cool component's luminosity class apears to be III. Then, using the calibration of Lee (1970), $MBOL_c = -1.9m$, $R = 56\ R_\odot$, $D = 0.5$ kpc. If, however, the luminosity class is II, $MBOL_c = -3.8m$, $R_c = 134\ R_\odot$, and $D = 1.24$ kpc, putting the star 0.6 kpc below the galactic plane. Note that even in this case the star remains substantially smaller than its critical Roche lobe (Keyes and Plavec, 1980).

When we first detected the hot continuum in the far ultraviolet spectrum of AG Peg with the IUE, we were very happy to be able to fit it by an atmospheric model with $T_{eff} \cong 30,000$ K. However, the Zanstra temperature derived from the He II lines is considerably higher, close to 10^5 K, and this probably is the actual temperature of the hot star in AG Peg; what we observe with the IUE must therefore be the Rayleigh-Jeans tail of the energy distribution. From the observed flux we conclude that the luminosity of the hot star is $L_h = 6.9 \times 10^3\ D^2\ (L_\odot)$, $MBOL_h = -4.8 - 5 \log D$, and its radius $R_h = 0.28$ D $(R_\odot)$. As we will explain in more detail elsewhere (Plavec and Keyes, in preparation), we prefer D =

0.6 kpc, in which case we have the following parameters of the system: $R_h = 0.16\ R_o$, $MBOL_h = -3.7m$, $R_c = 65\ R_o$, $MBOL_c = -4.4m$.

While we do not know how representative these stellar parameters are for the whole family of the symbiotics, we can at least be more definite about the spectral types of the components. The most homogeneous compilation of atmospheric model fitting to the optical and UV continua of the symbiotics was made by Slovak (1980). In five out of nine objects, he finds that the far ultraviolet continuum can be fitted by a B0 V star. This means that the hot components in these objects are rather similar to AG Peg as described above: while the observed slope of the continuum can be formally fitted by a Kurucz model atmosphere continuum for an effective temperature not far from 30,000 K, the presence of the He II emissions indicates that this continuum is most likely only the Rayleigh-Jeans tail of a much hotter object, whose actual effective temperature probably lies between 6×10^4 and 1.5×10^5 K. In three of the remaining four objects surveyed by Slovak, the continuum in the ultraviolet can be formally fitted by a stellar continuum corresponding to about spectral type A0 V. I think this fit only means that the UV continuum is essentially flat, without any definite trace of the hot component. But such a hot component must be present, since the observed emission lines are pretty much the same as in the previous five cases. The interpretation, already suggested by Slovak, is that the hot star is surrounded by a disk or an envelope which is sufficiently optically thick to produce its own continuum and to obscure the hot star inside. The last of the nine cases, AG Dra, is best fitted in the UV by a B5 V star, and this continuum may well be a superposition of a hot stellar continuum and a disk or envelope.

As to the cool components, Slovak's survey finds an M giant in seven out of nine cases, while in the remaining two symbiotics the cool component is K4. It is worth noticing that all nine cool giants are classified as of luminosity class III --a fact that is not easy to reconcile with evolutionary considerations, as we will see in Section 3. On the whole, we see that AG Peg is reasonably representative. In this object, too, there is an additional near-ultraviolet continuum due to circumstellar hydrogen. There probably exists a continuous transition between systems displaying the hot star continuum clearly, and those in which the circumstellar continuum dominates entirely. The observed range in the sources of the cool continuum is much narrower. Thus it is easier for us to start an attempt at modeling and evolutionary interpretation by a discussion of binaries that can harbor a cool giant.

3. BINARIES WITH A LATE-TYPE GIANT COMPONENT.

An old Latin proverb, applied to the birth of a baby, says: "Mater semper certa, sed pater incertus." We have a similar case in the symbiotics: one of the "parents" is known: the cool component in each symbiotic binary system is a late-type giant of spectral type M, K, or occasionally G. Therefore these components are located on either of the two giant branches of the stellar evolutionary tracks in the H-R diagram, and have deep convective envelopes. Moreover, the observed mass outflow from these stars (necessary to maintain the circumstellar/ circumbinary nebula, and even possibly to generate or stimulate the production of ionizing photons) probably demands that the cool giant either fill its critical Roche lobe, or be at least not very far from it (say its radius should be at least 50% of the critical Roche radius).

What kind of binary systems can contain a star satisfying these conditions? Consider first systems either in the phase of the first mass transfer, or approaching it (depending on whether the late giant fills its critical lobe or not). In the notation I used at the Trieste Colloquium No. 59 on mass loss from stars (Plavec, 1981), these binaries are of type Bc (hydrogen shell burning, contracting He core, deep convective envelope) with the cool star on the first giant branch, or of type Cc (He and H shell burning, CO core, deep convective envelope), with the cool star on the asymptotic branch. In both cases the cool star must be the more massive component.

In order to be more quantitative, assume now that the mass ratio is 3:1 (many binaries have mass ratios closer to 1:1; our choice has little effect on the numbers that follow). Assume that the mass of the primary component is $M_1 = 5\ M_\odot$. Then the primary star will reach its critical Roche surface on the first giant branch if the system's period is $30 \leqslant P \leqslant 55$ days, and on the asymptotic branch if $55\ d \leqslant P \leqslant 3$ years. On the first giant branch, before the mass loss starts, the star will be a G8 -K3 III giant. On the asymptotic branch, it will be late K to M star, with a luminosity class changing from II to Ib.

If the primary is, instead, a $3\ M_\odot$ star, the critical periods for the first giant branch will be $7d \leqslant P \leqslant 25\ d$, and for the asymptotic branch $25\ d \leqslant P \leqslant 16$ years. On the first giant branch, the star is again a G8-K3 III giant. In its long ascent to the point where it ignites carbon --and this point is identical with that for the $5\ M_\odot$ star! (Paczynski 1971a) --it changes all the way from K3 III to something like M5 II-Ib.

It is known (see, e.g., Webbink, 1979) that the probability for a given star to have a less massive companion revolving about it in a period P is approximately 0.14 per decade of the period. Imagine 100 binaries with a $5\ M_\odot$ primary and with periods less than 3 years. Out of this sample, we can expect that 40 will have periods corresponding to case Cc, 8 to case Bc, and 42 to Br (Roche lobe overflow from a radiative envelope, when the primary crosses the Hertzsprung gap). For systems with a $3\ M_\odot$ primary, the odds shift even more in favor of convective mass transfer: out of 100 systems with periods shorter than 11 years, 61 can be expected to start the mass transfer on the asymptotic branch, 15 on the first giant branch, and 24 in the Hertzsprung gap.

However, this does not mean that we will observe such binary systems with primary components located along their evolutionary tracks according to the above probability distribution. The lifetimes at the various phases will affect the observed distribution, and will heavily favor the inconspicous and uninteresting configuration in which the two stars form a wide detached main-sequence system. The odds are only 1 in 4 for the 5 M_o primary (1 in 3 for the 3 M_o primary) that it will be seen outside the Main Sequence band; and 1 in 19 (or 1 in 12, repectively) that it will be caught in one of the phases of rapid expansion. All the evolutionary phases outside the main sequence, particularly the giant phases are relatively very short, except for the core He-burning phase lying between the two giant branches. The question is, though, if we can have a symbiotic object with the primary in the core helium burning phase. A star of 5 M_o makes a loop in the H-R diagram all the way to spectral type about A4 III, where its radius is only 21 R_o. Such a star will not generate a sufficiently strong stellar wind to maintain the circumstellar/circumbinary nebula needed for a symbiotic object, even if we were willing to stretch out the criterion and accept the A giant as the late-type component in a symbiotic. However, stars of 3 M_o or less do not make this blue loop, and remain in the red giant region, although their radius does shrink temporarily when they start core helium burning. I think they remain candidates for symbiotics even at the phase of core helium burning, thereby increasing the probability of catching a binary system with a $\leq$ 3 M_o primary at the red giant stage.

It is better to intercompare the probabilities for the phases of a rapid expansion only. The chances that a system will be seen with the primary in the Hertzsprung gap are about the same as that we will see the primary on the second giant branch (better in favor of the latter for less massive stars). Compared to them, the odds of catching the primary on the first giant branch are 6 to 20 times lower.

In think that this is a most interesting result. It suggests that most of the observed symbiotics should be on the asymptotic branch, and indeed the observed orbital periods very strongly support this conclusion. Actually, no symbiotic is known to have an orbital period of the order of a few months, as it would be appropriate for the first giant branch. Where are the short-period symbiotics? My guess is that they do not look like symbiotics. I think that the dimensions indicated in AG Peg, namely of the order of several AU, are essential, since only this size of the nebulosity enables us to see the prominent forbidden lines so typical for the symbiotics. Also the rate of mass outflow by stellar wind will be considerably higher for the more luminous giant on the asymptotic branch. It is quite possible that the short-period relatives of the symbiotics are the W Serpentis stars, to be discussed later on; if it is so, then we must take this potential relationship into consideration when we discuss the models for both kinds of objects.

However, there is a puzzle here. The stars on the asymptotic branch have luminosities of class II if not Ib, whereas observationally, the symbiotic giants appear to be of lower luminosity class, namely III. The difference is about 2 magnitudes in the absolute visual magnitudes, not negligible! Assuming class III, we get reasonable distances from us as well as reasonable z-distances of the symbiotics from the galactic plane. If we postulate class II for example in T CrB, we get an improbable value of $z = 2.2$ kpc. On the other hand, we must admit that we seldom have a direct evidence from the spectrum, since the criteria

distinguishing classes II and III in M stars lie in the infrared and
are rarely accessible to most observers. It may be that Iben's evolutionary tracks I have been using for this discussion are not suitable,
since they assume conservation of mass; quite possibly the observed
symbiotic giants have already suffered such a large mass loss that they
are now evolving along different tracks, and the usual correlation
between luminosity and surface gravity is broken. In the models
we calculated some time ago for an initially 7 M_o mass-losing giant
(Plavec, Ulrich and Polidan, 1973), the star indeed deviated from the
"conservative" track": its effective temperature decreased so that its
spectral type changed from K to M, and its luminosity strongly depended
on the instantaneous rate of mass loss. It is worrisome to realize
that our estimates of distances, and therefore also of the luminosities
of the hot components, which are based on a luminosity classification
of the red giant, may not be fully reliable since the spectrum might
actually reflect the surface gravity rather than luminosity, and be
affected by the rate of mass outflow from the giant.

Incidentally, the above evolutionary calculations, and those by
Harmanec (1974) suggest that, following a short phase of truly devastating mass loss, a red giant on the first giant branch will embark
on a relatively long phase of quiet evolution during which the mass
loss from its atmosphere is still fairly high to begin with (10^{-4} M_o per
year) but declines quickly, to be replaced by a phase of no mass loss
through the Roche lobe overflow (stellar wind was not considered, but
must of course be present), when the star is burning helium in the core.
The common thing for these two stages is that the star remains in the
red giant region, but is now the less massive component! Such binaries
do indeed exist: the peculiar binary shell stars AX Monocerotis and 17
Leporis have the required properties. They are not symbiotics, however,
although they show mass transfer and interaction between the components.
The hotter components are obviously not hot enough to ionize and excite
the nebulosity. But one cannot exclude the existence of symbiotics of
this type.

The common property of all these systems discussed so far, i.e.
of systems before or at the first phase of mass transfer, is that the
other component is less advanced in its evolution, i.e. in all probability it is a Main Sequence star, and definitely not a degenerate star.
Such a star cannot have sufficiently high effective temperature to ionize the surrounding nebulosity. Therefore the only type of a symbiotic
object that can form in the first phase of mass transfer is the one
advocated by Bath (1977), namely what I will later call the <u>Algol symbiotic</u>: a non-degenerate star surrounded by an accretion disk which is
hot enough in its central regions to simulate a hot star and to produce
the required number of ionizing and exciting photons.

Most definitely, we can have symbiotics near the second stage of
mass transfer, by which term we mean the situation when the initially
less massive star has become the more massive one, and is now ascending
one of the giant branches. Perhaps due to some sort of cosmic justice,
it is now about to return some of the acquired matter back to its mate.

But is the other component ready to accept it? There exists a rather bewildering variety of possible evolutionary tracks leading to quite different configurations for this second phase of mass transfer; some almost unexplored alternatives still obtain, too. I would like to ask the interested reader to study the excellent review by Webbink (1979). Here I will concentrate on the much narrower problem of potential symbiotic systems.

Mass transfer occuring, in the first phase of mass transfer, by Roche lobe overflow when the primary component was crossing the Hertzsprung gap produces the semidetached systems of the well-known Algol type. The mass transfer ends when either helium is ignited in the loser and it detaches itself from the Roche lobe, or when the hydrogen-rich envelope is nearly completely exhausted and the remnant collapses on the helium-rich degenerate core. The latter case occurs for stars initially less massive than about 3.6 M_o, and directly leads to helium white dwarfs with masses below 0.46 M_o. Stars initially more massive than $\simeq$ 3.6 M_o will convert into helium-burning stars. As already mentioned above, helium-burning stars can also be obtained in case Bc. In both cases, hydrogen-rich envelope of a non-negligible mass still exists around the helium-burning core, so the star is not a pure helium star yet. Most unfortunately, practically no evolutionary calculations are available for this phase, so important for us. Paczynski (1971b) computed evolutionary models for pure helium stars, starting with equilibrium configurations on the zero-age helium main sequence. One interesting result is that pure helium models with masses in the range (about) 1 $M_o \leq M \leq 2.6\ M_o$ evolve into the red giant region again when helium burning is shifted into a shell. Thus they may initiate a second phase of mass transfer from the same star, with the other star being practically anywhere along its own evolutionary track -- on the Main Sequence or beyond. In any case, the companion is non-degenerate, so if a symbiotic system is to result, we will again have the formerly discussed Algol symbiotic type, nothing new.

Eventually, all the helium stars less massive than 2.6 M_o should become carbon-oxygen white dwarfs with masses below the Chandrasekhar limit. In view of the fact that probably the masses of the hot components in the symbiotics are not higher than, say, 2 M_o at most, it is not necessary to discuss more massive helium stars. Thus the final stage appears to be always a white dwarf, and in view of the longevity of the white dwarfs, models of symbiotics involving them must certainly be of importance. The white dwarf itself, although often quite hot, is of too low luminosity to produce the necessary quantity of ionizing photons; therefore it must be "rejuvenated" by accretion to such an extent that hydrogen and helium burning shells are ignited and maintained by fresh supply of hydrogen-rich material from the red giant. This is in essence the model proposed first by Tutukov and Yungelson (1976) and further developed by Paczynski and Rudak (1980). This model works on the same principal basis as the conventional model for nova outbursts. At first I thought that it could be called a cataclysmic symbiotic. However, this term may be misleading, since among the cataclysmic variables we include the dwarf novae, where the outbursts are caused by accretion disks, not by a sudden ignition of a nuclear fuel. Therefore, the proper term for these white dwarf symbiotics is novalike symbiotics.

But the helium remnants of Algol binaries should not be forgotten in our discussion. For a certain time, they exist on their own nuclear fuel, and are quite luminous; in fact, Paczynski's models for helium stars with masses of no more than about 2 M_o yield just the right order of luminosities postulated by observations. A critical question is the lifetime of these objects. Paczynski's pure helium models have lifetimes longer than those of red giants.

The relation between the binary orbital period and the relative expectation of finding the other star on the giant branches remains the same for the second phase of mass transfer, since it depends only on the masses. In general, the first phase of mass transfer lengthens the orbital period, so the chances are greater that indeed the other star will be approaching its Roche limit when on one of the giant branches -- more likely again on the asymptotic branch.

Finally, some of the observed symbiotics seem to have periods so long that a Roche lobe overflow is unlikely to occur at all. Then the initially more massive components, provided they are stars of lower to moderate mass, will lose most of their mass by stellar wind and then by a planetary nebula ejection. Such a wide system will be ready to become a symbiotic when the other star reaches its giant stage. The hot component is then a central star of a planetary nebula. It contains a highly degenerate carbon-oxygen core surrounded by an envelope with a helium and a hydrogen burning shells. Mass flows into the core as it burns in the shells, and the envelope is rapidly consumed (Paczynski, 1971c), especially for relatively large core masses (above 1 M_o). Soon the shells die out and the object cools off to the white dwarf stage, unless a steady supply of accreting matter keeps the shells alive. So, while again here is another track possibly leading to the white dwarf, novalike symbiotics, there exists a phase when the hot component is a subdwarf, intrinsically much more luminous than a white dwarf because of its own energy sources, independent of accretion.

4. POSSIBLE MODELS FOR THE SYMBIOTICS

The abovove discussion of the possible symbiotic models was based entirely on their combination spectra. However, actual theoretical models of the symbiotics have been developed with an additional postulate, namely to explain their photometric activity (flares and eruptions). This postulate is probably justified, since eruptive activity on a moderate scale appears to be endemic in the symbiotics. The question is only if the photometric activity is really an inevitable aspect of all symbiotics, or if we are misled by observational selection and overemphasize this characteristic. There exist the so-called BQ[] stars (Ciatti, D'Odorico, and Mammano, 1974), which may be quite similar to the symbiotics, yet do not display any conspicuous eruptive activity. I believe that systems with combination spectra need not be eruptive.

On the basis of the preceding discussion, it appears that we can have three physically distinct physical models for the symbiotics:

(1) A red giant combined with a main sequence star. The ionizing photons are produced in an accretion disk surrounding the main-sequence star. This is the model proposed and developed by Bath (1977), which I called the <u>Algol symbiotic</u>. Mass transfer between the components and accretion on a main-sequence star is the main characteristic of the semidetached binaries briefly called Algols. It is true that in typical Algols, the accretion rate is too low to generate a substantial disk of high central temperature, and that the luminosity generated by accretion is negligible compared to the intrinsic luminosity of the gainer. But all these "typical" Algols must have passed through a stage when the mass transfer was much more significant. We will eventually find some binaries in this rapid phase of mass transfer; indeed, we may already have identified them: β Lyrae is a very likely candidate, and others may be hiding under the label of the <u>W Serpentis stars</u> (Plavec, 1980).

Bath's primary concern were actually the nova-like eruptions. He modeled his type of symbiotics very much alike his model of novae: If the accretion rate becomes supercritical, the disk is disrupted, an induced stellar wind will create an optically thick stellar envelope which expands and gradually thins out. Thus, if the supercritical rates can indeed be accomplished in nature, we will have a "cataclysmic Algol". An important feature of Bath's model is an instability of the red giant component.

(2) A red giant combined with a white dwarf. The ionizing photons are available from the white dwarf since the hydrogen and helium nuclear burning shells have been re-ignited and are maintained by accretion of material coming from the red star. The difference from the previous case is not solely in the different nature of the gainer. In the above case, accretion must occur at a very high rate since it directly generates all the required "hot star" luminosity; in the present case, much lower rate of mass accretion is needed, since it only stimulates the nuclear energy production. As I explained before, these symbiotics may be called the <u>novalike symbiotics</u>.

(3) Finally, the possibility must be considered seriously that the companion to the red giant in a binary system is intrinsically hot and luminous enough, so that no accretion is needed. This would be the simplest, "natural" type of a symbiotic, in which the mass loss from the cool star is needed only for maintaining the nebulosity which is to be ionized by the hot star (and it is not excluded that here, as in the preceding two cases, the hot object may itself contribute to the formation of the nebula). This model would simply require a subdwarf, similar to the nuclei of planetary nebulae. Thus perhaps this type of a symbiotic object (if it indeed exists) should be called a <u>subdwarf symbiotic</u> or a <u>PN symbiotic</u>.

I would now like to discuss these three models in turn.

5. ALGOL SYMBIOTICS: MODELS WITH AN ACCRETING MAIN-SEQUENCE STAR

In an Algol symbiotic, the necessary flux of ionizing photons is produced by an accretion disk, more precisely it originates in the interior part of the accretion disk and predominantly in a hot transition zone in which the gas particles leave the Keplerian orbits and pass through a series of shocks, eventually landing on the surface of the accreting star. Pringle (1977) derived a formula for the temperature of the transition zone, which in solar units reads

$$T = 2.3 \times 10^6 \, \dot{M}^{6/19} \, M^{8/19} \, R^{-18/19} \quad (K) \qquad (1)$$

where the mass M and radius R of the accreting star are in solar units, and the accretion rate is in solar masses per year. The radiation has an approximately blackbody distribution. Now let us assume that the accreting star is a main-sequence object. This specification is sufficient for the crude estimate we need, since there exists a close correlation between M and R for main sequence stars. In other words, this assumption transforms equation (1) into a relation between T and $\dot{M}$ only, while the remaining terms on the right-hand side can be lumped together into a constant. Thus, for main-sequence accreting stars, equation (1) can be approximated by the relation

$$\dot{M} = 3 \times 10^{-20} \, T^{19/6} \qquad (2)$$

For $T = 10^5$ K postulated by the equivalent width of He II λ 164 nm in AG Pegasi and most other well-studied symbiotics, we find from equation (2) that the mass <u>accretion</u> rate must be on the order of 2×10^{-4} M_o/ year. For the Roche lobe overflow, we may equate the accretion rate on the gainer to the mass transfer rate from the loser. But if we assume mass loss from the loser via stellar wind rather than Roche lobe overflow, then we must postulate initial mass outflow rates from the loser to be at least by a factor of 10^2 higher, i.e. about 2×10^{-2} M_o/year. It is obvious that the Algol model of the symbiotics demands mass transfer from the cool star to its mate by means of a directed stream emanating from the first Lagrangian point and due to Roche lobe overflow (in analogy to Algols, which is another reason why I think that the term I am using is not bad).

Many binary systems must be of the type which makes an Algol symbiotic phase possible. We also know that once a giant star with a deep convective envelope reaches the Roche critical surface by its photosphere, an almost catastrophic mass transfer ensues. (Paczynski and Sienkiewicz, 1972; Plavec, Ulrich and Polidan, 1973). Tremendous amounts of gas will be transferred to the other star on a nearly dynamical time scale of the giant. This event will not create a quiescent symbiotic; perhaps it can explain outbursts often observed in symbiotics. Possibly the high mass transfer rates occur in spurts if Bath's model of the envelope instabilities in red giants is correct (Bath, 1972), the gainer forms an optically thick envelope like in Bath's models of novae outbursts (Bath, 1978), and this envelope takes a fairly long time to disperse entirely. Thus the slow-nova outburst of AG Peg could be explained. The small effective radiating area of the hot component in AG Peg (expressed implicitly in the finding that the radius of the hot star is only about 0.16 R_o) is in this model translated into the statement that the hot transition region of the disk is naturally small. The difficulty with the rather small mass of the hot object need not be serious: the mass is actually quite uncertain and can easily be 2 M_o if the giant is ~ 6 M_o; or we can assume that the gainer is a star on the lower part of the Main Sequence, which is almost a necessary postulate, since its intrinsic spectrum does not show. But a very serious objection is that the high photon flux must be maintained, i.e. the high rate of mass transfer must exist now --and there is no evidence of it.

It is most unlikely that this model can explain all symbiotics. If the symbiotics were interacting systems like the Algols, their galactic distribution would be similar, i.e. they would be young disk objects. But observational evidence shows that the distribution of the symbiotics is very much like that of the planetary nebulae, so that most of them must be old disk population objects --in other words, evolved systems probably in the second phase of mass transfer. However, even a single observed Algol symbiotic, positively identified, would be very interesting. Recently, Bath (1981) and Kenyon et al. (1981) suggested that the repetitive outbursts in CI Cygni are accretion-powered, although the nature of the central star is not clear. Another potential candidate for an Algol symbiotic is T CrB, which deserves a more detailed discussion.

6. T CORONAE BOREALIS: A CATACLYSMIC ALGOL-TYPE SYMBIOTIC?

T Coronae borealis is a well known recurrent nova, which erupted in 1866 and in 1946 in two apparently similar, extremely fast outbursts. Unlike typical novae, it is not a short-period binary system consisting of two dwarfs. Its orbital period is 227 days, and the late-type component is an M3 III giant. To further stress the difference from ordinary novae, the mass of the hotter star appears to be definitely above the Chandrasekhar limit. The spectroscopic observations by Kraft (1958) were rediscussed by Paczynski (1965) who obtained $M_h \geqslant 1.6\ M_o$, $M_c \geqslant 2.2\ M_o$.

According to our recent observations at Lick Observatory, this object probably does not now fully qualify as a symbiotic object, if we apply Merrill's criteria. The late-type continuum is there all right, but the optical emission lines are weak (except for Hα); in fact, there may be no He II or O III emissions present at all. At the time of Kraft's observations, i.e. 1956/57, He II λ 468.6 nm and O III λ 376.0 nm were "very feeble", and most likely weakened since then. But in the far UV, we do see the typical emission lines of the symbiotics, He II λ 164 nm and the various intercombination lines which are strong e.g. in

AG Peg and AR Pav. More importantly, the object definitely met Merrill's criteria between 1921 and its most recent outburst in 1946 (see, e.g., Swings and Struve, 1943).

T CrB attracted our attention when we studied the case of Roche lobe overflow in late-type giants (Plavec, 1973). Paczynski and Sienkiewicz (1972) found that Roche lobe overflow from a deep convective envelope leads to an extremely rapid mass loss on a timescale approaching the dynamical timescale. Subsequently, we studied the process in more detail (Plavec, Ulrich and Polidan, 1973), on a giant originally of 7 M_0. When the expanding giant reaches the Roche limit, a rapid adiabatic phase of mass loss sets in and the mass loss rate grows exponentially until it reaches rather unbelievable values, such as 0.1 solar masses per year. When the mass ratio in the system is reversed, the rate slows down considerably to $\dot{M} \sim 10^{-4}$ M_0/year. This phase is then followed by a stage of a still slower mass loss (by another 3 or 4 orders of magnitude), during which the devastated giant still adheres to the Roche lobe but only its outermost atmospheric layers exceed it. Only this phase may be relatively long: it is terminated when the core of the giant ignites helium and the star shrinks, and this occurs quite independently of the amount of mass previously lost from the envelope. Thus, if the giant reaches the Roche lobe near the bottom of the giant branch, the slow phase of mass loss will be as long as a single-star ascent to the red giant tip.

Our calculations were primarily intended to explain the system AX Monocerotis, where a K2 II giant supports a variable circumstellar shell around a B2 IV main-sequence star (Cowley, 1963). The period of the system is 232.5 days, and the mass ratio is $M_c/M_h \simeq 0.4$. Thus if the system ever followed anything like our scenario, it must be in the slow mass loss phase now. Searching for a counterpart in the rapid phase, we found T CrB where the late-type giant is not dissimilar to AX Mon, and the orbital period is identical, 227 days. The mass ratio in T CrB is in favor of the cool star, $M_c/M_h \simeq 1.4$, which must be so for the rapid phase. The two observed outbursts of course suggest rather intermittent mass transfer, perhaps triggered by an instability of the red giant as suggested by Bath (1972). Perhaps, as suggested by Webbink (1978), the system happens to be just in the very short evolutionary phase immediately preceding the onset of the catastrophic mass transfer.

Webbink (1976) studied the outbursts in considerable detail and concluded that the light curve almost demands an explanation in terms of accretion on a main-sequence star. Our recent optical scans, combined with IUE spectra, should enable us to check on these ideas. The IUE spectra show, in addition to a number of moderately strong emissions (C IV, N III], O III], N IV], Si III]), a continuum which would be rather flat if it were not for numerous very deep absorptions. The continuum can be formally fitted reasonably well by a Kurucz model atmosphere with Teff = 11,000 K and log g = 2. It is reasonable to assume that the cool component is a luminosity III giant. With an apparent visual magnitude of 10.15^m and E(B-V)= 0.08^m, the distance to the system is D $\simeq$ 1.35 kpc. Again, we use Kurucz model atmospheres (Kurucz, 1979). The Kurucz model fit for the hotter component then fixes its radius by means of the relation R/D = 0.91, which for the adopted distance yields R_c = 1.2 R_o, a little too small for a B9 star, but not by a large margin. But we cannot boost its radius by assuming a larger distance. If we assume that the red giant is of luminosity class II, that places it at a distance of 3.2 kpc; but with the high galactic latitude of the object, β = 47°, this would place it 2.2 kpc above the galactic plane. Yet the main arguments against the main-sequence interpretation of the FUV continuum are elsewhere. Firstly, we do not observe any Balmer jump which would increase the total light from the system longward of λ 365 nm; our Lick scans show only the light of the M star. Secondly, T_{eff} = 11,000 K is totally inadequate to explain the presence of the emission line λ 164 nm of He II which is definitely present, although considerably fainter than in AG Peg. The reasonably good fit of the FUV continuum by a 11,000 K atmosphere only means that the continuum is to a large degree flat. Most likely, this continuous radiation comes from a disk surrounding a much hotter but also much smaller object.

We can learn something about the nature of the object by using the He II line to determine its Zanstra temperature. The total flux in the line is f ~ 5.5 × 10^{-13} erg s^{-1} cm^{-2}, which translates to a total power emmitted in the line of 0.03 solar lumniosities. In AG Peg, the power is 9 L_o. Assuming T_{eff} = 10^5 K as in AG Peg, we find that in T CrB the hot object should have a radius of only 0.01 R_o as against the 0.16 R_o in AG Peg. Most likely the hot object is cooler than in AG Peg, but we cannot go below about 80,000 K, otherwise there would be no He II line. Taking all the uncertainties, it is possible to adopt a radius several times larger, but it will still remain in the domain of extremely small subdwarfs. It is therefore not surprising that we see no direct evidence of its light: what we observe in the far ultraviolet is only the light of a surrounding disk.

Our preliminary conclusion seems to be in favor of a subdwarf in T CrB, rather than a main-sequence star, as the gainer.

7. THE NOVALIKE SYMBIOTICS: BINARIES WITH REJUVENATED DEGENERATE DWARFS

This appears to be the most popular model nowadays, and you will hear very detailed accounts from Rudak and from Tutukov and Yungelson. I will make only a small remark. The theoreticians often talk about the hot component being a white dwarf in this model. This stirs a number of objections since the surface of a "naked" white dwarf lies in a deep potential well, and this in turn leads to very high temperatures for the accretion disks. Using formula (1) with M_h = 0.6 M_o and R_h = 0.0158 R_o, we find peak disk temperatures on the order of 10^6 K, and therefore we must expect X-rays coming from this gainer, as they indeed do in cataclysmic variables. But actually what the theoreticians are talking about are subdwarfs, namely objects with degenerate carbon-oxygen cores like genuine white dwarfs, but surrounded by a non-negligible hydrogen-rich envelope, which is large and dense enough to stop any infalling material high above the degenerate core, and make the star larger. For example

Tutukov and Yungelson (1976, p. 347) consider a "dwarf star" with a radius about 10^{10} cm, which is 0.14 R_o. This of course is no white dwarf! Thus it is important to realize that when, for example in this volume, the observers consistently talk about subdwarfs and the theoreticians equally consistently talk about white dwarfs, they actually mean the same thing: and the thing should properly be called a subdwarf. Consider a subdwarf with the dimensions we found above for AG Peg: $M_h = 1$ M_o, $R_h = 0.16\ R_o$. If the accretion rate is $\dot{M} = 10^{-7}\ M_o$/year as is often assumed for these objects, then the temperature in the inner parts of an accretion disk may be as high as 8×10^4 K, and the disk's luminosity about 20 L_o. While the disk temperature is not too different from the intrinsic temperature of the gainer, the luminosity of the disk is negligible compared to the gainer's intrinsic luminosity.

As explained by Paczynski and Rudak (1980 and this volume) and by Tutukov and Yungelson (1976 and this volume), the induced nuclear luminosity and photometric behavior of the model is a very sensitive function of the mass accretion rate. High mass influx will convert the hot star into a core of a supergiant; in a narrow range we get a fairly stable hot subdwarf in which eruptions must be due to a variable mass outflow rate of the giant (symbiotics of type I in the notation by Paczynski and Rudak); and for still smaller mass accretion rates, we get hydrogen flashes leading to slow nova outbursts (type II).

8. PN SYMBIOTICS, OR SUBDWARFS UNPOWERED BY ACCRETION: DO THEY EXIST?

But what if the accretion rate is negligibly small? I think such a case is also possible. The mass outflow rate due to stellar wind from late-type giants can be calculated e.g. by means of a formula by Reimers (1975):

$$\dot{M}_c = 4 \times 10^{-13}\ L/\ g\ R \qquad (3)$$

where all the quantities, including the surface gravity g , are in solar units, or by a similar formula given by Mullan (1978):

$$\dot{M}c = 1.6 \times 10^{-9}\ M\ R^{1/2}. \qquad (4)$$

Inserting the values for AG Peg again, $M_c = 1\ M_o$, $R_c = 65\ R_o$, $L_c = 600\ L_o$, we obtain $\dot{M}c = 1.6 \times 10^{-8}$, respectively $1.3 \times 10^{-8}\ M_o$/year. Of this amount, the hot star accretes only 0.7%, so that $\dot{M}_{acc} \cong 10^{-10}\ M_o$/year. This is too low to power the subdwarf in either way. One can argue that all our values for AG Peg are too low, but there certainly is no evidence for an accretion disk in it. Possibly also, the proximity to the Roche critical surface enhances the stellar wind blowing from the symbiotic giants, and the above formulae must be modified for binaries. On the other hand, there are indications that the orbital periods in some symbiotics are as long as 20 years. In such a case the fraction of the stellar wind accreted by the subdwarf is less than 10^{-3}, and even a fairly strong stellar wind will not be able to power the subdwarf.

Yet if such long-period symbiotics with undersize giants do exist, we will have to conclude that the subdwarf is intrinsically sufficiently hot and luminous to provide enough ionizing photons. After all, central stars of planetary nebulae do the job without accretion. And the spacial distribution of the symbiotics is surprisingly similar to that of planetary nebulae (Boyarchuk 1975, Wallerstein 1980). The hot component of AG Peg has the characteristics of a Wolf-Rayet nucleus of a planetary nebula (Keyes and Plavec, 1980). Or the hot component may be a core helium burning star, a remnant of an Algol subgiant. In any case, it is worth looking for these "natural" symbiotics. Their eruptions would not be easy to explain, but do we have to postulate eruptions in all cases? May be the BQ[] stars are of this type.

9. RELATED BINARY SYSTEMS THAT ARE NOT SYMBIOTICS

We can learn somethin about the symbiotics also if we study binary systems that have some similar properties, yet are not symbiotics. I already mentioned AX Monocerotis, in which a less massive K giant is combined with a B1-3 IV star. There is evidence of gas streaming due to Roche lobe overflow in spite of a long period of 232 days. An absorption shell spectrum signals the presence of circumstellar material around the hotter star, and an outflow from that region is evident from P Cygni profiles of the Balmer lines, but otherwise there are no emission lines. Why? Apparently the hotter component is not hot enough, and the mass transfer rate is not high enough for an Algol symbiotic. A related object is the shell star 17 Leporis, which has an appropriate period (260 days) and appropriate giant (M1 III) for a symbiotic, does indeed show the presence of a circumstellar envelope around the accreting star, and some mass outflow from its vicinity is indicated by their violet displacements, but no emission is seen except in one or two Balmer lines. The hotter star, an A6 III giant (Plavec et al., 1981) is again not hot enough, and accretion is insufficient.

The bright star δ Sagittae (Reimers and Kudritzki, 1980) is interesting in this context, because the authors find evidence of an accretion disk surrounding a late B star, whose companion is a luminous M2 II giant, and the system is unusually large, the period being 10 years.

Perhaps most interesting is the fact that the rather flat ultraviolet energy distribution in the eclipsing symbiotic AR Pavonis is very similar to what we see in the eclipsing system RX Cassiopeae (Plavec, Weiland, Dobias, and Koch, 1981). In the optical region, late-type giants dominate: M3 III in AR Pav, K1 III in RX Cas. The hot component is hidden is a disk or envelope, but must be there, since we observe high-ionization emission lines in the ultraviolet. The hot star may be cooler in RX Cas since we do not see He II, only He I in emission. But this may primarily be a density effect. The system of RX Cas, with its period of 32 days, is much more compact, and the absence of most intercombination indicates much higher density ($N_e \simeq 10^{12}$ cm^{-3}) than in AR Pav. RX Cas is in turn similar to other members of the W Serpentis group (Plavec 1980), for example to SX Cas, which contains a K3 III giant. Are these objects transition cases between the symbiotics and the "ordinary" Algols? Are they perhaps quasi-symbiotics with the giants on the first giant branch?

ACKNOWLEDGEMENTS

This work was supported by grants from NASA and NSF. My thanks are due to Dr. C. D. Keyes for collaboration and discussions.

REFERENCES

Bath, G.T.: 1972, Astrohpys. J. 173, 121.
Bath, G.T.: 1977, Mon. Not. R.A.S. 178, 203.
Bath, G.T.: 1978, Mon. Not. R.A.S. 182, 35.
Bath, G.T.: 1981, in Proc. North Amer. Workshop Symbiotic Stars, (ed. R. Stencel), JILA, 20.
Boyarchuk A.A.: 1975, in "Variable Stars and Stellar Evolution", (ed. V. E. Sherwood and L. Plaut), Reidel, 377.
Ciatti, F., D'Odorico, S., and Mammano, A.: 1974, Astron. Astrophys. 34, 181.
Cowley, A.P.: 1963, Astrophys. J. 139, 817.
Harmanec, P.: 1974, in "Late Stages of Stellar Evolution", (ed. R. Tayler), Reidel, 195.
Hutchings, J.B., Cowley, A.P., and Redman, R.O.: 1975, Astrophys. J. 201, 404.
Kenyon, S.J.: 1981, in Proc. North Amer. Workshop Symbiotic Stars, (ed. R. Stencel), JILA, 21.
Kenyon, S.J., Webbink, R.F., Gallagher, J.S., and Truran, J.W.: 1981, to be published in Astron. Astrophys.
Keyes, C.D., and Plavec, M.J.: 1980, in "Close Binary Stars: Observations and Interpretation", (ed. M.J. Plavec, D.M. Popper, and R.K. Ulrich), Reidel, 535.
Kraft, R.P.: 1958, Astrophys. J. 127, 625.
Kurucz, R.L.: 1979, Astrophys. J. Suppl. 40, 1.
Lee, T.A.: 1970, Astrophys. J. 162, 217.
Mullan, D.J.: 1978, Astrophys. J. 226, 151.
Paczynski, B.: 1965, Acta Astron. 15, 197.
Paczynski, B.: 1971a, Acta Astron. 21, 271.
Paczynski, B.: 1971b, Acta Astron. 21, 11.
Paczynski, B.: 1971c, Acta Astron. 21, 417.
Paczynski, B. and Sienkiewicz, R.: 1972, Acta Astr. 22, 73.
Paczynski, B. and Rudak, B.: 1980, Astron. Astrophys. 82, 349.
Plavec, M.J.: 1973, in "Extended Atmospheres etc." (ed. A.H. Batten), Reidel, 216.
Plavec, M.J.: 1980, in "Close Binary Stars: Observations and Interpretation", (ed. M.J. Plavec, D.M. Popper, and R.K. Ulrich), Reidel, 251.

Plavec, M.J.: 1981, in "Effects of Mass Loss on Stellar Evolution",
    (ed. C. Chiosi and R. Stalio), Reidel, 431.
Plavec, M.J., Ulrich, R.K., and Polidan, R.S.: 1973, Publ. Astron. Soc.
    Pacific 85, 769.
Plavec, M.J., Weiland, J.L., Dobias, J.J., and Koch, R.H.: 1981, Bull.
    Amer. Astron. Soc. 13, 523.
Plavec, M.J., Dobias, J.J., Weiland, J.L., and Stone, R.P.S.: 1981,
    published in "Be Stars", (ed. M. Jaschek and
    H.-G. Groth), Reidel; UCLA preprint No. 117.
Pringle, J.E.: 1977, Mon. Not. R.A.S. 178, 195.
Reimers, D.: 1975, Mem. Soc. Roy. Sci. Liege, 6th Serie 8, 369.
Reimers, D. and Kudritzki, R.P.: 1980, in "Second European IUE Confer-
    ence, Tübingen", 229.
Slovak, M.H.: 1980, Bull. Amer. Astron. Soc. 12, 868.
Swings, P. and Struve, O.: 1943, Astrophys. J. 98, 91.
Thackeray, A.D. and Hutchings, J.B.: 1974, Mon. Not. R.A.S. 167, 319.
Tutukov, A.V. and Yungelson, L.R.: 1976, Astrophysics, 12, 342.
Wallerstein, G.: 1980, preprint.
Webbink, R.F.: 1976, Nature 262, 271.
Webbink, R.F.: 1978, in "Changing Trends in Variable Star Research",
    (ed. F.M. Bateson, J. Smak, and I.H. Urch),
    Hamilton, N.Z.: Univ. Waikato, 102.
Webbink, R.F.: 1979, in "White Dwarfs and Variable Degenerate Stars",
    (ed. H.M. Van Horn and V. Weidemann), Rochester:
    Univ. of Rochester, 426.

DISCUSSION ON BINARITY

Kwok: I am not sure we should place too much emphasis on wind accretion. Observations of single-star mass loss show that M-giant winds are always accompanied by dust emission. The implication is that S-type symbiotics have no significant cool-star wind. This is consistent with the correlation of radio emission with D-type infrared excess.
If there are weak cool-star winds in S-type symbiotics, they could have been detected in the radio.

Plavec: I agree that the source of the material might be the hot star. But if it is a general rule, then: (1) why is always an M giant present? (2) Why don't we observe P Cygni emission profiles in a typical symbiotic star? I know only of AG Peg as a case that shows mass outflow from the hot star.

Andrillat: My question concerns the hot component. There exists a very small number of WR stars which are members of symbiotic stars and they are of the WN type. In my opinion it is possible to find also WC types, because among the nuclei of Planetary Nebulae we have both WN and WC types.

Plavec: I thought that the central stars of the PN, if they are of the WR type, tend to belong to the WC subclass. On the contrary, in AG Peg the hot component is WN. Since this is the only surely found WR star among the hot components of the symbiotics, I don't dare to predict what the rule is.

Kafatos: In all fairness to the accretion model of symbiotics by Bath, the high accretion rates that you mentioned ($\dot{M} \gtrsim 10^{-5} M_\odot yr^{-1}$) are only needed to provide the outbursting mechnism, not all the time.

Plavek: Yes. How long will then an object remain a symbiotic? In AG Peg, 120 years has elapsed since the outburst. This time is comparable to the dispersion time of the nebula as estimated by Tutukov and Yungelson.

Rudak: (1) Prof. Plavek mentioned among the arguments against Roche lobe outflow, that orbital periods would be in that case below 100 days. I should propose to cancel it, as in Case C evolution of binaries, Roche lobe outflow can take place.
I would rather emphasize the importance of a possible mass ratio greater than one, favouring the cool component, as in that case one would expect

very rapid mass transfer on an almost dynamical time scale for the giant's envelope.

(2) Let me make some comments on "Algol-type" symbiotic model, which have been developed by G. T. Bath. One should be very careful considering the qualitative picture of instabilities in mass transfer arising in the outer layers of giant's envelope filling its Roche lobe. As was indicated by Wood in 1977, the way in which mass transfer takes place strongly depends on surface conditions accepted on the Roche lobe. What Wood got, was a constant mass outflow in contradiction to Bath's episodic mass transfer with a periodicity of several hundred days.
The influence of the deep convective zone is also of great importance, as Osaki's work indicates.

Viotti: Concerning the binary as opposed to single star models, I would like to recall that there are some generally accepted criteria which may give direct evidence of binarity. They are:
(1) the simultaneous presence of two absorption "photospheric" spectra;
(2) the presence of a "photospheric" absorption spectrum with a periodically variable radial velocity; (3) a light curve characteristic of an eclipsing binary, with minima separated by constant time intervals;
(4) astrometric observations of the apparent orbit of the visible component. Only a few number of symbiotic objects satisfy one or two of these criteria (e.g. CI Cyg, AR Pav), while for other objects we have only indirect evidence for binarity, that in many cases is open to criticism.

Hack: Your binary model explains in a very natural way many symbiotic features. However, I think that at least in some cases a symbiotic spectrum can equally well be explained by a single star model.
An M giant in the transition stage to a planetary nebula can have an instability phase, possibly correlated with the occurrence of the Helium-flash, and the star can emit a shell sufficiently thick to produce a blue-UV continuum. This shell moving in the circumstellar envelope produced by slow wind of an M star will excite the gas by collision, thus producing low excitation features and permitted emission lines.

Cassatella: Is it a general trend for what you called Algol type symbiotics to show the presence (e.g. from line profiles) of an accretion disk?

Plavec: I think that the evidence for a large disk in the Algol symbiotics should indeed be rather obvious — and since I don't see much of it, this only corroborates my doubts about the broad applicability of the Algol concept to the symbiotic stars.

Slovak: The IUE spectra of symbiotic stars do not show absorption lines, with the exception of TX CVn. Thus there is no direct evidence in the UV for a main sequence companion, but argues for a hot subdwarf or a main sequence star obscured in an accretion disk.

Plavec: I agree, although I know now that absorption lines may also form in or outside a hydrogen circumstellar disk or envelope, or in front of a disk. So even if I do see the absorptions of say spectral type A, it still does not necessarily imply that a main-sequence (or any other) A star is indeed present.

MODELS FOR SYMBIOTIC STARS IN THE LIGHT OF THE DATA

Michael Friedjung
Institut d'Astrophysique (CNRS), Paris, France

ABSTRACT

Different single and binary models of symbiotic stars are examined. Single star models encounter a number of problems, and binary models are probable. There are however difficulties in the interpretation of radial velocities. Accretion disks play a role in some cases, but winds especially from the cool component must be taken into account in realistic models. There is some evidence of excess heating of the outer layers of the cool component. Outbursts may be related to sudden changes in the characteristics of the cool star wind.

I. INTRODUCTION

Doubts can be raised as to whether a class of "Symbiotic Stars" really exists, and it is far from clear that all stars so classified have the same physics. Therefore one can even ask whether it is justified to talk about models for these stars! In this review I shall mainly concentrate on the "Classical Symbiotic Stars", which do appear to have certain common features.

The objects upon which this review will especially be centred, possess in quiescence a composite spectrum, with a component apparently due to a cool giant usually combined with a hot continuum which tends to dominate at short wavelengths, and always emission lines including at some times some of very high excitation (NV, FeVII, etc.). In active phases brightening occurs, with the cool continuum tending to be veiled by a hotter one, while the high excitation emission lines disappear. Certain "Symbiotic Stars" such as V1016 Cyg, however may not completely satisfy this description.

Other features which are basic for any model, include the cyclic spectroscopic changes of period $10^2 - 10^4$ days, as well as the periodic photometric variations sometimes detected, which in a few cases (CI Cyg, and AR Pav at least) are best interpreted as due to eclipses. Different characteristics will be discussed later.

Various quite different models can be proposed for the "Classical Symbiotic Stars" as well as for related objects. These models will now be described and criticized.

II. SINGLE STAR MODELS

A number of attempts have been made to model symbiotics on the basis of single stars possessing outer layers having different regions with very varied physical conditions. Two general types of single star model exist:

(a) <u>Hot central object surrounded by cool envelope.</u>

Several suggestions of this nature for stars related to the symbiotic class have been made. For instance, Sobolev (1960) considered that cool stars with emission lines such as symbiotic stars could have a hot nucleus surrounded by an envelope with a significant optical thickness in some subordinate continua like the Balmer continuum, giving rise to the cool absorption spectrum. A similar model for V1016 Cyg has been recently proposed by Nussbaumer and Schild (1981), according to which it could be a young planetary nebula with the region where hydrogen is ionized having a mass of 3×10^{-4} $M_\odot$, surrounding a hot star with $T_* = 1.6 \times 10^5$ °K and $R_* = 0.06$ $R_\odot$. Other single star models for this object exist (for references to them see Nussbaumer and Schild).

The observed properties of classical objects do not appear to fit this type of model. A cool envelope could be cool regions of a wind from a hot subdwarf, but then a wind velocity of the order of the stellar escape velocity ($\sim 10^3$ km s^{-1}) of the type of star required would be needed. Neither emission lines with a corresponding width nor absorption lines with a corresponding blue shift are usually observed, AG Peg perhaps being an exception. In addition both line and continuum absorption of the hot continuum by the cool envelope might be expected unless there were large deviations from spherical symmetry. No sign of non interstellar excited neutral absorption lines has been reported in high dispersion ultraviolet spectra, though Johnson (1981) considers that continuum absorption by amorphous silicate smoke may occur in the ultraviolet of R Aqr. However as pointed out by Johnson this may be explained by a binary model, with absorption of hot component radiation by the cool star's wind. Even when low excitation lines are seen in the near-ultraviolet, they do not appear to be associated with the spectrum resembling that of a cool star. For instance Faraggiana and Hack (1971) found that the M6III star type absorption spectrum of CH Cyg was veiled by the blue continuum present in 1967.

Absorption of the hot continuum might be less important if there were deviations from spherical symmetry, such as in the model proposed by Menzel (1969) for certain stars, with a cool ring formed by a magnetic field around the hot star. However the cool component seen in symbiotic stars appears fairly normal, and often non variable as in the case of Z And (Altamore et al. 1979). Also no magnetic fields were detected by Slovak (1978) for symbiotic stars.

These objections would be less strong for models like that of Nussbaumer and Schild for V1016 Cyg and similar stars. An envelope could have been ejected from a previously existing red giant with a lower escape velocity, the giant having become a hot subdwarf following formation of the nebula as in the mechanism of Kwok et al. (1978). In the model the hot star does not give rise to more than half the ultraviolet continuum except below 1600 A.

(b) Cool central object surrounded by hot envelope.

Most normal stars appear to be surrounded by chromospheres/coronae, which are hotter than the visible photosphere. It is therefore attractive to consider an enhancement of this process, which would lead to emission lines formed in these layers being seen in the visual region, and not only in the far ultraviolet and beyond. Models of this type have been popular in recent years among certain astrophysicists inspired by far UV and X-ray observations of normal stars, and by theories of chromospheres and coronae. Such models were first proposed by Aller (1954) and considered in more detail by Gauzit (1955). One precise suggestion of this kind was made by Wood (1974), who considered relaxation oscillations of an asymptotic branch star producing shock fronts, which dissipate energy in the expanding envelope.

A major problem of such models, is to produce a hot optically thick region responsible for the hot continuum. Symbiotic star spectral energy distributions have been decomposed into a hot optically thick component, a hot gas component, a cool stellar component, and sometimes dust emission, starting with the classical work in the visual region by Boyarchuk (1966, 1967, 1968) on AG Dra, AG Peg and Z And. Since then hot continua have been found in the ultraviolet for various symbiotic stars by for instance Hack (1979), Gallagher et al. (1979), Keyes and Plavec (1980), Johnson (1980), Kafatos et al. (1980), Slovak (1981), and by Altamore et al. (1981). The last authors show that the gas producing the high excitation emission lines seen in the ultraviolet was probably not responsible for the hot continuum.

One could always imagine a hot continuum formed in heated optically thick spots of the stellar photosphere resembling solar faculae, but such a hypothesis seems somewhat artificial. A model of this sort involving magnetic heating was proposed for CH Cyg by Wdowiak (1977). However upper limits to coherent magnetic fields of 200 G were found for CH Cyg, AG Peg and EG And by Slovak (1978).

In addition the radial velocities of AG Peg characteristic of a double lined spectroscopic binary, and what are probably eclipses seen for other objects, are hard to explain with a single star model. To summarize it is hard to rigorously disprove the hot single star envelope model for all symbiotic objects, but it offers a physically less fruitful and more arbitrary approach than the models to be now discussed. However as will be seen certain features of the models just considered, probably need to be combined with the binary approach.

III. BINARY MODELS

None of the partecipants at the "North American Workshop on Symbiotic Stars" held in June 1981 in Boulder, defended a single star model for any symbiotic object. The binary star conception seemed to be unanimously accepted!

Binary models were as far as I am aware first proposed by Berman (1932) for several stars and by Hogg (1934) for Z And. It is clear that certain features of symbiotic stars, such as the composite spectral energy distribution, the cyclic spectral and photometric variations, the latter in some cases been very probably eclipses, very strongly indicate that these stars are binary. Binary models are for instance described in the reviews of Boyarchuk (1969, 1974). Many often complex processes are known to occur in interacting binaries, and the different properties of binaries which may be relevant will now be considered.

(a) Radial velocity variations.

Radial velocity variations usually differ considerably from those of a non interacting binary, and their interpretation poses perhaps the greatest difficulty for binary models. Let us consider some specific stars.

As shown by Cowley and Stencel (1973), AG Peg is a double lined spectroscopic binary. Visual region emission lines of highly ionized ions come according to these authors from regions near the hot star, but other emission lines come from other regions, and can in some cases be perturbed by violet shifted absorption. A more detailed model for this star involving mass flow from the hot to the cool component was proposed by Hutchings et al. (1975), but the physics of such a situation if it really occurs would be hard to understand.

The radial velocities of AR Pav were interpreted by Thackeray and Hutchings (1974) as those of a single lined spectroscopic binary. Gas streams were also detected for this star, passing as one expects from the cool to the hot component. The star also appears to have eclipses, which agree in phase with what one expects from the radial velocity solutions.

The radial velocity variations of V1329 Cyg measured by Grygar et al. (1979) and by Iijima et al. (1981) were interpreted by both groups of authors in the single lined spectroscopic binary framework. Photometric minimum seems to occur at the phase expected for eclipses from the radial velocity curve, though it is not at all certain that classical eclipses are involved. The radial velocities of BF Cyg, RW Hya and R Aqr also show periodic variations, which could be orbital according to Boyarchuk (1969).

The situation is however not so clear for other symbiotic stars. Radial velocity variations are observed for Z And, but as shown by Boyarchuk (1968), the radial velocities instead of depending on an orbital phase, are correlated with photographic brightness. Orbital variations cannot be more than 5 km s^{-1}, and Boyarchuk supposing that the lines were formed around the secondary having a mass of only 1/9 of that of the primary, concluded that the inclination of the system could not be more than 10°. This is a somewhat improbable situation with a probability of only 1/66, and it appears that the only likely binary models of this star require emission line formation either between the components, or around the primary, generally taken as the cool star.

A similar problem exists for AG Dra according to Smith and Bopp (1981), who found no clear evidence for radial velocity phase dependence. These authors suggested that the red component had pulsations; one could indeed imagine that a combination of pulsations and orbital motion produces complex radial velocity variations. Pulsations have also been suggested by Smith (1980) to explain the radial velocities of EG And; it may however not be a classical symbiotic star. Pulsations were detected by Faraggiana and Hack (1971) for CH Cyg; a complex situation exists for it whose explanation in the framework of binary hypothesis is not clear, though these problems may be overcome by invoking gas streams and a wind to be discussed later.

A more serious problem may exist for RR Tel, if one uses the data compiled by Thackeray (1977). No evidence for periodic radial velocity changes was found. Though small systematic differences between different emission lines were seen, interpreted as due to blending with weak P Cygni absorption components, mean radial velocities for several ions were determined having standard errors between 1.9 and 0.5 km s^{-1}! It might be useful to re-examine the data. Formation of lines around a massive primary, combined with small pulsations having another period (periodic light variations occurred before outburts), might make it difficult to detect periodic variations. Thackeray himself suggested that some lines might be circumstellar.

To summarize, while the radial velocity variations of some symbiotic stars agree well with the binary conception, difficulties exist for several of the stars discussed. More studies of these stars and others not mentioned, are required.

(b) <u>Accretion</u>.

The various phenomena of accretion (Roche lobe overflow, bright spots, accretion disks, boundary layers, accretion columns, etc.) are basic to the physics of cataclysmic binaries, to which symbiotic stars are often related. As we shall see these phenomena are probably present, but all characteristics of symbiotic stars probably cannot be explained by them.

There is evidence in some cases that the red component is best not considered as filling its Roche lobe. Hutchings et al. (1975) found that the red component of AG Peg was 4 to 6 times smaller than the Roche lobe polar radius. Using the red star radius conditions of Keyes and Plavec (1980), the star would have to be at a distance of 2 kpc to fill the Roche lobe of Hutchings et al., and hence at a distance of 1 kpc from the galactic plane. AG Peg is however not a high velocity star. A similar situation may exist for Z And if the 680 day period is orbital. The size of the Roche lobe can be estimated to be of the order of 2×10^{13} cm, while the red star radius given by Altamore et al. is 9×10^{12} cm. It is not clear however whether the difference is significant. On the other hand Thackeray and Hutchings (1974) found that the red component of AR Pav could fill its Roche lobe, and the same applies to CI Cyg according to Stencel et al. (1981). In cases where the Roche lobe is not filled accretion could of course be from a wind.

Examination of emission line profiles clearly suggests the existence of an accretion disk for AR Pav. Thackeray (1959) considered that the emission lines could come from a thick rotating ring, but had difficulty explaining all observations. It might be interesting to re-examine them, taking account of possible bright spot eclipses by the red star and its wind.

The emission line profiles of other symbiotic stars are less easily interpreted in this framework. For instance as pointed out by Altamore et al. (1981), the half intensity widths of the emission lines of Z And observed with IUE correspond to Doppler velocities of not more than 60 km s^{-1}. If formed in a disk such lines would unless an improbable geometry is assumed, have to be at at least 3×10^{12} cm from the central star. Even lines of highly ionized species such as NV, which might be expected to be formed near the disk center, could only be formed near the outer edge of a disk not much smaller than the Roche lobe. In that paper it was also found that the NIII] lines come from a region of low radiation density at at least 1×10^{12} cm from the hot component, while somewhat less rigorous reasoning led to a maximum line of sight thickness and suggested line formation in a region with a large extension compared with its thickness.

It may be noted that line formation in an outer "excretion" disk such as that which probably existed during the 1978 outburst of the unusual dwarf nova WZ Sge, might also be possible. Such disks with a smaller rotation velocity, would give rise to narrow lines. The nature and conditions for their existence are however badly known.

A preliminary examination of IUE spectra of AG Dra suggests emission line half intensity widths of the same order as those of Z And, and this also is the case for CI Cyg. The latter and perhaps both AG Dra and CI Cyg have eclipses, so it would be hard to explain line narrowness by orientation effects. The eclipses of CI Cyg moreover can be used to test for the region of narrow line formation. An eclipse was observed with IUE by Viotti et al. (1980) and by Stencel et al. (1981). Different lines are eclipsed to different extents, indicating formation in somewhat different regions. In particular NV appears hardly to be eclipsed, and this casts doubt on the occurrence of the most ionized regions close to the hot star, especially as the continuum eclipse appears to be 0.35 dex in the far UV (Baratta et al. at this meeting). This suggests that these unlike less ionized regions, occur rather around the eclipsing star than in an accretion disk. It should also be added that the eclipsed lines need not necessarily be formed in a disk; some could be partly formed on the side of the eclipsing star facing its companion, and partly in gas streams and or parts of an eclipsing star wind deflected by the latter.

The presence of an accretion disk would also affect the continuous energy distribution of a symbiotic star. Webbink (1981) analysed the light curve of the 1975 eclipse of CI Cyg, and considering the effect of the emission lines was not large, concluded that one needed to assume the total eclipse of an accretion disk to explain the form of the light curve. A large mass accretion rate of at least 7×10^{-4} M$_\odot$yr^{-1} is needed for Webbink's model; it should be noted however that CI Cyg was active at the time of eclipse, and that the situation may be quite different in quiescence. In any case the calculation should be done with new data at different wavelengths for different epochs.

The spectral energy distribution can be compared with disk models, which can also be used to predict the expected photoionization of gas in various regions. Kafatos (1981) considered the ionization of "nebular" gas by a disk plus a hot boundary layer, and was able to explain emission line fluxes. In addition Kenyon (1981) has attempted to fit observed energy distributions to disk plus boundary layer models, and obtained fits for all stars considered except BF Cyg. The accreting star according to him could be a main sequence star or a subdwarf. However it is not sure that the boundary layer temperatures of his models would be high enough to produce the states of highest ionization in all cases by photoionization. In particular his model gives a boundary layer temperature of $\lesssim$ 35000°K for Z And, not very different from the black body temperature of 43000°K found by Altamore et al. (1981). The flux of photons able to doubly ionize helium would be a factor of 10^3 too small to produce the observed 1640 A flux of Z And for the latter temperature; the temperature required in fact would be near 80000°K. The situation is probably even worse for the production of the ground state of NV. It must be noted however that if as reported by Viotti at this meeting, there was a very hot component to the continuum of Z And, this argument could not be used.

To conclude this section, accretion from disks almost certainly occurs in some cases at least; it is however not clear whether accretion disks are present all the time for all stars, particularly in the quiescent state in which Z And now is.

(c) Stellar winds from each component

Winds can occur from both components of a binary, and their properties are probably important for understanding symbiotic stars.

Winds from the hot stellar component seem to play a major role in some cases. Such a wind could be expected to have a velocity of the order of the stellar escape velocity, that is near 10^3 km s^{-1}. The broad P Cygni part of AG Peg line profiles mentioned by Keyes and Plavec (1980) probably comes from such a wind, especially as Hutchings et al. (1975) found from ground based data that gas was apparently leaving the hot star. The mass loss rate of 10^{-6} M$_\odot$yr^{-1} found by Gregory et al. (1977) using radio observations is very large, and presumably associated with continuing activity since the outburst of AG Peg in the 19th century.

If this mass loss occurred from the hot star, and the distance of AG Peg is 0.5 kpc, one finds that the wind would have an optical depth of unity for Thomson scattering at about two solar radii. The photosphere defined as the level above which a photon in its random walk between scatterings would cross a free-free plus free-bound optical thickness of one would then be near a radius of 5×10^{10} cm at 2000 A. This is not far from the hot star radius given by Keyes and Plavec, which then would have a false photosphere formed by an optically thick wind.

The Wolf-Rayet features of RR Tel observed by Thackeray and Webster (1974) between 1951 and 1960 with a width of the order of 2000 km s^{-1} at some epochs, are also most easily interpreted as due to wind from the hot component. The P Cygni absorption component of RX Pup at velocity near 1000 km s^{-1} observed by Swings and Klutz (1976) may have a similar explanation.

The winds arising from the cool stellar component probably play a more fundamental role for many symbiotic stars. The compact "nebula" of R Aqr could be the result of mass loss from the cool star according to Michalitsianos et al. (1980). Such a wind would have a low velocity, and could be where the narrow emission lines seen in many stars are formed. Information about normal cool giant velocities is given by Reimers (1980), where one sees that velocities can be of the order of 10 km s^{-1}, much less than the stellar surface escape velocity. Using the Altamore et al. (1981) data, one finds that for the HeII 1640 A line to be formed in a wind, a mass loss rate of about $3-8 \times 10^{-7}$ $M_\odot yr^{-1}$ is required. Taking the Kudritzki and Reimers (1978) constant for the red giant mass loss formula, the expected mass loss rate of the cool star of Z And if it was single is $3 \times 10^{-8} M_\odot yr^{-1}$, when the star is supposed to have a mass of 3 $M_\odot$. In view of the uncertainties, this is surprisingly close. It should be noted that there is some doubt concerning the mass loss rate formula; while a conversation with Kwok at this meeting suggests that a mass loss rate below 10^{-7} $M_\odot yr^{-1}$ would be hard to detect in the radio region.

Thackeray (1977) suggested formation of the lines of RR Tel in an accelerating wind, with ionization increasing outwards. This is almost certainly the opposite situation to that of a classical nova in its decline after maximum. It is probably easiest to suppose formation of the non Wolf-Rayet lines in the cool star wind, even though the terminal velocity of the order of 100 km s^{-1} at least at certain times, is high for a normal red giant. In any case line profiles are perhaps more easily explained by a wind than by rotating disk.

Feast et al. (1977) related the presence of strong winds to the cool star being a Mira variable and the presence of dust. In such a case they suggested that accretion was from a Mira wind, while in other cases it was due to Roche lobe overflow. This division into two classes parallels the division of emission line stars including symbiotic ones by Allen and Glass (1974) into S (stellar) and D (dust) types, according to the nature of the infrared energy distribution. Z And belongs to the former, and RR Tel to the latter type. The study of Mira variables in D type symbiotic stars has been extended by Willson (1981). According to her the Mira variables that occur are fairly normal, though the period distribution is different perhaps because large mass loss rates for wind accretion are associated with long period Miras. From evolutionary considerations a critical orbital period was defined, which determined whether a binary will become an S or D type symbiotic star.

From the previous discussion it appears that cool star winds can also be important in S type cases like Z And. In addition in a very recent study with R.E. Stencel and R. Viotti (paper in preparation), it was found that the high excitation permitted far ultraviolet resonance lines in all symbiotic stars for which we have data, are red shifted with respect to the wavelength system defined by the intercombination lines. The former lines can be expected to be optically thick, and such a shift could be produced either by the presence of weak P Cygni absorption or by a radiative transfer effect, which occurs when photons are scattered many times in a slowly expanding medium. Both explanations suggest either line formation in a slow wind or at the base of the wind where expansion has begun, but not in a rotating disk.

Collisions between the winds from each component might be expected to play an important role in some cases. For instance it might using such concepts be possible to explain the observations of AG Peg in a physically more probable way than that of Hutchings et al. (1975). Wallerstein (1981) has described colliding wind calculations of Willson that can explain the peculiar changing profiles of V1016 Cyg and HM Sge. Such a picture does not necessarily contradict the existence of an outer nebular shell, which exists in the model of Nussbaumer and Schild (1981). These problems have also been considered by Kwok (1981).

(d) Enhanced outer layer heating of the cool component

The study of Z And by Altamore et al. (1981) as well as new additional arguments of mine given here for Z And and CI Cyg, suggest formation of high ionization ultraviolet lines far from the hot component of stars of this type, and can most easily be interpreted by formation around the cool component. It may be noted that emission lines of singly ionized metals also may be produced by photoionization in the atmosphere of the cool component, as for instance suggested by Boyarchuk (1967) for AG Peg.

There may nevertheless be problems of explaining the most ionized states by photoionization, so line formation in a hot wind or in a hot region at the base of the wind, having a large extension compared with its thickness, is suggested.

However since the work of Linsky and Haisch (1979) it is known that single red giants resembling the cool components of most symbiotic stars, seem neither to possess hot outer layers, nor hot winds. Binaries nevertheless can have enhanced activity; in particular the RS Cvn stars which rotate faster than single stars of the same class. In the recent study with R.E. Stencel and R. Viotti already mentioned, we suggest that rotation could also be associated with enhanced heating of the outer layers of the cool components of some symbiotic stars. Normal red giants are expected to have rotational velocities of the order of several cm s^{-1}, while red components of symbiotic stars could be tidally spun up to velocities of several km s^{-1}. Details still need to be considered, including reasons why not many more stars are symbiotic.

The clearest test of such concepts would be to measure the electron temperatures of the regions where the states of highest ionization occur. Electron temperatures are notoriously more difficult to determine than electron densities; the most suitable method is perhaps that of Stickland et al. (1981) where fluxes of collisionally excited lines are compared with those of the next lower stage of ionization formed by recombination in the same region. The method is unfortunately not sensitive at high temperatures. An approximate attempt to measure the temperature of the NV region of Z And from the 1718/1240 A flux ratio gives a minimum electron temperature of 18000°K, which at least is not inconsistent with the model proposed here.

When the method was applied however to RR Tel by Penston et al. (1981), an electron temperature of only 15000°K was found, not much higher than that corresponding to the other line ratios formed in less ionized regions. RR Tel is thought however to have dust in its wind, and winds from D type symbiotic stars are presumably cool. The results presented by Mussbaumer at this meeting, suggesting possibly a high electron temperature for RR Tel indicates that such reasoning could be nevertheless too simplistic.

It may be noted that the "yellow" symbiotic stars defined by Glass and Webster (1973), whose cool components are less cool, including AG Dra, may have a somewhat different behaviour. Their cool components if single would according to the terminology of Dupree (1981) be perhaps "hybrid", having "warm" coronae with temperatures of at least 2×10^5°K. Excess heating might be less necessary for these stars. In any case the winds and outer layers of symbiotic stars in general are probably not very hot (i.e. do not have temperatures of more than a few times 10^5 °K), as X-ray emission was not detected in many cases.

IV. CAUSES OF SYMBIOTIC ACTIVE PHASES

I shall only briefly touch on this subject. Nuclear burning in particular will be considered by other speakers.

Bath (1977) has proposed that accretion powers optically thick winds, accelerated by the radiation pressure associated with a luminosity close to and occasionally in excess of the Eddington limit. He considered that the visual brightness changes are due to mass loss rate changes affecting the radius to which the wind is optically thick, and hence the flux per unit area and the effective temperature of the photosphere. A major difficulty is that there is no evidence of such large luminosities of the hot component. The luminosity of that of Z And according to Altamore et al. (1981) is 7.7×10^{35} erg s^{-1} or 5×10^{-3} of the Eddington limit of a one solar mass star. To bring it up to the Eddington limit, one would have to suppose the star at 17 kpc or 3.6 kpc above the Galactic plane. The distance estimate of Altamore et al. is based on the uncertain luminosity of the cool component, but increasing the estimated distance to 17 kpc, would make this component bigger than the size of the orbit of the hot component, while one would expect it to be a high velocity star for which there is no evidence.

Using the results of Keyes and Plavec (1980), AG Peg would have to be at a distance of 8 kpc and 4 kpc above the Galactic plane for the hot component to be at the 1 $M_\odot$ Eddington limit. Similar problems arise as those for Z And. The situation is here however puzzling, because as we saw, there is probably an optically thick hot component wind, and it is not clear how it can be driven if AG Peg is far below the Eddington limit.

It should also be noted that according to a recent study of mine (Friedjung 1981), probable solutions for supercritical winds should satisfy a condition relating the radiative flux to that of kinetic energy. Classical novae which probably have such winds after outburst, approximatly satisfy the condition, but Z And does not.

Somewhat speculatively I would like to suggest that active phases of at least S type symbiotic stars are related to changes in the chromosphere, corona, and wind of the cool star. A warm wind might become a cool one of low velocity, and perhaps also a slight expansion of the chromosphere might enable Roche lobe overflow to take place. The accretion rate would increase, as accretion from a wind is very sensitive to its velocity, varying as the inverse fourth power of it. At the same time lines of high ionization species would tend to disappear in the spectrum. An accretion disk might form, or grow considerably if already present. AG Peg seems to require something more; perhaps nuclear burning of accreted gas by the hot component.

V. CONCLUSIONS

Symbiotic stars are probably binary, though it is hard to regorously disprove models of a cool central object surrounded by a hot envelope. The interpretation of the radial velocity measurements also poses problems for certain stars.

A number of different physical processes seem to play important roles in symbiotic stars. Accretion disks exist at least in some cases, but it is not clear whether they are always present in quiescence. Winds from both components, and particualrly from the cooler one, are important, while there is evidence that excess heating of the outer layers of the cool component occurs at least for some symbiotic stars.

The nature of outbursts was considered. Bath's model of optically thick winds driven by radiation pressure encounters difficulties. A new suggestion was made, connected with a change of the cool star's wind.

AKNOWLEDGEMENTS

I would like to thank Scott Kenyon for communicating the results of his calculations concerning the fit of the observed energy distributions to accretion disk plus boundary layer models, and Bob Stencel for communicating results on CI Cyg before publication. Mike Seaton helped also to elucidate some points about the electron temperature calculation of Stckland et al. (1981).

REFERNCES

Allen, D.A.; 1980, Mon. Not. R. astr. Soc. $\underline{192}$, 521.
Allen, D.A., Glass, I.S.: 1974, Mon. Not. R. astr. Soc. $\underline{167}$, 357.
Aller, L.H.: 1954, in "Astrophysics - Nuclear Transformations, Stellar Interiors, and Nebulae", The Ronald Press Company, New York, p.180.
Altamore, A., Baratta, G.B., Viotti, R.: 1979, Inf. Bull. Var. Stars, No. 1636.
Altamore, A., Baratta, G.B., Cassatella, A., Friedjung, M., Giangrande, A., Ricciardi, O., Viotti, R.: 1981, Astrophys. J. $\underline{245}$, 630.
Bath, G.T.: 1977, Mon. Not. R astr. Soc. $\underline{178}$, 203.
Berman, L.: 1932, Publ. Astr. Soc. Pacific $\underline{44}$, 318.

Boyarchuk, A.A.: 1966, Astrofizica 2, 101.
Boyarchuk, A.A.: 1967, Astron. Zh. 43, 976.
Boyarchuk, A.A.: 1968, Astron. Zh. 44, 1016.
Boyarchuk, A.A.: 1969, in "Non Periodic Phenomena in Variable Stars", Academic Press, Budapest, p.395.
Boyarchuk, A.A.: 1975, in "Variable Stars and Stellar Evolution", IAU Symposium No.67, V.E. Sherwood and L. Plaut (eds.), D. Reidel, Dordrecht, p.377.
Cowley, A., Stencel, R.E.: 1973, Astrophys. J. 184, 687.
Dupree, A.K.: 1981, in "Effects of Mass Loss on Stellar Evolution", IAU Colloquium No.59, C. Chiosi and R. Stalio (eds.), ρ
Faraggiana, R., Hack, M.: 1971, Astron. Astrophys. 15, 55.
Feast, M.W., Robertson, B.S.C., Catchpole, R.M.: 1977, Mon. Not. R. astr. Soc. 179, 499.
Friedjung, M.: 1981, Acta Astronomica in press.
Gallagher, J.S., Holm, A.N., Anderson, C.M., Webbink, R.F.: 1979, Astrophys. J. 229, 994.
Gauzit, J.: 1955, Ann. Astrophys. 18, 354.
Glass, I.S., Webster, B.L.: 1973, Mon. Not. R. astr. Soc. 165, 77.
Gregory, P.C., Kwok, S., Seaquist, E.R.: 1977, Astrophys. J. 211, 429.
Grygar, J., Hric, L., Chochol, D., Mammano, A., 1979, Bull. astr. Inst. Czech. 30, 308.
Hack, M.: 1979, Nature 279, 305.
Hogg, F.S.: 1934, Publ. Am. astr. Soc. 8, 14.
Hutchings, J.B., Cowley, A.P., Redman, R.O.; 1975, Astrophys. J. 201, 404.
Iijima, T., Mammano, A., Margoni, R.: 1981, Astrophys. Space Sci. 75, 237.
Johnson, H.M.: 1980, Astrophys. J. 237, 840.
Johnson, H.M.: 1981, talk at North American Workshop on Symbiotic Stars.
Kafatos, M.: 1981, talk at North American Workshop on Symbiotic Stars.
Kafatos, M., Michalitsianos, A.G., Hobbs, R.W.: 1980, Astrophys.J. 240, 114.
Kenyon, S.J.: 1981, private communication.
Keyes, C.D., Plavec, M.J.: 1980, in "The Universe at Ultraviolet Wavelengths - Two Years of IUE", NASA, p. 443.
Kudritzki, R.P., Reimers, D.: 1978, Astr. Astrophys. 70, 227.
Kwok, S.: 1981, talk at North American Workshop on Symbiotic Stars.
Kwok, S., Purton, C.R., Fitzgerald, P.M.: 1978, Astrophys.J. 219, L125.
Linsky, J.L., Haisch, B.M.: 1979, Astrophys. J. 229, L27.
Menzel, D.H.: 1969, in "Les Transitions Interdites dans les Spectres des Astres", 15me Coll. Astrophys. de Liège, p.341.
Michalitsianos, A.G., Kafatos, M., Hobbs, R.W.:1980, Astrophys.J. 237, 506.
Nussbaumer, H., Schild, H.: 1981, Astr. Astrophys. 101, 118.
Penston, M.V., Benvenuti, P., Cassatella, A., Heck, A., Selvelli, P.L., Beeckmans, F., Macchetto, F., Ponz, D., Jordan, C., Cramer, M., Rufener

P., Manfroid, J.: 1981, Mon. Not. R. astr. Soc. in press.
Reimers, D.: 1980, in "Second European IUE Conference", ESA SP-157, page xxxiii.
Slovak, M.H.: 1978, Bull. Am. astr. Soc. 10, 609.
Slovak, M.H.: 1981, paper presented at the 157th meeting of the AAS.
Smith, S.E.: 1980, Astrophys. J. 237, 831.
Smith, S.E., Bopp, B.W.: 1981, Mon. Not. R. astr. Soc. 195, 733.
Sobolev, V.V. 1960, in "Moving Envelopes of Stars", Harvard University Press, Cambridge, Mass., p.82.
Stencel, R.E., Michalitsianos, A.G., Kafatos, M., Boyarchuk, A.A.:1981, in preparation.
Stickland, D.J., Penn, C.J., Seaton, M.J., Snijders, MA.J., Storey, P.J.: 1981, Mon. Not. R. astr. Soc. 197, 107.
Swings, J.P., Klutz, M.: 1976, Astr. Astrophys. 46, 303.
Thackeray, A.D.: 1959, Mon. Not. R. astr. Soc. 119, 629.
Thackeray, A.D.: 1977, Mem. R. astr. Soc. 83, 1.
Thackeray, A.D., Hutchings, J.B.: 1974, Mon. Not. R. astr. Soc. 167, 319.
Thackeray, A.D., Webster, B.L.: 1974, Mon. Not. R. astr. Soc. 168, 101.
Viotti, R., Giangrande, A., Altamore, A., Baratta, G.B., Cassatella, A., Ponz, D., Friedjung, M., Muratorio, G.: 1980, IAU Circ. No. 3518.
Wallerstein, G.: 1981, talk at North American Workshop on Symbiotic Stars.
Wdowiak, T.J.: 1977, Publ. Astr. Soc. Pacific 89, 569.
Webbink, R.F.: 1981, talk at North American Workshop on Symbiotic Stars.
Willson, L.A.: 1981, talk at North American Workshop on Symbiotic Stars.
Wood, P.R.: 1974, Astrophys. J. 190, 609.

SYMBIOTIC STAR UV EMISSION AND THEORETICAL MODELS

M. Kafatos
Department of Physics, George Mason University, Fairfax, Va.,
U.S.A.

Observations of symbiotic stars in the far UV have provided important information on the nature of these objects. The canonical spectrum of a symbiotic star, e.g. RW Hya, Z And, AG Peg, is dominated by strong allowed and semiforbidden lines of a variety of at least twice ionized elements. Weaker emission from neutral and singly ionized species is also present. The Mg II doublet is usually very strong and may be associated with the M giant primary. A continuum may or may not be present in the 1200 - 2000 A range but is generally present in the range 2000 - 3200 A range, the latter arising from free-free and bound-free emission in the same nebula that is responsible for the UV line emission (CI Cyg, RW Hya, RX Pup). The suspected hot subdwarf continuum is seen in some cases in the range 1200 - 2000 A (RW Hya, AG Peg, SY Mus). High resolution observations of lines are important because they yield information on densities, temperatures and sizes of the line emitting region(s); in general, however, such observations are difficult and time-consuming to obtain with "IUE". Densities are found to range from a low of $\sim 10^6$ cm^{-3} in R Aqr and V1016 Cyg through typical values of $10^8 - 10^9$ cm^{-3} in RW Hya to a high of $\sim 10^{11}$ cm^{-3} in Z And. Sizes range from $\sim 10^{11}$ in the resonance line emitting region in Z And to $\sim 10^{14} - 10^{15}$ cm in the more extended regions of R Aqr and V1016 Cyg. Temperatures are generally $\lesssim$ 20,000 K. High resolution profiles generally show single component nebular emission (RW Hya, SY Mus, AG Peg, V1016 Cyg). Complex profiles showing multiple velocity structure present in rings and/or streamers have been detected in RX Pup. Continua, which very often are flat, are harder to interpret but it seems that line blanketed models of B, A and F-type stars generally fail. A combination of different sources of continua seems ro be required: nebular emission (particularly for $\lambda \gtrsim 2000$ A); hot subdwarf continuum; and/or continuum arising in an accretion disk. The presence of an accretion disk is difficult to demonstrate and to this date the best candidate for accretion to a main sequence star remains CI Cyg. A number of equations have been derived by the author that can yield the accretion parameters from the observable quantities. Boundary layer temperatures $\sim 10^5$ K and accretion rates $\gtrsim 10^{-5}$ M$_\odot$/yr are required for accreting main sequence companions. To this date, though, most of the symbiotics may only require the presence of a $\sim 10^5$ K hot subdwarf.

DISCUSSION ON THEORETICAL MODELS

<u>Chairman</u>: I want to recall the audience that we must decide <u>if the Symbiotic Stars are binary or not</u>, otherwise we cannot go on to Session IV which is devoted to the binary evolution.

<u>Slovak</u>: CH Cygni = HD 182917 has been extensively studied by Hack, Wallerstein and many others, but there is <u>no</u> convincing evidence for the binary nature of this system. Admittedly, it is an extreme example of a symbiotic star (very low excitation, latest spectral type secondary, displays rapid variations), but it still remains a troublesome case which cannot be ignored.

<u>Boyarchuk</u>: I have some arguments against the single star hypothesis of the nature of symbiotic stars.
First if we propose that a symbiotic star is a cool star with a hot corona or chromosphere, in this case we must remember that the luminosity of the corona would have to be 10^6 times larger than the solar one in order to observe it in total light in the visual spectral region.
It means that the bolometric luminosity of such a corona will be even more intense than the bolometric luminosity of the photospheric radiation. Under such conditions it is impossible to keep the surface temperature cool enough. As a result we will not have a symbiotic star.

Secondly, if we propose that a symbiotic star is a hot star with a hot nebula, and that TiO-bands and other absorption features are formed in other parts of this nebula, we should note that in the spectra of many symbiotic stars we observe the absorption line of CaI λ 4227. This line has very extended "wings" which is normal for a cool star spectrum. But, if we calculate the column density which is needed to produce such wings in a nebula, and multiply by the surface of the nebula which is huge, then we will obtain the mass of the absorption envelope that is equal to several solar masses. It is difficult to understand how such envelope could exist.

The main argument in favour that the symbiotic stars are single stars is a lack of strong evidence for the binary nature of some symbiotic stars. But this argument is not sufficient. It is necessary to give the arguments that a single star can produce the symbiotic phenomenon.

<u>Hack</u>: The model you discussed (hot star and cool envelope) cannot work in the case of CH Cygni. In fact this star in a period of relative quiescence lasting several years, had a normal M 6 III spectrum.

The blue-UV continuum appearing during outburst is due to a semi-transparent layer, because the TiO bands are always visible (also in the blue part of the spectrum). The temperature indicated by the blue continuous energy distribution is not very high (less than about 10^{4}°K), and the blue continuum lasts a few months or years, and is surely not sufficent to heat the photosphere of the M giant.

Slowak: The lack of magnetic fields, at least as determined by conventional Zeeman studies in symbiotic stars, argues against the single star model proposed for a red star, which is analogue of the flare star model.

Michalitsianos: A kilogauss magnetic field measured on an average magnetic field for a symbiotic star would imply local field strengths of enormeous values. It is likely that high energy processes would occur under such circumstances, which is contrary to observations; for example soft X-ray emission is not present.

Kwok: One important element in our understanding of the nature of symbiotic stars is their division into type S and type D (Allen 1979), which correlates well with the type I and type II of Paczynski and Rudak (1980). Since only type D (or type II) objects have clear evidence for an M-star wind, one might wonder whether the presence of such a wind is in fact the cause of the slow nova phenomenon, through their abilities to transfer mass in spite of a wide binary separation.

Rudak: As I understood, Bath applied geometrically thick ($z/r=0.5$) accretion disk in his calculations to reproduce the observed light curve. I don't know how the distribution of surface temperature on the disk was calculated. The way to do that in the case of an α-disk, which gives us a very bright boundary layer dominating the disk and a temperature of disk decreasing outwards, is not proper for considerably thick disks.
And the reason is that in geometrically thick disks, the energy generated due to accretion is not transferred entirely to the point on the surface which lies directly above the point inside the disk, where this generation take place. We will expect the effective influence of energy generation in one place giving surface temperature in other places. Similar problems are considered for example in thick accretion disks around massive black holes and give some QSO-characteristics.

Kafatos: About the theoretical attempts to fit the optical continua (by Bath) or the more recent ones to fit the IUE UV continua (by S. Kenyon) they go, in my opinion, in the wrong direction or are - at least - incomplete. The reason for this is that we have a lot of information about the nature of the ionizing radiation from the UV lines, and that

information has to be incorporated in any self-consistent disk model.
The continua in the long wavelength region of IUE seem in most cases to
arise from the same optically thin nebuale that give rise to the UV lines
and there is no need to invoke disks.
The only continua that are left are in the short wavelength UV region of
IUE (1200-2000 Å). This is, in my opinion, too small a region to try to
fit accretion disk continua which are in any case highly uncertain.
Until someone has looked at the ionizing radiation responsible for the
line forming region, how that radiation arises in a self-consistent disk
model, and how this radiation escapes from the inner regions of the geometrically thick disks of Bath, the disk models of symbiotics remain I
think incomplete.

Hack: I wish to say that also IUE observations indicate that probably the physical conditions in chromosphere-corona of late dwarf stars
are different from those in giants and supergiants. Only the dwarfs show
hot chromospheres, while cool giants and supergiants present low excitation features.

Kwok: From the paper of Dr. Kafatos, it seems that the characteri-
stic densities of D-type objects are lower than that of S-type objects.
Again, it points to D-type objects as having wind-like nebulae, whereas
S-types do not.

SESSION IV

EVOLUTIONARY CONSIDERATIONS

Chairman: Y. Andrillat

Introductory reports on:

EVOLUTIONARY STATUS   (B. Rudak)

MODEL OF SYMBIOTIC STARS   (A.V. Tutukov and L.R. Yungelson)

Symbiotic Stars: Accretion onto a Companion?

THE EVOLUTIONARY STATUS OF SYMBIOTIC STARS

Bronislaw Rudak
Copernicus Astronomical Center
Warsaw, Poland

Abstract: The evolutionary relations between symbiotic stars and cataclysmic variables are presented. The symbiotic stars are assumed to be long period detached binaries containing a carbon-oxygen degenerate primary and a red giant losing its mass through a spherically symmetric wind. Such systems can be obtained in Case C evolution, provided a common envelope during a rapid mass transfer phase was not formed. The same way recurrent novae containing a red giant as a secondary component may be produced. The factors influencing the differences between symbiotic stars and nova-type stars are discussed.

1. INTRODUCTION

Considering the problem of where symbiotic stars belong among the different kinds of stellar objects, their evolutionary stage and their possible natural linkages with other types of systems, we face real difficulties. Some of them seem to be trivial, but yet they are not, especially when one is looking through the observational data published to date. Even such a necessary fact as their binary or single star nature, which is desirable to begin any theoretical considerations, is not well established. Although the latest observations seem to favour the former possibility in more and more cases /e.g. Ciatti et al. 1979, Kenyon and Cahn 1979, Slovak 1981/, for a significant number of symbiotic stars one cannot definitely say what is their morphological structure. The problem is partly derived from the non-precise definition of "symbioticity". Regardless of three or four necessary conditions to be fulfiled, it is still a very general term, and as such, leaves room for some objects which have nothing to do with "classical" symbiotic stars. Thus the heterogeneity of the set of objects classified as symbiotic ought to be treated seriously.

In this review, however, we intend to focus our attention on those stars which are binary systems. We believe this group to be rather homogeneous, since one definite model can explain its behaviour, allowing for all particular observed differences between the member stars.
This resembles the situation of novae some fifteen years ago, when the wealth of observational data was at last satisfactorily explained.

2. GENERAL REMARKS

The idea of binary nature of Z And type stars was originally undertaken by Hogg in 1934, and then developed by Merrill, Payne-Gaposchkin, Struve, Swings, Kuiper and others. In the middle of seventies an interesting model was proposed by Bath /1977 and references therein/. Looking for tight couplings between symbiotic stars and cataclysmic binaries, Bath adopted his model of the latter group to explain the symbiotic phenomena. In that model, the binary consists of a main sequence dwarf or subdwarf and a red giant filling its Roche lobe. The activity of the system is due to instabilities arising in the outer layers of the giant, which in turn force the episodic mass transfer with a periodicity of few hundred days. The enhanced amount of matter falling onto the primary's surface at a super-Eddington rate via the accretion disk, causes a radiatively driven expansion of its photosphere. Although such a model reproduces the general features of the behaviour of the above systems, there are observational indications which do not confirm Baths idea.

The binary nature of symbiotic stars was essentially deduced from variations in radial velocity curves, representing the movement of each component. The examination of these curves /Cowley and Stancel 1973, Boyarchuk 1975/ leads to the conclusion that the giant is the more massive component, with $\mathfrak{M} \approx 3 - 4 \mathfrak{M}_\odot$. In this case, the mass loss from the more massive star will cause a shrinkage of its Roche radius and a subsequent decrease of the orbital period. But after removing some material from the outer layer of the convective envelope, the new equilibrum radius cannot be fitted with a new smaller radius /Paczynski 1970/. This equilibrium radius is larger than the new Roche radius, so that a considerable amount of matter can be freely removed on a dynamical time scale for the giant. In effect, the majority of the giant's envelope will be quickly striped off revealing the degenerate core, and the mass ratio of the system will reverse. In other words, the life-time of symbiotic stars would be comparable to the dynamical time scale for giants.

That is why we doubt the model of red giants filling their Roche lobes in symbiotic binaries.
More careful examination of shapes of emission lines arising from the circumstellar matter, and observations of light changes on short time scales would show us if an accretion disks in symbiotic stars are really absent, as suggested by present-day observations. CI Cyg and CH Cyg seem to be the exceptions.
As for the main sequence component, one cannot rule out its presence, but new observations favour the primary to be a much more compact object /e.g. Allen 1980/.
Therefore through the rest of the paper we shall assume that, as in cataclysmic variable stars /Robinson 1976/, the hot components are white dwarfs.
So finally, we are left with a detached system containing a white dwarf and a red giant, as a compatible model for significant number of symbiotic stars.

3. THE PLACE OF SYMBIOTIC STARS AMONG EVOLVED BINARIES

The fate of an initially detached system of two main sequence stars depends mainly on their separation at that stage and the total mass of the system.
In principle, low-mass and massive binaries have to be described separately. For the latter group, usually the explosive carbon ignition inside the initially more massive component takes place at advanced stages of its evolution, giving rise to X-ray binaries with neutron stars or black holes. The critical mass which separates this group from the former one is not precisely established, but we can accept it to be near to 5 $M_\odot$. Therefore, all symbiotic systems will fall safely into the low-mass group. As far as we know, there is only one object, GX 1+4, belonging to the massive group, which sometimes is classified as a possible member of the symbiotic stars. Nevertheless, the spectroscopy of this star was examined too poorly to treat this classification seriously. In any event its structure, with a probable neutron star component, does not fit the rest of the symbiotic stars.
From now on, when using the expression "binary system", we shall mean low-mass systems only.
Let us now repeat the most characteristic features of the different stages in binary evolution as a function of initial separation.
Consider first the system whose initial orbital period is larger than 1 day but shorter than 100 days. This will produce the so called Case B evolution. The initialy more massive component /primary/ attains its Roche lobe radius as a subgiant with a helium core and a hydrogen burning

shell. The Roche lobe overflow causes rapid mass transfer through the inner Lagrangian point onto the secondary, a more or less evolved main sequence star. This effect proceeds on very short time scale compared to the stars life-time. For stars with radiative envelopes it is of the order of the thermal time scale, and for stars with developed convective envelopes it may be even shorter, approaching the dynamical time scale /Paczyński 1970/. This mass transfer forces a decrease in stellar separation, reaching its minimal value for $m_1 \approx m_2$, and then its subsequent increase. After the mass ratio is reversed, further mass loss goes on at a nuclear time scale for the hydrogen burning shell. This semi-detached phase lasts as long as there is a fresh hydrogen supply for burning in the shell. The resulting detached system, with a more massive secondary and a degenerate helium white dwarf primary will unavoidably form the semi-detached phase once again in its future: when the secondary manage to fill its Roche lobe, the mass transfer in the reversed direction will take place. If the mass ratio of the system is extreme enough before the primary overflows its Roche lobe, the rapid mass transfer can lead to the formation of dense common envelope around both stars. The drag, which will arise between this envelope and two orbiting stars embedded in it, is effective enough to transfer a considerable amount of angular momentum out of the binary system. This mechanism decreases the orbital period to the values well below 1 day. The binaries of Case B evolution, which underwent this stage, form as a result the class of cataclysmic variables.

Naturally, the mechanism which generates the activity of cataclysmic variables can work also for objects with periods already comparable to those of symbiotic stars /e.g. T CrB/. These exceptional objects managed to avoid the common envelope phase or were formed via Case C evolution, and the latter possibility explains naturally, why these stars are so similar to symbiotic objects. This point will be discussed later.

Because typical orbital periods for symbiotic binaries are longer than 100 days by a factor of 3 - 10, it is natural to seek their history in Case C evolution. This was independently considered by many authors /Paczyński 1980 and references therein/.

Case C evolution of a binary starts with separation wide enough to enable the ignition of the helium core before the primary reaches its Roche lobe. The initial mass of the primary in proto-symbiotic system is probably larger than $2 m_\odot$, so that the helium ignition proceeds in the form of calm burning. As a result, a massive, degenerate carbon-oxygen core is formed. It is surrounded by an extended supergiant-type envelope expanding at the rate of nuclear burning in the hydrogen shell. After the star touches its Roche lobe, the rapid mass transfer onto the secondary can take place. As the existence of well developed burning shells enables the envelope to become deeply convective, the mass loss phase will last no longer than primary's dynamical time scale /for reasons already mentioned/. In effect, all hydrogen-rich matter of the envelope is rapidly transfered to the main sequence companion, leaving a degenerate C-O core which eventually can radiate at the expense of its own cooling only.

If the analogous to Case B common envelope is not formed during this mass transfer, the resulting orbital period will not be changed significantly, comparing to its initial value. The object is ready now to undergo the symbiotic phase of its evolution. The more massive secondary component developes now into a red giant, but still remains detached from its Roche lobe.

The spherically symmetric mass loss from its surface, enhanced by the vicinity of the Roche lobe can lead to a significant amount of accretion of this hydrogen-rich matter onto the white dwarf component. This will lead to the ignition of thermonuclear burning in the hydrogen and helium shells. Such explanation of activity of symbiotic stars was proposed by Tutukov and Yungelson /1976/, Paczyński and Żytkow /1978/ and Paczyński and Rudak /1980/. The advantage of such scenario lies in that, it naturally leads to the increase of primary's luminosity to the level at which it can compete with giant component's luminosity.

So, essentially the nature of symbiotic variability is identical to nova-type activity. Different are only the values of the characteristic parameters governing the quantitative picture of events. That is why the symbiotic stars of type II /in Paczyński and Rudak's notation/ resemble so much some of the recurrent and slow novae in their behaviour.

As main parameters we would mention here three of them, namely:

- accretion rate onto the primary $\dot{M}_{ac}$
- mass of the primary M_{core}
- CNO abundance of accreted matter.

Let us briefly consider the significance of each of these factors successively.

$\dot{M}_{ac}$ for classical and recurrent novae are relatively low and range from 10^{-9} to $10^{-7} M_\odot/yr$, though the secondary component fills its Roche lobe. In effect, the hydrogen in the shell surrounding the degenerate core burns in the form of flashes only. A similar situation arises with type II symbiotic stars. But if $\dot{M}_{ac}$ exceeds the critical value $\dot{M}_{cr}$, stable burning takes place /Fig 1./. For details· see Paczyński and Rudak /1980/.

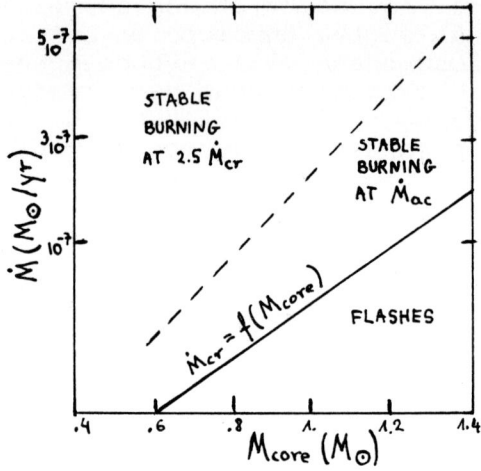

Fig 1. The ranges of hydrogen-burning character in shell for given core mass and accretion rate are shown. The falling matter has hydrogen content X=0.7, and shell burns on C-O core.

The second factor, M_{core}, plays a major role in determining the time scale of dynamical events on the surface of primary. This can be easily deduced from Fig 2., adopted from Paczyński /1971/. It represents the set of static envelopes built for different values of M_{core}.

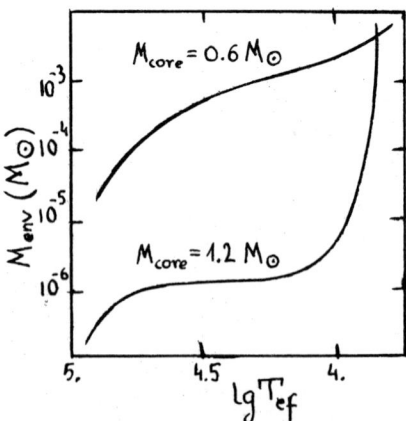

Fig 2. The dependance of static envelope mass surrounding the degenerate C-O core, on the effective temperature of the envelope /Paczyński 1971/.

We may see how sensitive the mass of envelope is to the value of M_{core}, and how sensitive the effective temperature of the primary is to any slight changes in M_{env} for the most massive cores.

The third factor governs the energetics of an outburst. It was shown by Starrfield et al. /1974/ that at temperatures exceeding 10^8 K in the flashing shell, the rate of decay of β^+ -unstable nuclei cannot be treated anymore to take place instantanously in the CNO cycle. Therefore the 10 - 100 folding enhancement of proton capturing nuclei in the burning matter is necessary to enable the explosive ejection of a massive envelope in fast-nova event. However, for slow nova outburst, the normal abundance is already sufficient to push out a considerable amount of the envelope by radiation pressure /Sparks et al. 1977, Prialnik et al. 1977/.

The outbursts observed for type II symbiotic stars resemble those of slow novae stars. This means that essentially, one would expect the normal CNO abundances in their ejecta.

The careful abundance analysis of both symbiotic stars and recurrent slow novae is necessary to find out if any differences in the CNO content can lead to the independent division between these two types of objects.

4. CONCLUSIONS

There is no doubt that symbiotic stars, whatever their character /single or binary stars/, represent advanced stages of stellar evolution.

Their distribution, though the sample is still not numerous, coincides with the old disk population /Boyarchuk 1975, Wallerstein 1980/.

The question, to what extent the symbiotic binaries constitute a common family with the cataclysmic binaries and double-core planetary nebulae cannot be answered before new observations are carried out. Though in few cases confusion arises in the proper classification of a given object as either symbiotic or e.g. recurrent nova, there is no doubt yet that the formation of binaries with initial orbital periods well over 100 days will favour their development to the symbiotic phase, provided the common envelope during the rapid mass transfer phase is not formed.

To understand the detailed processes which take place in symbiotic stars simultaneous observations in optical, UV and X-ray ranges during their activity are necessary. The abundance analysis, particularly of CNO is also of great importance.

We should also expect positive results in looking for

possible content of s-process products. If the cool component is really a well evolved red giant with deep convection and effective mass loss, those elements will be abundant enough in the environment of symbiotic stars to be observed.

Acknowledgments: We would like to thank Prof. B. Paczyński for his many stimulating discussions and comments on the subject of symbiotic stars. Also many thanks to Dr. R. Sienkiewicz for his critical comments, and to Drs. J. Ziółkowski, W. Krzemiński and R. P. Olowin for careful reading of the manuscript.

REFERENCES

Allen D.A., 1980, M.N.R.A.S., 190,75.
Bath G.T., 1977, M.N.R.A.S., 178,203.
Boyarchuk A.A., 1975, in Variable Stars and Stellar Evolution, ed. Sherwood and Plaut.
Ciatti F., Mammano A., Vittone A., 1979, Asiago Astrophysical Observ. preprint.
Cowley A., Stencel R., 1973, Astrophys. J., 184,687.
Kenyon S., Cahn J. H., 1979, IAP 79-40.
Paczyński B., 1970, Proceedings of the IAU Coll. No 6.
Paczyński B., 1971, Acta Astr., 21,417.
Paczyński B., 1980, Highlights of Astronomy, p.27.
Paczyński B., Rudak B., 1980, Astr. Astrophys., 82,349.
Paczyński B., Żytkow A., 1978, Astrophys. J., 222, 604.
Prialnik D., Shara M.M., Shaviv G., 1977, in Novae and Related Stars, Reidel, p.220.
Robinson E.L., 1976, Ann. Rev. Astr. Astrophys., p.119.
Slovak M.H., 1981, paper presented at the 157th meeting of the AAS.
Sparks W.M., Starrfield S.G., Truran J.W., 1977, in Novae and Related Stars, Reidel, p.219.
Starrfield S., Sparks W.M., Truran J.W., 1974, Astrophys. J. Suppl., 28,247.
Wallerstein G., 1980, preprint.

ON THE MODEL OF SYMBIOTIC STARS

A. V. Tutukov and L. R. Yungelson
Astronomical Council, USSR Ac. of Sci.
(read by A. V. Fiederova)

Abstract. We discuss conditions necessary for appearance and discovery of the symbiotic star phenomenon within the model of a binary consisting of a red (super)giant 3 $M_\odot$ not filling the Roche lobe and of an accreting hot degenerate CO-dwarf 0.8 $M_\odot$. Within this model "classical" symbiotic stars may exist only within a narrow region of mass accretion rates and separations of components: $10^{-7} \lesssim \dot{M} \lesssim 3 \cdot 10^{-7}$ $M_\odot$/y and $3 \cdot 10^{13} \lesssim a \lesssim 2 \cdot 10^{14}$ cm. The evolutionary status of symbiotic stars and related objects and the mechanisms of their variability are discussed.

Introduction

Symbiotic stars are relatively rare irregular variable stars with spectra where molecular absorption bands are combined with high-excitation emission lines. The most complete review of their properties was given by Boyarchuk (1981). The best explanation of the features of symbiotic stars is provided by a model of a binary with one component a cold (super)giant (G-M), and the second - a hot star (Boyarchuk 1969, 1981). The binary nature of some symbiotic stars is directly indicated by variations of their radial velocities and light curves. The measured orbital periods of symbiotic stars are from ~ 1 year to 26.7 years. Both components are immersed into an H II region with a radius of about $10^{15} - 10^{16}$ cm, $n_e \gtrsim 10^6 - 10^7$ cm^{-3}, T $\sim 10^4$ K. Wright and Allen (1978) who studied the radioemission of circumstellar matter found that the rate of mass inflow into the nebula is $10^{-6} - 10^{-5}$ $M_\odot$/y. This is close to the outflow rates for Miras. The matter for the nebula is probably provided by stellar wind from cold star. It is possible that also a hot star ejects some matter into the nebula.

Tutukov and Yungelson (1976) have shown that it is possible that hot components of symbiotic stars are degenerate carbon-oxygen dwarfs, the luminosity of which is provided by shell hydrogen burning of accreted stellar wind matter. The temperatures of carbon-oxygen dwarfs may reach values up to $\sim 10^6$ K (Pottash 1981). Temperatures of the same order are determined for the sources of excitation in some symbiotic stars (e.g. AG Dra, Oliversen et al. 1980). The ultraviolet emission of hot dwarf ionizes the matter of the nebula. A degenerate dwarf in a <u>wide</u> system is formed after the loss by the initially more massive component of its extended hydrogen-rich envelope.

Slovak and Africano (1978) have discovered that the symbiotic star CH Cyg is oscillating on the time-scale of about 5 minutes with an amplitude $0.^m02 - 0.^m04$. This indicates the presence of a compact energy source in this system. Allen (1980a) who studied line-widths in the spectra of some symbiotic stars has shown that for their hot components the value of $(M/M_\odot) / (R/R_\odot)$ is not less than 5, but for HM Sge it even exceeds 25. The data of Slovak and Africano and of Allen point also to the possibility of the presence of degenerate carbon-oxygen dwarfs in symbiotic stars.

There exists a number of objects with features close to those of "classical" symbiotic stars: stars with quite the same optical, IR and radio features, but without detected variations of visual magnitude. Those objects are sometimes called BQ[ ] -stars (Ciatti et al. 1974). Moreover, there are so-called slow novae, for which only one outburst was detected so far. Their spectra before the outbursts were similar to the spectra of symbiotic stars. There does not exist any unique classification of symbiotic and related stars. E.g. the same star AG Peg is named symbiotic (Boyarchuk 1970) and a slow nova (Allen 1980b) star. Both types of systems contain K or M (super)giants or Mira variables (Allen 1980b). It is possible that all objects with spectra in which features of a cold giant are combined with emission lines differ only in the frequency and amplitude of the outbursts.

Paczyński and Rudak (1980) suggested the classification of symbiotic stars based on differences in the kind of their activity and its sources. Symbiotic stars of type I have quasiperiodic variability in the time-scale of several months with amplitudes not greater than 4^m ; the high-excitation emission features are observed together with late giant features in the minima of brightness. The high excitation lines disappear when the brightness increases.

On the basis of results of computations of hydrogen burning in the envelopes of accreting carbon-oxygen dwarfs Paczyński and Rudak have supposed that hydrogen burning in shell sources of hot components of type-I symbiotic stars is stationary, and that their optical variability is caused by variations of accretion rate which change the effective temperature of a dwarf almost without changing their luminosity. A typical representative of these stars is Z And.

In the spectra of type-II symbiotic stars low-excitation emission lines are observed, but the degree of excitation rapidly increases after the outbursts when the visual magnitude grows up to $\Delta m \approx 5^m$. During the outburst ejection of matter from a hot component is possible, and it may possess the stellar wind. The list of type-II symbiotic stars compiled by Paczyński and Rudak includes some stars named as slow novae by other students (e.g. V 1016 Cyg). Paczyński and Rudak have supposed that the activity of type-II stars is caused by nonstationary hydrogen burning in the envelopes of accreting dwarf components. Typical for type-II symbiotic stars is HM Sge.

The Model

The aim of the present paper is to study conditions necessary for the phenomenon of symbiotic stars. We shall study the model of a binary star that contains a red (super)giant accompanied by a hot carbon-oxygen dwarf. We take the mass of the dwarf equal to 0.8 $M_\odot$, that of the (super)giant - 3 $M_\odot$. These values of masses are quite typical because masses of carbon-oxygen dwarfs are 0.5 - 1.4 $M_\odot$, and the mass ratio in symbiotic stars is 3 to 4, while the lower limit of estimated masses of observed cold components is 3 $M_\odot$ (Boyarchuk 1970).

The cold component loses matter by stellar wind. Part α of this wind is captured by a dwarf:

$$\alpha \approx \frac{r^2}{4a^2} \approx \frac{G^2 m^2}{v^4 a^2} \approx \frac{m^2 R^2}{4 M^2 a^2} . \qquad (1)$$

Here a is the distance between components, m - the dwarf mass, $r = 2Gm/v^2$ - the radius of capture by the white dwarf, $v = (2GM/R)^{1/2}$ - the velocity of stellar wind assumed to be equal to the escape one, M - the mass of the giant, R - its radius. The accretion onto the white dwarf supports the activity of the nuclear burning shell. The luminosity of the shell is limited by the Paczyński-Uus limit:

$L/L_\odot \lesssim 6 \cdot 10^4 \, (m/m_\odot - 0.5)$.

The system under study may be described by two parameters: the mass loss rate by the (super)giant $\dot{M}_G$ and the distance between components (Fig. 1). Employing (1)

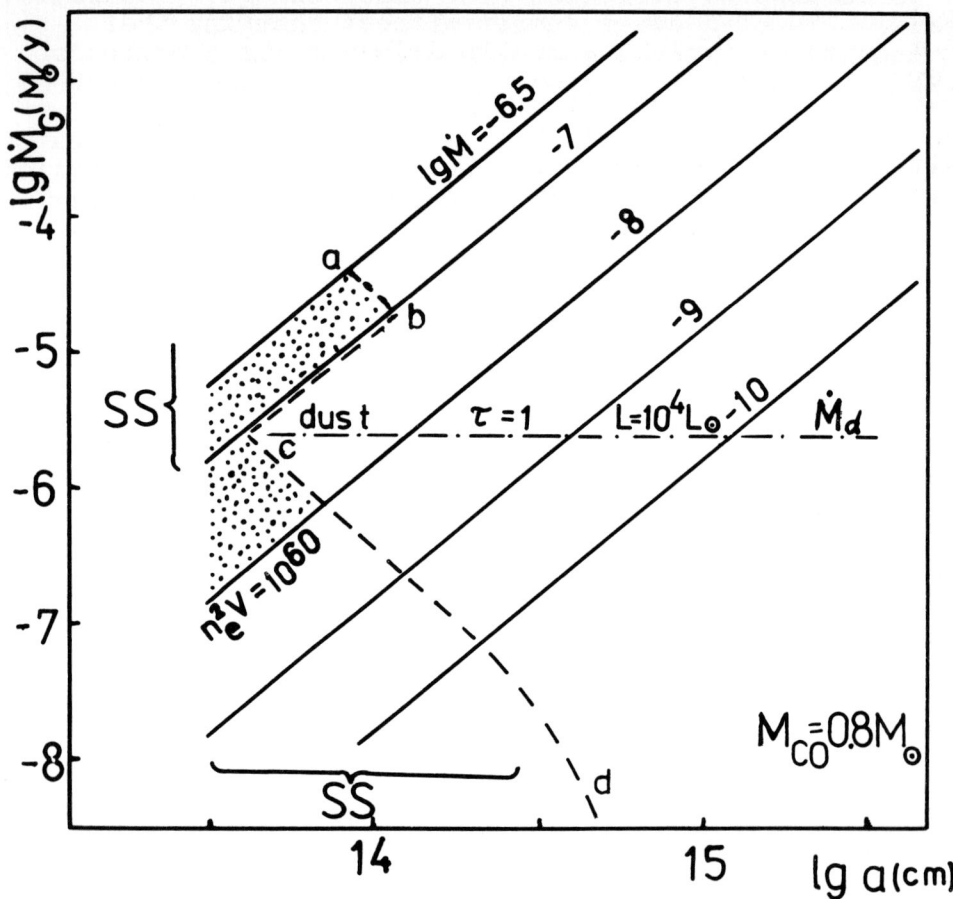

Figure 1. The relation between the separation of components and mass loss rate by giant 3 $M_\odot$ for wide binary systems. The lines of constant mass accretion rate by 0.8 $M_\odot$ carbon-oxygen degenerate dwarf are marked by values of lg $\dot{M}$ in $M_\odot$/y. Lines a-b and c-d limit the region where the system is not screened by dust from the (super)-giant envelope. Type-I symbiotic stars exist in the dotted region where $\dot{M} \gtrsim 10^{-7}$ $M_\odot$/y, in the region of lower $\dot{M}$ type-II symbiotic stars exist. The values of a and $\dot{M}_G$ observed in symbiotic stars are indicated. $\dot{M}_d$ is the mass loss rate for which the optical thickness of the dust in the envelope is 1 (for $L_G = 10^4$ $L_\odot$).

we may plot in Fig. 1 the lines of constant accretion rate $\dot{M} = -\dot{M}_G$. (We assume for definiteness that $R = 3 \cdot 10^{13}$ cm.) Paczyński and Żytkow (1978) have shown that hydrogen burning in the thin shell sources of accreting degenerate carbon-oxygen dwarf 0.8 $M_\odot$ is stationary if $10^{-7} \lesssim \dot{M} \lesssim 3 \cdot 10^{-7}$ $M_\odot$/y. There does not exist any systematic study of conditions of stationary burning for other values of m. However, Sienkiewicz (1980) has shown that the thin sources are thermally stable only for $\dot{M}$ close to maximal, i.e. $\sim 10^{-7}$ $M_\odot$/y for all dwarfs 0.6 $M_\odot \lesssim M < 1.4$ $M_\odot$. If the accretion rate exceeds about $3 \cdot 10^{-7}$ $M_\odot$/y, the accreted matter does not have enough time to burn out, the dwarf's envelope expands, cools out, its emission stops to ionize the outflowing matter and the symbiotic star phenomenon disappears. If the accretion rate is lower than about 10^{-7} $M_\odot$/y, hydrogen burning is nonstationary. Hydrogen burns in outbursts. Between the outbursts the matter is accumulated in the envelope, on the base of which the temperature is not high enough for hydrogen burning. If $\dot{M} \approx 10^{-10} - 10^{-7}$ $M_\odot$/y, the interval between outbursts is $10^2 - 10^7$ years (Tutukov and Ergma 1980). During the outbursts the radius and luminosity increase for $(10 - 10^3)$ years, the effective temperature decreases, and if the white dwarf does not have stellar wind strong enough to prevent its transformation into a red giant, the symbiotic star phenomenon disappears. Thus, stationary hydrogen burning necessary in the Paczyński-Rudak type-I symbiotic star model is possible in an extremely narrow interval of $\dot{M}$. If the conditions of stationary burning are not fulfilled ($\dot{M} < 10^{-7}$ $M_\odot$/y), the energy liberation during accretion determines the luminosity of the dwarf.

High-excitation emission lines typical for spectra of symbiotic stars appear in the compact H II regions around them. For the existence of a stationary H II region the flux of ionizing quanta must be high enough for primary ionization of all incoming from the giant matter and for ionization of hydrogen after recombination. We may write down this condition as

$$\frac{\dot{M}_G}{m_H} + \int_{R_d}^{R_{H\bar{II}}} \left(\frac{\dot{M}_G}{4\pi x^2 V m_H}\right)^2 \alpha(T) 4\pi x^2 dx \lesssim \frac{\dot{M}_G r^2}{4a^2} \frac{\varepsilon}{E} \cdot \quad (2)$$

Here m_H is the mass of a hydrogen atom, $\alpha(T) = 2.2 \cdot 10^{-13}$ cm^3/s for $T = 10^4$ K - recombination coefficient of hydrogen, ε - energy liberated by the burning of 1 g of hydrogen or by accretion of 1 g of it, $E = 2.2 \cdot 10^{-11}$ erg - energy of L_α -quantum. The integration of (2) must be performed from the dust formation radius R_d in the stellar wind, because if the border of H II region were beyond R_d, dust would screen out the giant, as the dust incoming into the H II zone does not evaporate. Taking the dust formation temperature $T_d \approx 10^3$ K, we find $R_d \approx 10^{12.3} (L_G/L_\odot)^{1/2}$ cm. Substituting $v^d = 5.15 \cdot 10^7$ cm/sec for $M_G = 3 M_\odot$, $R_G = 3 \cdot 10^{13}$ cm, and integrating (2) we obtain

$$\dot{M}_G \lesssim 10^{-22.4} R_d \left(\frac{10^{12.7} \varepsilon}{a^2} - 1 \right) M_\odot/y . \qquad (3)$$

The value of ε is $10^{18.8}$ erg/g for nuclear energy generation, and $\varepsilon = 10^{17}$ erg/g for accretion. In Fig. 1 the curves a-b and c-d correspond to condition (3) (we assumed that $L_G = 10^4 L_\odot$ for determination of R_d). According to Boyarchuk (1970) for the parts of nebulae that radiate in the continuum the relation $n_e^2 V \gtrsim 10^{60}$ cm^{-3} is valid (V is the volume). Taking into account that most of the ionizing quanta are spent on ionization of recombined atoms we may write down:

$$\alpha(T) \cdot 10^{60} \lesssim \dot{M} \frac{\varepsilon}{E} \quad \text{or,} \quad \dot{M} \gtrsim \frac{10^{70.8}}{\varepsilon} . \qquad (4)$$

This means that if the dwarf radiation is due to stationary shell-hydrogen burning, the condition of the discovery of continuum radiation of nebula $n_e^2 V \gtrsim 10^{60}$ cm^{-3} is always fulfilled. However, it is not fulfilled if the luminosity of the dwarf is due to accretion. In this case we may take the discovery of hydrogen recombination lines, e.g. the H_β -line, on the continuous background as the condition for discovery of a symbiotic star. If we require that the same amount of energy per 1 Å must be emitted by nebula in the H_β -line of width $\Delta \lambda$ as emitted by the cold component in the same spectrum region, we obtain:

$$L_{H_\beta} \approx \alpha_{42} n_e^2 V \frac{1.97 \cdot 10^{-8}}{\lambda \Delta \lambda} \gtrsim L_{mon} .$$

Substituting $\alpha_{42}(10^4 K, 10^6$ cm$^{-3}) \approx 3.1 \cdot 10^{-14}$ cm^6/sec (Brocklehurst 1971), $L_{mon} = 2.2 \cdot 10^{31}$ erg/sec/Å, $\lambda = 4861$ Å we obtain

$$n_e^2 V \gtrsim 1.8 \cdot 10^{56} \Delta \lambda \text{ cm}^{-3} .$$

As for stars under consideration $\Delta \lambda$ is of the order of several tens of Å, we may take $n_e^2 V \gtrsim 10^{58}$ cm^{-3}. Then the nebula is detectable if $\dot{M} \gtrsim 10^{-8} M_\odot/y$.

One more limitation on the emergence of the symbiotic star

phenomenon is due to the formation of dust in the matter outflowing from the (super)giant. If the H II region is not developed enough and does not stretch out up to the dust formation radius R_d, then the star will be observed as a symbiotic star only if the optical thickness of the dust is lower than 1. Taking $\mathcal{X}d \approx 10^2$ cm^2/g we may estimate that the dust does not screen the star if

$$\dot{M}_G \gtrsim \dot{M}_d \approx 10^{-7.6} (L_G/L_\odot)^{1/2}. \qquad (5)$$

Discussion

Using expressions (3)-(5) we may distinguish the region of symbiotic stars in the M_G-a diagram (Fig. 1). From the side of small a it is limited by the line $a = R_G$. Symbiotic stars with stationary shell sources of hydrogen burning the variability of which is due to the variability of accretion rate may exist in the region limited by the lines $\dot{M} \approx 10^{-7}$ $M_\odot$/y and $\dot{M} \approx 3 \cdot 10^{-7}$ $M_\odot$/y, and by the line a-b corresponding to the condition of the stationarity of the H II region without dust. If a and M_G are greater than those corresponding to the line a-b, the star is screened out by dust and becomes an infrared object which is possibly variable.

In the region $\dot{M} \lesssim 10^{-7}$ $M_\odot$/y the existence is possible of objects with hot components, the luminosity of which is fed by liberation of energy at accretion. Dwarf components of these stars must be cooler, and the sizes of the emitting regions of their nebulae smaller than in symbiotic stars with stationary hydrogen burning. If in such binaries M_G and a exceed the values corresponding to the line c-d, but $\dot{M}_G \lesssim 10^{-6} - 10^{-5.5}$ $M_\odot$/y, they will possess considerable infrared excesses. In those stars hydrogen burns out in outbursts separated by $10^2 - 10^7$ years. During the outbursts the luminosity of a star increases by several orders of magnitude, remains quite constant for several tens - several thousands of years (if the dwarf does not lose mass) and afterwards decreases to the initial state (Paczyński and Żytkow 1978). In "low" state one may probably identify these objects with BQ[ ]-stars, in "high" state - with symbiotic stars of type II or with slow novae. If $M_G \gtrsim 10^{-5.5}$ $M_\odot$/y, the dust completely reprocesses the optical radiation of stars into the infrared one, transforming them into infrared sources. The position of the region of symbiotic stars depends on the masses of components (see (1)).

Observed symbiotic stars have the separations of components $a \lesssim 2 \cdot 10^{14}$ cm ($P \lesssim 10^{4}$ days), and mass loss rates $\dot{M}_G \gtrsim 10^{-6}$ $M_\odot$/y (Wright and Allen 1978). These values of a and $\dot{M}_G$ cover the interval of parameters for which the excitation emission spectra of symbiotic stars may be explained by stationary shell hydrogen burning. Note that for given masses of components the interval of a is very narrow - only about 0.5 in lg a (see Fig. 1).

The energy generation by nuclear burning or by accretion can explain the source of emission spectrum excitation, but the source of observed variability still remains unclear. Accretion occurs through the disk. It is possible that the matter firstly accumulates in the ring and afterwards is accreted through the disk. Such a model was suggested by Osaki (1974) for explanation of large outbursts of U Gem stars that occur with the quasiperiod from tens to hundreds of days and have amplitudes up to $4^m - 5^m$. The accumulation of matter in the disk may be modulated by variations in the mass outflow rate from the cold component. The mass outflow rate may be modulated by e.g. filling now and again the Roche lobe by the variable cold component (CH Cyg, see Luud 1980) or by the instability of a Roche-lobe-filling star with a deep convective envelope (Bath 1978). The disk may also be unstable and/or have a nonhomogeneous structure. Probably the observed diverse variability of symbiotic stars may be explained by overlapping of several mechanisms.

In Fig. 1 the evolutionary track of a binary system passes from below to the top. The main parameter that determines the evolution is the mass loss rate by the cold component. As M_G grows the system may successively pass through the stage of BQ[ ] -star, which after the outburst may be identified with type-II symbiotic stars or slow novae, and afterwards - through the stage of a type-I symbiotic star. In systems which are wider than $(1.5 - 2) \cdot 10^{14}$ cm the symbiotic star phenomenon probably is not observed because at $\dot{M}_G \gtrsim 10^{-6}$ $M_\odot$/y they are completely screened by dust and transform into infrared sources.

If a (super)giant has a companion, it may lead to the concentration of outflowing matter in the orbital plane of the system, and formation of an optically and geometrically thick gaseous disk. The outflow of matter from the (super)-giant ceases when the mass of its hydrogen envelope decreases to $\sim 10^{-3}$ $M_\odot$ and the envelope begins to contract. In this stage, if the disk-like envelope of the system becomes transparent in polar directions, then due to radiation scattering of the central star one will observe a bipolar nebula, like Roberts-22, V 645 Cyg, CRL 618, IV Zw 67, M1-92 (Allen et al. 1980). In these nebulae the central star is an O-F supergiant. In some ($10^3 - 10^4$) years the supergiant will transform into a carbon-oxygen dwarf - a usual nucleus of planetary nebula which is also able to lose matter. When the radius of the expanding lost envelope reaches $\sim 3 \cdot 10^{17}$ cm, it will be ionized by the nucleus radiation and become a usual planetary nebula. Ciatti et al. (1978) have noted the similarity between the nebulae around HM Sge and V 1016 Cyg, which are classified as type-II symbiotic stars and slow novae, and the most compact planetary nebulae. The bipolar structure may be a peculiar feature of the planetary nebulae with binary cores. According to Seaton (1968) a considerable part of planetary nebulae has a bipolar structure. It is an extremely hard task to discover the binary nature of a planetary nebula that was formed from BQ[ ] or a symbiotic star due to the long orbital period of the system and to large difference in the luminosities of components, because the "old" dwarf without accretion rapidly cools.

Let us estimate the number of symbiotic stars in the Galaxy. The mass function of binary stars is

$$\frac{d^2 N}{d\lg a \; d\lg(M/M_\odot)} = \frac{1}{6}\left(\frac{M}{M_\odot}\right)^{-2.5}$$

According to Kraitcheva et al. (1978) the number of double stars per unity of lg a is constant for $10^{14} \lesssim a \lesssim 10^{17}$ cm. The lifetime of the red (super)giant is $\tau \approx M_G/\dot{M}_G \approx 10^5$ years for typical $\dot{M}_G \approx 10^{-5} M_\odot/y$ (see Fig.1). The mass interval of cold components is 3 to 10 $M_\odot$ (Boyarchuk 1970). Taking all these data we obtain:

$$N_{SS} \approx \frac{1}{12} \int_3^{10} \frac{10^5}{\left(\frac{M}{M_\odot}\right)^{2.5}} d\left(\frac{M}{M_\odot}\right) \approx 10^3 \;.$$

This number satisfactorily agrees with the estimate of the number of symbiotic stars in the Galaxy by Boyarchuk (1970) $\sim 10^3$, which is based on the counts of these stars in the vicinity of the Sun. We may increase our estimate if the symbiotic star phenomenon may also appear in systems with masses of components lower than 3 $M_\odot$, because the carbon-oxygen white dwarf may form in all binaries that are wide enough. The reason for the absence of stars with $M_G < 3 \; M_\odot$ among symbiotic stars with estimated giant component mass is still obscure. One of possible explanations is short duration of the mass loss stage with $\dot{M}_G \gtrsim 10^{-6}$ $M_\odot$/y by stars with $M_G \sim M_\odot$, and their lower mean luminosity. Besides, usual estimation of masses M_G according to their spectral type makes them unreliable.

According to Sienkiewicz (1980) the maximal effective temperature of carbon-oxygen dwarfs with hydrogen burning in the envelopes strongly depends on their mass. One may approximate this dependence by $\lg T_e \approx 4.7 + M/M_\odot$ and use it for the estimation of the lower limit of the mass of the components of symbiotic stars. Our analysis is relevant if the system contains a carbon-oxygen dwarf. In principle the existence of systems where a cold supergiant is accompanied by a helium dwarf or a neutron star or a black hole is possible. A degenerate helium dwarf may form in a binary system from a star with a mass lower than 2.5 - 3 $M_\odot$ due to mass exchange. Hydrogen burning in the shell source of an accreting helium dwarf is unstable, the time-scale of instability is not less than $\sim 10^2$ years (Sienkiewicz 1980). Between the outbursts the luminosity of the dwarf may be supplied by accretion with $\xi \sim 10^{17}$ erg/g and by cooling. Such a system may also be observed as a star with emission lines in the spectra. It is also possible for a carbon-oxygen dwarf in the symbiotic star to be accompanied by a giant with a helium core, if the mass loss rate by a giant exceeds $10^{-8} - 10^{-7}$ $M_\odot$/y.

If the companion of a (super)giant is a neutron star or a black hole, then the supercritical accretion ($\dot{M} \gtrsim 10^{-8}$ $M_\odot$/y) may lead to formation of an optically thick envelope around the compact object, in which the X-ray emission may be reprocessed into ultraviolet or optical emission. X-ray and ultraviolet emission may ionize the stellar wind matter. The degree of ionization and the temperature may be extremely high. It is possible that GX 1+4 is an example of such systems. It is a hard X-ray source in the optical spectrum of which high-excitation emission lines are discovered (Davidsen et al. 1977). In the case of supercritical accretion the formation of jets in polar directions, like those observed in SS 433, is possible.

Tutukov and Yungelson (1976) have mentioned that accretion onto a carbon-oxygen dwarf in a symbiotic star may increase its mass to the Chandrasekhar limit and change the thermal conditions of its interiors in such a way that at the instance when the mass of the dwarf reaches 1.39 $M_\odot$, its central density exceeds that in the 1.39 $M_\odot$ carbon-oxygen core of a single star. If $\rho_c \gtrsim 5 \cdot 10^9$ g/cm^3, the formation of a neutron star is possible after the Supernova explosion which is due to explosive carbon ignition.

To sum up, we may conclude that conditions necessary for phenomenon of a "classical" symbiotic star are fulfilled in a rather narrow interval of mass loss rates by cold components and of distances between components. The "region of existence" of symbiotic stars is limited by conditions for discovery of the emitting region and by conditions of the system screening by the dust. Outside this region, if mass loss rates by giant components are $\dot{M}_G \approx 10^{-7} - 10^{-6}$ $M_\odot$/y and a $\lesssim 10^{14}$ cm, the existence of stars with emission lines in spectra whose activity, however, differs from that of symbiotic stars is possible.

References

Allen, D.A.: 1980a M.N.R.A.S. 190, p.75.
Allen, D.A.: 1980b, M.N.R.A.S. 192, p.521.
Allen, D.A., Hyland, A.R., Caswell, J.L.: 1980, M.N.R.A.S. 192, p.505.
Bath, G.T.: 1978, M.N.R.A.S. 178, p.203.
Boyarchuk, A.A.: 1970, In: Eruptivnyje zvezdy, eds. A.A. Boyarchuk and R.E. Gershberg, Nauka, Moscow, p.113.
Boyarchuk, A.A.: 1981, Soviet Sci. Revs. - Astrophys. and Space Phys., ed. R. Syunyaev, in press.
Brocklehurst, M.: 1971, M.N.R.A.S. 153, p.471.
Ciatti, F., D'Odorico, S., Mammano, A.: 1974, Astron. Astrophys. 34, p.181.
Ciatti, F., Mammano, A., Vittone, A.: 1978, Astron. Astrophys. 68, p. 251.
Davidsen, A., Malina, R., Bowyer, S.: 1977, Astrophys. J. 211, p.866.
Kraitcheva, Z.T., Popova, E.I., Tutukov, A.V., and Yungelson, L.R.: 1978, Astron. Zh. 55, p.1176.
Luud, L.S.: 1980, Astrophysics 16, p.443.
Oliversen, N.A., Anderson, C.M., Cassinelli, J.P.: 1981, Bull. Amer. Astron. Soc. 12, p.819.
Osaki, Y.: 1974, Publ. Astron. Soc. Japan 31, p.429.
Paczynski, B., Rudak, B.: 1980, Astron. Astrophys. 82, p.349.
Paczynski, B., Zytkow, A.N.: 1978, Astrophys. J. 222, p. 604.
Pottash, S.R.: 1981, Astron. Astrophys. 94, L13.
Seaton, M.J.: 1968, In: IAU Symp. No. 34 "Planetary Nebulae", eds. D.E. Osterbrock and C.R. O'Dell, Reidel, Dordrecht, p.1.
Sienkiewicz, R.: 1980, Astron. Astrophys. 85, p.295.
Slovak, M.H., Africano, I.: 1978, M.N.R.A.S. 185, p.591.

Tutukov, A.V., Ergma, E.V.: 1980, Pisma Astron. Zh. 5, p.531.
Tutukov, A.V., Yungelson, L.R.: 1976, Astrophysics 12, p.521.
Wright, A.E., Allen, D.A.: 1978, M.N.R.A.S. 184, p.893.

DISCUSSION ON EVOLUTION

Kwok: It is interesting to note that a radiative-driven wind from the C,O core not only explains some of the nova characteristics, but it also prevents the C,O core to evolve backwards to become a red giant.

Rudak: That is true. Such a wind will tend to decrease the value of $\dot{M}_{acc}$. On the other hand, however, one would hardly expect the spherically symmetric accretion rate for these wide systems to be significantly larger than a maximal possible H-burning rate in the shell on a C-O core. For example, for the C-O core of 1 $M_\odot$, the accretion rate has to exceed 3×10^{-7} $M_\odot yr^{-1}$. Only in these pathological cases, the hot star would look like a giant or a supergiant, wuth its colour depending on the specific enthropy of accreted matter.

Cassatella: Is it expected from theory that during the thermonuclear burning on the surface of the hot primary, enough neutrons (rapid or slow) are liberated so that r or s-process elements are formed? Actually, Audouze et al. (1981 Astr. Ap. 93, 1) do observe s-process elements to be overabundant in the symbiotic star CI Cyg.
More in general, is it expected to see anomalies in the chemical abundances in symbiotic stars like is observed for example in classical novae?

Rudak: The reason for which the enhanced abundances of s-process elements are observed in some symbiotic objects lies not necessarily in the presence of a degenerate hot component burning shells on its surface. Well developed convection in the red giant component evolves both the origin of a slab with H, C and O, where slow neutrons are produced, and the effective transport of s-process elements onto the giant's surface.

Plavec: There seems to be a difference in vocabulary between the theorists and the observers, which may lead to misunderstandings and possible gaps - - apparent on real. The observers often talk about sub-dwarfs, meaning objects that lie between the MS and the white degenerate dwarfs, or rather to the left of the upper main sequence and above the white dwarfs (in fact, to the left of most of them, as well as above). The subdwarfs are believed to be evolving to the WD stage, but still not completely degenerate. Or they can be "helium stars" lying on a helium main sequence — with anevolutionary past and future which is rather obscure. It is these objects we seem to find most frequently in the symbiotics. I wonder if Dr. Rudak can comment on their position and function in his models.

Rudak: The hot subdwarfs of O or B spectral types cannot be expected to be the members of symbiotic stars with shell nuclear burning, because of too low gravity (log g ~ 6.0). I should not treat them also to be the progenitors of white dwarf components in symbiotics. The reason for this is that, according to Schönberner's calculations, they can become to be the helium-rich white dwarfs with relatively low masses well below 1 $M_\odot$. In nuclear burning models, very massive white dwarfs are necessary, if their envelopes are to be sensitive for any small changes in M_{env}.

Kwok: In Bath's model, he only needs a UV continuum source and surface nuclear burning is entirely unnecessary. Would you comment on that?

Rudak: If I understand you correctly, you were considering nuclear burning models to be artificially too much complex for symbiotic phenomena? I do not share your opinion. I should like to emphasize that in "simpler" models, with mass transfer via accretion disk onto a main sequence star, there is one crucial problem which still has not been overcome: the proper treatment of the physical phenomena in the envelope of red giant filling its Roche lobe, to find out how the mass transfer from this star looks like.

Nussbaumer: Your reply still leaves Kwok's question open. You say that you consider Bath's accretion disk model as reasonable. I guess that you also believe in your model. Does this imply that both these models would result in the same symbiotic phenomena?

Rudak: Generally yes. Let me remind you the similar situation with classical novae stars, where Bath's model competes with the model of Starrfield and his collaborators.

Kafatos: The accretion model for symbiotic stars requires, I think, main sequence stars of ~ 1 $M_\odot$. If the secondary is considerably smaller than a one solar mass main sequence star, then the inner regions are too hot and should give off soft X-rays.

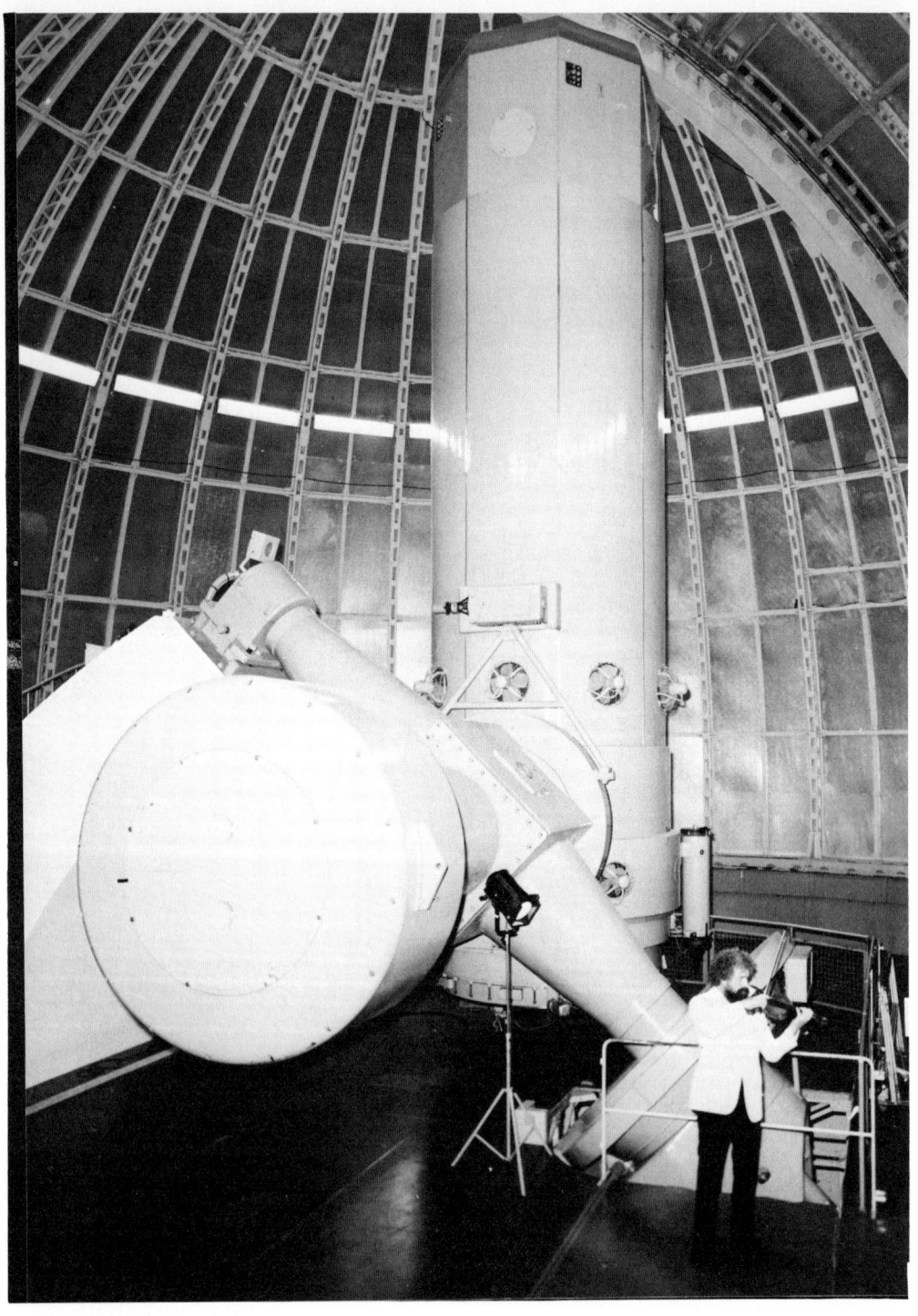

CONCLUDING REMARKS

J.P. Swings
Institut d'Astrophysique, Université de Liège, Belgium

What have we been talking about during the last three days? According to Boyarchuk and Plavec, it is a small class of celestial sources: old disk population objects, just like planetary nebulae. Their number in catalogs has increased from 21 (Boyarchuk's compilation) to about 100 (Allen's catalog(s)), and possibly newer catalogs could contain up to 10^3 members: this is still debatable, however, due to uncertainties on distance estimates. But, in any case, we have been considering a wide, and wild, series of targets, and if we wish to give the oath of ignorance Nussbaumer mentioned, we can get as big a number we wish: so let's leave this meeting with the impression that we tackled a vital astrophysical problem.

How can we define "Symbiotic Stars"? We are getting into deep water right away, but maybe we can try to agree on a few characteristics:
- presence of high excitation emission lines;
- presence of low temperature absorption features (even if they are not really seen!);
- presence of very inhomogeneous regions (gas and/or dust; optically thin and/or optically thick);
- conspicuous variations in the spectrum and in the light curve (stressed by several of you, Boyarchuk, Ciatti, Viotti, etc.; see types I and II of Paczynski and Rudak, mentioned again by Fiederova and by Boyarchuk e.g.), sometimes quiescent phases, smooth variations (with or without periodicity, depending on the wavelength region we are observing at), erratic variations, flickerings, i.e. just about any variability between very rapid changes (5 min or so) to periods of several hundred days, and up to 35 years. Of course, objects such as spectroscopic binaries with true eclipses ought to be easier to tackle on the theoretical side, and in that respect an object like AR Pav, as mentioned by Slovak, is a good candidate for more study;

— from IR data (see Allen's report, and references therein), there are at least two classes of symbiotics: S-type (75%; cool star colors in the 1-4 μ region), and D-type (20-25%; dust at $T \leq 800-1000°K$), and perhaps a third class, D', containing hotter stars (G-type e.g.) with much cooler dust. In addition, the IR variability leads Whitelock to define two classes, those objects with large amplitude (Mira-type), and those with small amplitude (normal late-type giants).

One thing is at least certain: we have added a lot to our knowledge, and/or to our confusion, by going to wavelength regions outside the visible: I shall try below to point out a few essential characteristics thus deduced for symbiotic stars.

a) The near infrared may give us the spectral type of the cool component, and is therefore very useful for classification purposes, as noted by Andrillat. It may also distinguish between objects of two types: those with poor and those with rich nebular spectra.

b) The infrared data and classes (S, D, D') correlate with dust temperature and variability, and give us very interesting knowledge of the presence of Mira variables (with periods such as 176, 387, 431, 580 days quoted by Whitelock).

c) The radio, where only about 10% of the objects are yet detected, probes different regions, and the power law index gives us some clue as to whether the object has an expanding shell, a wind or comes from a nova; I personally find fascinating the results obtained by V.L.A. techniques, e.g. the asymmetric structure + halo of V1016 Cyg and HM Sge that were reported by Kwok.

d) The ultraviolet data can be used at least in two ways:
(i) the first one is to know the continuum of the hot component, and in combination to visible and IR data, to deconvolve the energy curves and get the composition of the binary structure, such as B0 + M2 giant, as shown by Slovak and others; there may still remain a problem as to whether one deduces the true temperature of the star, or that of a disk: one has also to be sure not to try to extrapolate things too far from a very small wavelength region (Viotti, Keyes, Cassatella...).
(ii) the second is to use the lines in order to perform a diagnosis of densities, temperatures, extinction, abundances, and also, very interestingly, of the dimensions of some emitting regions: if the parameters are well known, models may be then derived. These lines analysis have been shown here by Nussbaumer and Kafatos e.g. One has to remain very careful, however, in applying the curves giving e.g. N_e or T_e from intensity ratios, and make sure that one is not trying

to achieve an unreasonable accuracy, but essentially obtain ranges
of values; or that one is really using the right lines for the right
zones and for deducing the appropriate astrophysical parameters in
the sort of nebula surrounding the objects we have been talking about.
The International Ultraviolet Explorer has obviously made a major
breakthrough in our observations (if not knowledge) of symbiotics,
and almost all the objects we considered at length in the sessions
on individual stars have been repeatedly observed with that satellite.
It is clear, however, that IUE does not solve all the problems, and,
as pointed out by Plavec, that data in the 900-1200 A are badly needed in order to convincingly get to know the nature of the hot component.

e) the X-ray (i.e. from the Einstein satellite) is of course the new domain, and we heard from Oliversen on the result of Anderson et al. on
AG Dra (and their conclusion on the presence of a white dwarf or a very
small compact object), and on the survey by Allen. The latter found only
three objects with X-ray fluxes, and these turn out to be three objects
that suffered slow-nova outburst(s): HM Sge, V1016 Cyg, RR Tel listed
here in order of decreasing flux, also in order of increasing coolness,
also in a normal sequence if one simply looks at the epoch of the (last)
outburst (if an e-folding time really exists). It is also very interesting to note that the X-ray emission may come from the interacting winds'
region in the models developed by Kwok.

f) So far I said nothing concerning visible data, although we heard an
interesting review by Ciatti (insisting on the fact that symbiotics are
variable, telling us about typical evolution of these stars, warning us
about the use of radial velocity curves to deduce without any doubt that
the objects are binaries, saying that the "flickering" I mentioned earlier is not necessary characteristic of symbiotics, mentioning that the
λ 6830 feature still remains unidentified,...). This fac
This was followed by Oliversen's report on H_α observations of a series
of stars to which we (i.e. Mrs. Andrillat and I) have added that proto-
planetaries, B[e] s, variables and symbiotics had been observed in the
same region in order to "test" the interacting winds theory for the formation of planetary nebulae.

If we now try to define symbiotic stars on the basis of Boyarchuk's
talk and the discussion that followed, I guess we can take two approaches
and say:
1 — A symbiotic star is a composite object, that suffers cyclic variat--
   ions, and that looks like a planetary nebula in the UV and like a
   late-type giant in the IR (Houziaux);
or try to be a bit more specific, and say:

2 - A symbiotic star in its quiescent phase comprises:
- the G-band, or absorption bands of TiO, H_2O, CO,..., and
- emission lines of HeII, or [NeIII], or [OIII], or of higher excitation, and varies on fairly large time scales.

It seems likely that the majority os symbiotic stars are binaries, although the situation is not crystal clear in all cases, as shown by Friedjung's talk and by this morning discussion. A wise word of caution was given by Plavec in the sense that one should not use criteria that are either too large or too narrow, so as not to get completly stuck.

Next, if I come to all the individual stars we discussed yesterday, according to an alphabetical order, so as not to put any bias on false groupings of objects, you will hopefully agree that it is impossible for me to summarize all the data that were thrown at us, although they were necessary to learn about. The objects actually included what several consider as the "prototype" of symbiotics, i.e. Z And, "fashionable" symbiotics (CI Cyg, AG Peg, RR Tel, HBV 475, and its counterpart in the LMC, HD 269227 (WN + M5) to become fashionable?), symbiotics with no forbidden lines in the visible (imagine the faces of Merrill, Struve, Thackeray, etc.!), but fortunately detected in the UV, such as AG Dra, what some of us include in the category of protoplanetary nebulae, like V1016 Cyg and HM Sge, weird beasts like RX Pup, marginal symbiotics such as CH Cyg, less fashionable symbiotics although quite interesting like YY Her, SY Mus, and even a "symbiotic or no symbiotic", in any case a nova, V4049 Sgr, and an object going from an F star to an M star (with [OIII] 5007 emission) to an F star again, PU Vul. I am pretty sure that each of you has his or her pet star, and even if recommendations were made as what star to observe in priority, nobody would agree (except maybe in the case of HBV 475 where an outburst is announced in 1982). I must personally say that I was impressed by the observations (essentially in the UV, but also in the visible) of several objects during eclipse(s), and to learn that e.g. permitted lines could be formed in a wind or in a shock, semi-forbidden lines in a low density envelope surrounding the system and the continuum in an accretion disk, keeping in mind that two very different things were to be considered: the eclipse, and the excitation mechanism(s). Also the fact that when a Mira exists in the systems, it keeps on displaying essentially the same IR light curve, whatever the outbursts in the visible and/or the modification(s) of the visible spectrum may be, is really quite impressive. The UV line profiles, within a same object, such as e. g. AG Peg or RX Pup, or from one to the other, as in the cases of "similar" objects like V1016 Cyg and HM Sge are truly something to try to understand, as well as the changes in the line ratios or the frequent case of CIV where the two components of the doublet at 1550 A have almost exactly the opposite ratio compared to what they "should" have.

CONCLUDING REMARKS

The next step is of course to attempt to build models of the symbiotics, although "Nature has not been able of producing simple symbiotics, but only unnecessarily complicated objects" (Plavec dixit). He gave us a very enthousiastic report about how to form natural symbiotics, what he calls "Planetary Nebula" symbiotics, by combining a late giant and a subdwarf. He told us that we needed sources of late-type, of nebular material, and of excitation and ionization. The same ingredients were also considered by Fiederova.

We also heard from Fiederova and from Rudak how the initial conditions in the binary system can be very important to create different types of symbiotics. Rudak showed us for instance that whether or not there existed a formation of a common envelope during the rapid mass transfer phase, one would obtain some types of cataclysmic variables or, maybe, what Plavec called a "cataclysmic symbiotic". The problem of CNO abundances was also stressed and shown to have important consequences on the amount of energy released, i.e. whether the object is a type II symbiotic, a slow nova or a fast nova. I will give you a bit later a recipe about to form your own "pet symbiotic", not to be quoted in astrophysical literature. In any case, Kafatos gave strong arguments in favour of the presence of a hot subdwarf in the system, on the basis of temperatures, radii, gravity, etc. obtained via IUE data, whereas accretion disks don't seem to be the most plausible explanation, although one of them seems to have been observed in AR Pav. In any case, one has to be careful about the temperatures involved, especially since the X-rays that would be expected are not observed.

To end up with a "summary", Recipe 1 tells you how to build up your typical symbiotic star.

In conclusion[1], I do think that more observations are needed: for instance very few polarimetric measurements have been made and many additional ones would probably be of great help in refining some models; more monitoring of interesting targets ought to be performed, in both hemispheres. I thus suggest that we reconvene in, say, 2 or 3 years to discuss all the new data we will have obtained in all spectral regions, as well as to hear, and probably criticize, the new esoteric, and/or exotic, theories that will certainly be imagined in the meanwhile. The cordial welcome we received here, the environment, the wheather, ... were so pleasant that I can think of no better meeting place to suggest than the Observatoire de Haute Provence. Thank you.

(1) not delivered at the meeting because, as in cithara concerts in India, the audience applauded afetr Recipe 1, indicating that it had heard enough...!

Recipe 1

HOW TO MAKE YOUR "TYPICAL" SYMBIOTIC STAR

<u>Start with</u>: a hot source
   and/or
   a cool source
   some gas (with abundances within a factor 10 of cosmic abundances)
   some dust

<u>Add</u> (according to personal taste)
some: black-body radiation(s); free-free emission; two photon emission; bound-free emission; thermal emission or re-radiation; collisions; chromospheric or coronal activity; line fluorescence;.....
a little wind(s) and/or shock(s)

<u>Mix well</u>; let expansion take place, and, if necessary, the Zanstra temperature become $\geqslant$ 100 000°K.

<u>Presentation</u>: make sure to distribute the right species onto an accretion disk, and don't worry about hot spots, inhomogeneities, Roche lobe overflows, variations,....

<u>Monitor</u>: line intensities with powerful rheostat

<u>RESULT</u>: Don't call it "SYMBIOTIC"

   that's "ignorance"

   but: <u>observe it as often as possible</u>
        in X-ray, UV, visible, IR, radio...

SUBJECT INDEX*

Abundance, chemical composition   40, 45, 70, <u>88-92</u>, 97, 205, 209, 294, 298, 301.
Accretion disk   11, 18, 93, 111, 112, 138, 142, 143, 156, 206, 237, 239, 240, 241, 245, 246, 250, 251, 258, 259, 265, 271, 272, 276, 277, 295, 300, 301.
Algol type symbiotic star   169, 237-242, 246, 247, 249, 250.

Be, BQ[]   4, 47-49, 67, 134, 208, 225, 229, 239, 246, 284, 289-291, 299.
Binary model, system, binarity   4, 5, 11, 13, 17, 38, 50, <u>65-66</u>, 69, 70, 103, 104, 127, 131, 136, 138, 142, 159, 166, 167, 169, 176, 189, 195, 199, 201, 206, 221, <u>231-251</u>, 254, <u>256-265</u>, <u>270</u>, 275-294, 299, 300.
Black hole   271, 277, 292.

Cataclysmic variable   103, 238, 242, 244, 258, 275-277, 281, 301.
Chromosphere   1, 2, 88, 181, 188, 255, 264, 270, 272.
Classification   3, 6, 17, 28, 186, 197, <u>225-230</u>, 253, 275, 284.
Cool, late-type component   3, 4, 11, 14, 31, 40, 43, 44, 47, 49, 52, 64, 147, 166, 171, 173, 197, 209, 226, 227, 233, 234, 239, 243, 253, 255, 261-265, 282, 283, 291, 293, 298.
Corona, coronal lines   3, 61, 67, 88, 96, 118, 131, 230, 255, 263, 264, 270, 272.

Disk   40, 65, 67, 128, 144, 234, 259-262, 290, 298.
D-type symbiotic star   18, 24, 26, <u>27-40</u>, 44, 104, 219, 226, 249, 262, 263, 271, 272, 298.
Dust, silicate grains   6, 18, <u>27-40</u>, 45, 52, 68, 112, 127, 140, 158, 163, 207, 209, 216, 226, 249, 254, 261, 263, 287-289, 297, 298.

Eclipse, eclipsing binary   65, 66, <u>141-148</u>, 152, <u>153-156</u>, 165, 166, 179, 192-194, 196-199, 254, 256, 258-260, 297, 300.
Eddington limit   18, 144, 264, 276.
Electron density, temperature   5, 12, 52, 84, <u>86-88</u>, 92, 93, 96-98, 101, 111, 112, 116, 126, 129, 130, 142, 145, 147, 155, 158, 162, 163, 183, 191, 197, 202, 205, 209, 217, 226, 263, 269, 283, 298.
Emission measure   101, 117-120.

* The <u>underlined</u> pages are expecially devoted to the argument.

Evolutionary model, stellar evolution   5, 61, 109, 230, 232, 235, 236,
        237, 249, <u>275-295</u>.
Excretion disk   259.
EXOSAT   159.

Fluorescence   66, 134, 135, 136, 142, 185, 204, 206.
Free-free, bound-free   18, 19, 38, 112, 113, 126, 129, 141, 189, 200,
        205,
Giant component, star   4, 5, 21, 27-40, 49-52, 61-63, 67, 108, 126,
        131, 136, 140, 156, 159, 162, 176, 193, 195, 199, 210, 227,
        229, 234, 269, 272, 279, 292, 294, 299.
Hα 26, 50, 51, 57, 58, <u>71-82</u>, 84, 128, 133, 134, 152, <u>153-155</u>, 157,
        177, 178, 210, 219, 221, 229, 242, 299.
Hot component, star   4, 5, 11, 40, 50, 52, 64, 65, 84, 126, 163, 167,
        173, 183, 191-193, 195, 201, 233, 237, 238, 249, 254, 256,
        259-262, 264, 265, 270, 277, 283, 285, 294, 299.
Hot spot   138, 181, 256, 258.

Infrared   6, <u>27-56</u>, 67, 69, 112, 127, 158, 162, 201, <u>207</u>, 208, 209, 215,
        216, 222, 226, 228, 229, 262, 284, 289, 290, 298.
Interstellar extinction, lines, 2200 A band   10, 38, 39, 44, 98, 105,
        126, 143, 145, 147, 153, 158, 183, 192, 197, 205, 208, 221,
        298.
IUE   6, 10, 11, 26, 44, 85, 93, 98, 103, 104, 107, 110, 112, 113, 121,
        126, 131, 134, 141, 145, 158, 163, 175, 176, 183, 189, 197,
        201, 204, 213, 217, 218, 222, 228, 233, 243, 251, 259, 269,
        271, 272, 299.

Light curve   23, 125, 141, 157, 158, 165, 179, 180, 183, 184, 188, 189,
        195, 196, 201, 207, 209, 214, 215, 226, 250, 271, 300.
Line profile   51, 71-82, <u>97-98</u>, 112, 135, 158, 166, 192, 193, 196, 201,
        207, 209, 214, 215, 226, 250, 271, 300.
Luminosity   29, 143, 169, 192, 209, 237, 239, 245, 264, 270, 285, 292.

Magellanic Clouds   4, 38, 39, 173.
Magnetic field   255, 271.
Mass accretion, exchange, transfer, accretion rate   4, 12, 24, 29, 68,
        69, 70, 109, 118, 142, 143, 167, 193, 206, 210, 215, 230,
        235, 238, 239, 241, 243, 245, 249, 257, 259, 262, 264, 269,
        276, 277, 279, 280, 283, 285-289, 292, 294, 295, 301.
Mass loss, outflow   20, 21, 23, 26, 131, 136, 158, 210, 227, 230, 237,
        243, 260, 261, 264, 276, 282, 283, 286, 289, 290, 292, 293.
Mira type, long period variable   2, 29, 30, 40, 62, 64, 68, 158, 159,
        163, 203, 207, 215, 216, 219, 225, 228, 261, 262, 283, 284,
        298, 300.

SUBJECT INDEX

Molecular bands, TiO   1, 2, 45, <u>48-54</u>, 61, 62, 83, 126, 152, 155, 157, 162, 173, 196, 207, 209, 215, 225, 228, 232, 270, 271, 283, 300.
M-type spectrum, star   2, 14, 26, 47, 48, 92, 125, 162, 165, 173, 199, 203, 209, 221, 226, 228, 232, 237, 242, 249, 270.
Near-infrared   <u>47-60</u>, 128, 165, <u>173</u>, 298.
Neutron star   277, 292.
NGC 7027   40, 67.
Nova, nova-like   3, 4, 18-24, 26, 40, 48, 49, 62-64, 66-69, 116, 138, 157, 167, 169, 215, 225, 228, 229, 238, 239, 242, 244, 264, 271, 275, 279, 281, 284, 285, 289, 290, 294, 295, 299, 301.
Nuclear processes   6, 149, 159, 235, 237, 264, 265, 277, 282, 293-295.

OAO-2   103, 108, 201.
Outburst   5, 19, 23, 43, 51, 61-63, 68-70, 104, 121, 125, 128, 131-133, 138, 141, 143, 157, 158, 167, 169, 171, 176, 179, 183, 210, 215, 216, 227, 238, 241, 242, 260, 270, 281, 284, 286, 289, 290.

P Cygni line profile, spectrum   26, 64, 107, 126, 132, 137, 166, 189, 203, 205, 246, 249, 258, 260, 262.
Planetary nebula   4, 5, 49, 63-67, 69, 84, 98, 101, 136, 157, 163, 192, 206, 209, 210, 228, 229, 232, 239, 240, 242, 243, 246, 249, 250, 254, 281, 291, 297, 299, 301.
Polarimetry   <u>139-140</u>, 221, 301.

Quiescence   104, 125, 227.
QSO   271.

Radial velocity   2, 11, 13, 14, 128, 132-135, 140, 145, 149, 155, 161, 166, 167, 171, 176, 183, 185-189, 196, 197, 250, 256, 258, 265, 276, 299.
Radio   13, <u>17-26</u>, 28, 158, 162, 163, 165, 202, 210, 210, 226, 283, 284, 298.
Roche lobe   4, 65, 69, 138, 143, 156, 166, 167, 171, 210, 233, 235, 237-239, 241, 243, 246, 249, 250, 258, 262, 264, 276-279, 290, 295.
Rotation   179, 181, 218, 258, 261-263.

Single star model   12, <u>67</u>, 158, 159, 167, 232, 250, <u>254-256</u>, <u>270</u>, 271, 275, 281.
Space Telescope   112, 129.
Spectral type   29, 50-53, 125, 132, 149, 157, 173, 177, 185, 201, 214, 221, 222, 233, 234, 236, 243, 246, 255.

S-type symbiotic star   18, <u>27-40</u>, 44, 45, 104, 112, 197, 226, 228, 249, 262, 264, 271, 272, 298.
Subdwarf   69, 111, 131, 143, 193, 215, 239, 244-246, 251, 260, 276, 294, 295, 301.
Sun   3, 96.
Supergiant   40, 49, 62, 186, 227, 245, 272, 291, 294.
Symbiotic phenomenon   210, <u>227</u>, 283, 291.
Symbiotic star, classification, model   <u>11-14</u>; 61, 63-64, 98-99, <u>225-230</u>, 239-246, <u>253-272</u>, <u>283-293</u>, <u>297-301</u>.

Thomson scattering   218, 261.
Type I, II symbiotic star   6, 279, 281, 284, 285, 291, 297.
Two photon emission   111, 126.

Ultraviolet spectrum   6, 10, 84, <u>85-113</u>, 126, 128, 134, 135, <u>141-146</u>, 158, <u>175</u>, <u>182-184</u>, <u>191-193</u>, 197-199, 201, <u>203-206</u>, 210, <u>213-214</u>, <u>217-218</u>, 226, 228, 229, 233, 254, 255, 298, 300.

White dwarf, degenerate dwarf   12, 93, 118, 144, 167, 171, 210, 238-240, 243, 277-279, 284-287, 291, 294, 295, 299.
Wind   13, 18-24, 26, 40, 66, 84, 128, 142, 143, 162, 166, 177, 209, 219, 232, 235, 236, 241, 245, 249, 250, 254, 257-265, 271, 285-287, 294, 299, 300.
Wolf-Rayet   47, 52, 64, 106, 173, 201, 209, 214, 218, 219, 246, 249, 261.

X-ray   68, 96, <u>115-121</u>, 129, 158, 177, 202, 210, 226, 244, 255, 264, 271, 277, 281, 292, 295, 299, 301.

Yellow symbiotic star   <u>30</u>, 40, 62, 226, 263, 298.

Zanstra temperature   110, 113, 121, 233, 244.

STAR INDEX

ANDROMEDA
Z   1, 4, 5, 9, 11, 13, 30, 43, 57, 62, 65, 69, 72, 73, 82, 83, 86, 87,
    88, 92, 96, 97, 98, 100, 101, 103, 106, 110, 111, 112, 113, 125-130,
    143, 145, 192, 226, 227, 255-264, 269, 276, 285, 300.
EG 31, 50, 72, 73, 74, 256.

AQUARIUS
R   1, 13, 17, 24, 30, 31, 38, 65, 72, 73, 93, 96, 205, 226, 254, 257,
    261, 269.

ARA
AE 116.

AURIGA
UV 31, 40, 63, 72, 73, 74.

CAMELOPADUS
Z   62, 69.

CANES VENATICI
RS 263.
TX 58, 251.

CARINA
η 202, 203.

CASSIOPEIA
RX 246.
SX 232, 247.

CEPHEUS
VV 62, 227, 232.

CETUS
o   113 (see also Mira-type variables)

CORONA AUSTRALIS
Y   116.

CORONA BOREALIS
T   1, 3, 4, 58, 61, 62, 65, 227, 233, 236, 242-244, 278.

CRUX
BI 31.

CYGNUS

BF 1, 11, 13, 50, <u>58</u>, 65, 72, 73, <u>77</u>, 103, 106, 110, 257, 260.
CH 30, 31, 38, 50, <u>51</u>, 56, 66, 72, 73, <u>77</u>, 86, 93, 103, 104, 106, <u>107</u>, <u>131-140</u>, 255-257, 270, 277, 284, 290, 300.
CI 1, 4, 9, 11, 13, 43, <u>51</u>, <u>58</u>, 64, 72, 73, <u>77</u>, 103, 106, 116, <u>141-156</u>, 206, 233, 242, 250, 253, 258, 259, 260, 262, 265, 269, 277, 294, 300.
V1016 4, 9, 17, 18, <u>20-21</u>, 26, 29, 38, 43, <u>53</u>, 61, 67, 68, 72, 73, <u>77</u>, <u>82</u>, 83, 84, 86, 88, 92, 93, 97, 98, 100, 101, 112, 113, 115, 116, 121, 145, <u>157-163</u>, 167, 205, 214, 253, 254, 255, 262, 269, 285, 291, 299, 300.
V1329 (HBV 475) 17, <u>23</u>, 43, 50, <u>53-54</u>, 68, 72, 73, <u>82</u>, 97, <u>165-176</u>, 300.
V1500 (nova 1978) 19, 20.

DRACO

AG 9, 13, 43, 50, <u>51-52</u>, 72, 73, <u>74</u>, 104, 109, 110, 113, <u>117-121</u>, 130, 145, 151, <u>177-190</u>, 218, 227, 234, 255, 257, 259, 263, 284, 299, 300.

GEMINI

U 62, 63, 69, 225, 290.
WY 4.

HERCULES

YY 93, 110, 300.

HYDRA

RW 1, 13, 65, 86, 87, 88, 96, 97, 98, 101, 110, 192, 205, 257, 269.

LEPUS

17 4, 237, 246.

LYRA

  4, 200, 240.

MONOCEROS

AX 4, 237, <u>243-244</u>, 246.
BX 93.

MUSCA

SY 93, 110, <u>191-194</u>, 269, 300.

OPHIUCUS

RS 3, 61, 62, 72, 73.

PAVO

AR 4, 11, 44, 65, 72, 73, 103, 104, 106, 156, <u>195-200</u>, 233, 243, 246, 250, 254, 257, 258, 297, 301.

STAR INDEX

PEGASUS
AG 4, 13, 23, 26, 44, 50, 52, 57, 65, 72, 73, 82, 86, 103, 104, 105–107, 109, 110, 112, 173, 189, 201–202, 218, 226, 233, 236, 241, 242, 243–246, 249, 254–256, 258, 260–265, 269, 284, 300.

PERSEUS
AX 1, 4, 11, 29, 58, 65, 72, 73, 74, 103, 104, 116.

PUPPIS
RX 18, 26, 27, 29, 40, 52–53, 56, 173, 203–208, 261, 269, 300.

SAGITTA
δ 246.
FG 149.
HM 4, 18, 22, 26, 38, 43, 50, 53, 67, 68, 72, 73, 83, 96, 97, 115, 116, 121, 162, 163, 167, 209–214, 262, 284, 291, 299, 300.
WZ 259.

SAGITTARIUS
V4049 219, 300.

SCORPIUS
CL 93.
HK 72, 73.
V455 116.

SCUTUM
FR 72.
RY 50, 72, 73.
V373 (nova 1975) 48.

SERPENS
W 200, 236, 240, 246.
RT 4, 68, 167.

TAURUS
T 225, 229.

TELESCOPIUM
RR 3, 4, 5, 9, 18, 24, 29, 30, 61, 68, 72, 73, 86, 88, 93, 96, 98, 113, 115, 116, 121, 130, 145, 157, 167, 215–219, 257, 261, 263, 299, 300.

VELA
WY 4.

VULPECULA
PU 221–222, 300.

AS 295B   72, 73, 116, 167.
GX 1+4   167, 277, 292.
H 1-36   29, 116.
HD 330036   40, 116.
HD 229227   173, 300.
He 2-38   29, 30, 116, 216.
MWC 603   72, 73, <u>77</u>.
SS 433   229, 292.

ASTROPHYSICS AND SPACE SCIENCE LIBRARY

Edited by

J. E. Blamont, R. L. F. Boyd, L. Goldberg, C. de Jager, Z. Kopal, G. H. Ludwig, R. Lüst,
B. M. McCormac, H. E. Newell, L. I. Sedov, Z. Švestka, and W. de Graaff

1. C. de Jager (ed.), *The Solar Spectrum, Proceedings of the Symposium held at the University of Utrecht, 26–31 August, 1963.* 1965, XIV + 417 pp.
2. J. Orthner and H. Maseland (eds.), *Introduction to Solar Terrestrial Relations, Proceedings of the Summer School in Space Physics held in Alpbach, Austria, July 15–August 10, 1963 and Organized by the European Preparatory Commission for Space Research.* 1965, IX + 506 pp.
3. C. C. Chang and S. S. Huang (eds.), *Proceedings of the Plasma Space Science Symposium, held at the Catholic University of America, Washington, D.C., June 11–14, 1963.* 1965, IX + 377 pp.
4. Zdeněk Kopal, *An Introduction to the Study of the Moon.* 1966, XII + 464 pp.
5. B. M. McCormac (ed.), *Radiation Trapped in the Earth's Magnetic Field. Proceedings of the Advanced Study Institute, held at the Chr. Michelsen Institute, Bergen, Norway, August 16–September 3, 1965.* 1966, XII + 901 pp.
6. A. B. Underhill, *The Early Type Stars.* 1966, XII + 282 pp.
7. Jean Kovalevsky, *Introduction to Celestial Mechanics.* 1967, VIII + 427 pp.
8. Zdeněk Kopal and Constantine L. Goudas (eds.), *Measure of the Moon. Proceedings of the 2nd International Conference on Selenodesy and Lunar Topography, held in the University of Manchester, England, May 30–June 4, 1966.* 1967, XVIII + 479 pp.
9. J. G. Emming (ed.), *Electromagnetic Radiation in Space. Proceedings of the 3rd ESRO Summer School in Space Physics, held in Alpbach, Austria, from 19 July to 13 August, 1965.* 1968, VIII + 307 pp.
10. R. L. Carovillano, John F. McClay, and Henry R. Radoski (eds.), *Physics of the Magnetosphere, Based upon the Proceedings of the Conference held at Boston College, June 19–28, 1967.* 1968, X + 686 pp.
11. Syun-Ichi Akasofu, *Polar and Magnetospheric Substorms.* 1968, XVIII + 280 pp.
12. Peter M. Millman (ed.), *Meteorite Research. Proceedings of a Symposium on Meteorite Research, held in Vienna, Austria, 7–13 August, 1968.* 1969, XV + 941 pp.
13. Margherita Hack (ed.), *Mass Loss from Stars. Proceedings of the 2nd Trieste Colloquium on Astrophysics, 12–17 September, 1968.* 1969, XII + 345 pp.
14. N. D'Angelo (ed.), *Low-Frequency Waves and Irregularities in the Ionosphere. Proceedings of the 2nd ESRIN-ESLAB Symposium, held in Frascati, Italy, 23–27 September, 1968.* 1969, VII + 218 pp.
15. G. A. Partel (ed.), *Space Engineering. Proceedings of the 2nd International Conference on Space Engineering, held at the Fondazione Giorgio Cini, Isola di San Giorgio, Venice, Italy, May 7–10, 1969.* 1970, XI + 728 pp.
16. S. Fred Singer (ed.), *Manned Laboratories in Space. Second International Orbital Laboratory Symposium.* 1969, XIII + 133 pp.
17. B. M. McCormac (ed.), *Particles and Fields in the Magnetosphere. Symposium Organized by the Summer Advanced Study Institute, held at the University of California, Santa Barbara, Calif., August 4–15, 1969.* 1970, XI + 450 pp.
18. Jean-Claude Pecker, *Experimental Astronomy.* 1970, X + 105 pp.
19. V. Manno and D. E. Page (eds.), *Intercorrelated Satellite Observations related to Solar Events. Proceedings of the 3rd ESLAB/ESRIN Symposium held in Noordwijk, The Netherlands, September 16–19, 1969.* 1970, XVI + 627 pp.
20. L. Mansinha, D. E. Smylie, and A. E. Beck, *Earthquake Displacement Fields and the Rotation of the Earth, A NATO Advanced Study Institute Conference Organized by the Department of Geophysics, University of Western Ontario, London, Canada, June 22–28, 1969.* 1970, XI + 308 pp.
21. Jean-Claude Pecker, *Space Observatories.* 1970, XI + 120 pp.
22. L. N. Mavridis (ed.), *Structure and Evolution of the Galaxy. Proceedings of the NATO Advanced Study Institute, held in Athens, September 8–19, 1969.* 1971, VII + 312 pp.

23. A. Muller (ed.), *The Magellanic Clouds. A European Southern Observatory Presentation: Principal Prospects, Current Observational and Theoretical Approaches, and Prospects for Future Research*, Based on the Symposium on the Magellanic Clouds, held in Santiago de Chile, March 1969, on the Occasion of the Dedication of the European Southern Observatory. 1971, XII + 189 pp.
24. B. M. McCormac (ed.), *The Radiating Atmosphere*. Proceedings of a Symposium Organized by the Summer Advanced Study Institute, held at Queen's University, Kingston, Ontario, August 3–14, 1970. 1971, XI + 455 pp.
25. G. Fiocco (ed.), *Mesospheric Models and Related Experiments*. Proceedings of the 4th ESRIN-ESLAB Symposium, held at Frascati, Italy, July 6–10, 1970. 1971, VIII + 298 pp.
26. I. Atanasijević, *Selected Exercises in Galactic Astronomy*. 1971, XII + 144 pp.
27. C. J. Macris (ed.), *Physics of the Solar Corona*. Proceedings of the NATO Advanced Study Institute on Physics of the Solar Corona, held at Cavouri-Vouliagmeni, Athens, Greece, 6–17 September 1970. 1971, XII + 345 pp.
28. F. Delobeau, *The Environment of the Earth*. 1971, IX + 113 pp.
29. E. R. Dyer (general ed.), *Solar-Terrestrial Physics/1970*. Proceedings of the International Symposium on Solar-Terrestrial Physics, held in Leningrad, U.S.S.R., 12–19 May 1970. 1972, VIII + 938 pp.
30. V. Manno and J. Ring (eds.), *Infrared Detection Techniques for Space Research*. Proceedings of the 5th ESLAB-ESRIN Symposium, held in Noordwijk, The Netherlands, June 8–11, 1971. 1972, XII + 344 pp.
31. M. Lecar (ed.), *Gravitational N-Body Problem*. Proceedings of IAU Colloquium No. 10, held in Cambridge, England, August 12–15, 1970. 1972, XI + 441 pp.
32. B. M. McCormac (ed.), *Earth's Magnetospheric Processes*. Proceedings of a Symposium Organized by the Summer Advanced Study Institute and Ninth ESRO Summer School, held in Cortina, Italy, August 30–September 10, 1971. 1972, VIII + 417 pp.
33. Antonin Rükl, *Maps of Lunar Hemispheres*. 1972, V + 24 pp.
34. V. Kourganoff, *Introduction to the Physics of Stellar Interiors*. 1973, XI + 115 pp.
35. B. M. McCormac (ed.), *Physics and Chemistry of Upper Atmospheres*. Proceedings of a Symposium Organized by the Summer Advanced Study Institute, held at the University of Orléans, France, July 31–August 11, 1972. 1973, VIII + 389 pp.
36. J. D. Fernie (ed.), *Variable Stars in Globular Clusters and in Related Systems*. Proceedings of the IAU Colloquium No. 21, held at the University of Toronto, Toronto, Canada, August 29–31, 1972. 1973, IX + 234 pp.
37. R. J. L. Grard (ed.), *Photon and Particle Interaction with Surfaces in Space*. Proceedings of the 6th ESLAB Symposium, held at Noordwijk, The Netherlands, 26–29 September, 1972. 1973, XV + 577 pp.
38. Werner Israel (ed.), *Relativity, Astrophysics and Cosmology*. Proceedings of the Summer School, held 14–26 August, 1972, at the BANFF Centre, BANFF, Alberta, Canada. 1973, IX + 323 pp.
39. B. D. Tapley and V. Szebehely (eds.), *Recent Advances in Dynamical Astronomy*. Proceedings of the NATO Advanced Study Institute in Dynamical Astronomy, held in Cortina d'Ampezzo, Italy, August 9–12, 1972. 1973, XIII + 468 pp.
40. A. G. W. Cameron (ed.), *Cosmochemistry*. Proceedings of the Symposium on Cosmochemistry, held at the Smithsonian Astrophysical Observatory, Cambridge, Mass., August 14–16, 1972. 1973, X + 173 pp.
41. M. Golay, *Introduction to Astronomical Photometry*. 1974, IX + 364 pp.
42. D. E. Page (ed.), *Correlated Interplanetary and Magnetospheric Observations*. Proceedings of the 7th ESLAB Symposium, held at Saulgau, W. Germany, 22–25 May, 1973. 1974, XIV + 662 pp.
43. Riccardo Giacconi and Herbert Gursky (eds.), *X-Ray Astronomy*. 1974, X + 450 pp.
44. B. M. McCormac (ed.), *Magnetospheric Physics*. Proceedings of the Advanced Summer Institute, held in Sheffield, U.K., August 1973. 1974, VII + 399 pp.
45. C. B. Cosmovici (ed.), *Supernovae and Supernova Remnants*. Proceedings of the International Conference on Supernovae, held in Lecce, Italy, May 7–11, 1973. 1974, XVII + 387 pp.
46. A. P. Mitra, *Ionospheric Effects of Solar Flares*. 1974, XI + 294 pp.
47. S.-I. Akasofu, *Physics of Magnetospheric Substorms*. 1977, XVIII + 599 pp.

48. H. Gursky and R. Ruffini (eds.), *Neutron Stars, Black Holes and Binary X-Ray Sources*. 1975, XII + 441 pp.
49. Z. Švestka and P. Simon (eds.), *Catalog of Solar Particle Events 1955–1969. Prepared under the Auspices of Working Group 2 of the Inter-Union Commission on Solar-Terrestrial Physics*. 1975, IX + 428 pp.
50. Zdeněk Kopal and Robert W. Carder, *Mapping of the Moon*. 1974, VIII + 237 pp.
51. B. M. McCormac (ed.), *Atmospheres of Earth and the Planets. Proceedings of the Summer Advanced Study Institute, held at the University of Liège, Belgium, July 29–August 8, 1974*. 1975, VII + 454 pp.
52. V. Formisano (ed.), *The Magnetospheres of the Earth and Jupiter. Proceedings of the Neil Brice Memorial Symposium, held in Frascati, May 28–June 1, 1974*. 1975, XI + 485 pp.
53. R. Grant Athay, *The Solar Chromosphere and Corona: Quiet Sun*. 1976, XI + 504 pp.
54. C. de Jager and H. Nieuwenhuijzen (eds.), *Image Processing Techniques in Astronomy. Proceedings of a Conference, held in Utrecht on March 25–27, 1975*. XI + 418 pp.
55. N. C. Wickramasinghe and D. J. Morgan (eds.), *Solid State Astrophysics. Proceedings of a Symposium, held at the University College, Cardiff, Wales, 9–12 July 1974*. 1976, XII + 314 pp.
56. John Meaburn, *Detection and Spectrometry of Faint Light*. 1976, IX + 270 pp.
57. K. Knott and B. Battrick (eds.), *The Scientific Satellite Programme during the International Magnetospheric Study. Proceedings of the 10th ESLAB Symposium, held at Vienna, Austria, 10–13 June 1975*. 1976, XV + 464 pp.
58. B. M. McCormac (ed.), *Magnetospheric Particles and Fields. Proceedings of the Summer Advanced Study School, held in Graz, Austria, August 4–15, 1975*. 1976, VII + 331 pp.
59. B. S. P. Shen and M. Merker (eds.), *Spallation Nuclear Reactions and Their Applications*. 1976, VIII + 235 pp.
60. Walter S. Fitch (ed.), *Multiple Periodic Variable Stars. Proceedings of the International Astronomical Union Colloquium No. 29, held at Budapest, Hungary, 1–5 September 1976*. 1976, XIV + 348 pp.
61. J. J. Burger, A. Pedersen, and B. Battrick (eds.), *Atmospheric Physics from Spacelab. Proceedings of the 11th ESLAB Symposium, Organized by the Space Science Department of the European Space Agency, held at Frascati, Italy, 11–14 May 1976*. 1976, XX + 409 pp.
62. J. Derral Mulholland (ed.), *Scientific Applications of Lunar Laser Ranging. Proceedings of a Symposium held in Austin, Tex., U.S.A., 8–10 June, 1976*. 1977, XVII + 302 pp.
63. Giovanni G. Fazio (ed.), *Infrared and Submillimeter Astronomy. Proceedings of a Symposium held in Philadelphia, Penn., U.S.A., 8–10 June, 1976*. 1977, X + 226 pp.
64. C. Jaschek and G. A. Wilkins (eds.), *Compilation, Critical Evaluation and Distribution of Stellar Data. Proceedings of the International Astronomical Union Colloquium No. 35, held at Strasbourg, France, 19–21 August, 1976*. 1977, XIV + 316 pp.
65. M. Friedjung (ed.), *Novae and Related Stars. Proceedings of an International Conference held by the Institut d'Astrophysique, Paris, France, 7–9 September, 1976*. 1977, XIV + 228 pp.
66. David N. Schramm (ed.), *Supernovae. Proceedings of a Special IAU-Session on Supernovae held in Grenoble, France, 1 September, 1976*. 1977, X + 192 pp.
67. Jean Audouze (ed.), *CNO Isotopes in Astrophysics. Proceedings of a Special IAU Session held in Grenoble, France, 30 August, 1976*. 1977, XIII + 195 pp.
68. Z. Kopal, *Dynamics of Close Binary Systems*, XIII + 510 pp.
69. A. Bruzek and C. J. Durrant (eds.), *Illustrated Glossary for Solar and Solar-Terrestrial Physics*. 1977, XVIII + 204 pp.
70. H. van Woerden (ed.), *Topics in Interstellar Matter*. 1977, VIII + 295 pp.
71. M. A. Shea, D. F. Smart, and T. S. Wu (eds.), *Study of Travelling Interplanetary Phenomena*. 1977, XII + 439 pp.
72. V. Szebehely (ed.), *Dynamics of Planets and Satellites and Theories of Their Motion. Proceedings of IAU Colloquium No. 41, held in Cambridge, England, 17–19 August 1976*. 1978, XII + 375 pp.
73. James R. Wertz (ed.), *Spacecraft Attitude Determination and Control*. 1978, XVI + 858 pp.

74. Peter J. Palmadesso and K. Papadopoulos (eds.), *Wave Instabilities in Space Plasmas. Proceedings of a Symposium Organized Within the XIX URSI General Assembly held in Helsinki, Finland, July 31–August 8, 1978.* 1979, VII + 309 pp.
75. Bengt E. Westerlund (ed.), *Stars and Star Systems. Proceedings of the Fourth European Regional Meeting in Astronomy held in Uppsala, Sweden, 7–12 August, 1978.* 1979, XVIII + 264 pp.
76. Cornelis van Schooneveld (ed.), *Image Formation from Coherence Functions in Astronomy. Proceedings of IAU Colloquium No. 49 on the Formation of Images from Spatial Coherence Functions in Astronomy, held at Groningen, The Netherlands, 10–12 August 1978.* 1979, XII + 338 pp.
77. Zdeněk Kopal, *Language of the Stars. A Discourse on the Theory of the Light Changes of Eclipsing Variables.* 1979, VIII + 280 pp.
78. S.-I. Akasofu (ed.), *Dynamics of the Magnetosphere. Proceedings of the A.G.U. Chapman Conference 'Magnetospheric Substorms and Related Plasma Processes' held at Los Alamos Scientific Laboratory, N.M., U.S.A., October 9–13, 1978.* 1980, XII + 658 pp.
79. Paul S. Wesson, *Gravity, Particles, and Astrophysics. A Review of Modern Theories of Gravity and G-variability, and their Relation to Elementary Particle Physics and Astrophysics.* 1980, VIII + 188 pp.
80. Peter A. Shaver (ed.), *Radio Recombination Lines. Proceedings of a Workshop held in Ottawa, Ontario, Canada, August 24–25, 1979.* 1980, X + 284 pp.
81. Pier Luigi Bernacca and Remo Ruffini (eds.), *Astrophysics from Spacelab*, 1980, XI + 664 pp.
82. Hannes Alfvén, *Cosmic Plasma*, 1981, X + 160 pp.
83. Michael D. Papagiannis (ed.), *Strategies for the Search for Life in the Universe*, 1980, XVI + 254 pp.
84. H. Kikuchi (ed.), *Relation between Laboratory and Space Plasmas*, 1981, XII + 386 pp.
85. Peter van der Kamp, *Stellar Paths*, 1981, xxii + 155 pp.
86. E. M. Gaposchkin and B. Kołaczek (eds.), *Reference Coordinate Systems for Earth Dynamics*, 1981, XIV + 396 pp.
87. R. Giacconi (ed.), *X-Ray Astronomy with the Einstein Satellite. Proceedings of the High Energy Astrophysics Division of the American Astronomical Society Meeting on X-Ray Astronomy held at the Harvard-Smithsonian Center for Astrophysics, Cambridge, Mass., U.S.A., January 28–30, 1980.* 1981, VII + 330 pp.
88. Icko Iben Jr. and Alvio Renzini (eds.), *Physical Processes in Red Giants. Proceedings of the Second Workshop, held at the Ettore Majorana Centre for Scientific Culture, Advanced School of Agronomy, in Erice, Sicily, Italy, September 3–13, 1980.* 1981, XV + 488 pp.
89. C. Chiosi and R. Stalio (eds.), *Effect of Mass Loss on Stellar Evolution. IAU Colloquium No. 59 held in Miramare, Trieste, Italy, September 15–19, 1980.* 1981, XXII + 532 pp.
90. C. Goudis, *The Orion Complex: A Case Study of Interstellar Matter*, 1982 (forthcoming).
91. F. D. Kahn (ed.), *Investigating the Universe. Papers Presented to Zdeněk Kopal on the Occasion of his retirement, September 1981.* 1981, X + 458 pp.
92. C. M. Humphries (ed.), *Instrumentation for Astronomy with Large Optical Telescopes, Proceedings of IAU Colloquium No. 67.* 1982 (forthcoming).
93. R. S. Roger and P. E. Dewdney (eds.), *Regions of Recent Star Formation, Proceedings of the Symposium on "Neutral Clouds Near HII Regions – Dynamics and Photochemistry", held in Penticton, B.C., June 24–26, 1981.* 1982, XVI + 496 pp.
94. O. Calame (ed.), *High-Precision Earth Rotation and Earth-Moon Dynamics. Lunar Distances and Related Observations,* 1982, xx + 354 pp.